# AGROCHEMICAL AND PESTICIDE SAFETY HANDBOOK

# AGROCHEMICAL AND PESTICIDE SAFETY HANDBOOK

Michael F. Waxman

CRC Press is an imprint of the
Taylor & Francis Group, an **informa** business

CRC Press
Taylor & Francis Group
6000 Broken Sound Parkway NW, Suite 300
Boca Raton, FL 3487-2742

First issued in paperback 2020



ISBN-13: 978-0-367-57930-2 (pbk)
ISBN-13: 978-1-56670-296-6 (hbk)



**Visit the Taylor & Francis Web site at**
**http://www.taylorandfrancis.com**

**and the CRC Press Web site at**
**http://www.crcpress.com**

**Library of Congress Cataloging-in-Publication Data**

Waxman, Michael F., 1942–
Agrochemical and pesticide safety handbook / Michael F. Waxman.
p. cm.
Includes bibliographical references (p. ) and index.
ISBN 1-56670-296-8 (alk. paper)
1. Pesticides--Handbooks, manuals, etc. 2. Agricultural chemicals--Handbooks, manuals, etc. 3. Pesticides--Safety measures--Handbooks, manuals, etc. 4. Agricultural chemicals--Safety measures--Handbooks, manuals, etc. I. Title.
SB951.W396 1998
632′.95—dc21 98-5485
CIP

Library of Congress Card Number 98-5485

# PREFACE

During the past ten years several important new and revised regulations have been issued. These regulations have mandated new certifications and training for those involved directly or indirectly with the application of pesticides. In addition, in 1992, EPA issued the Worker Protection Standard which mandated training for all agricultural workers and pesticide handlers and detailed the content of such training.

The *Agrochemical and Pesticide Safety Handbook* was designed to provide pesticide users, manufacturers, and formulators with additional "reader-friendly" knowledge on the safe use of agrochemicals and pesticides and information and recommendations which provide the reader with preventive measures and guidance during the initial stages of incidents involving agrochemicals and pesticides, when the correct, rapid responses can prevent a major problem. The Pesticide and Chemical Guides contain important information on more than 500 pesticides and 100 agrochemicals.

The handbook is composed of two parts. Part One provides the reader with valuable information on all aspects of pesticides. This section helps the reader recognize that pesticides and certain agrochemicals are essential tools in the control of most pests and that the safe use of these chemicals is an essential element in this control.

Part Two provides the reader with the necessary information needed to apply, store, and dispose of specific pesticides and other hazardous agrochemicals properly. Emergency guidelines and first aid procedures are also provided in the context of handling incidents such as fires, spills and clean-up.

Part One:

Chapter 1 discusses the market for pesticides. It reviews the current market status and its future trends.

Chapter 2 reviews the pertinent government regulations issued by the regulating agencies (such as U.S. EPA, OSHA, and the Department of Agriculture) and how these regulations impact various industries and their personnel.

The different types or groups of pests are defined and the various pest control methods for each group are discussed in Chapter 3.

Chapter 4 describes the different formulations used to prepare the finished pesticide and the beneficial qualities imparted by using different additives in the formulation. This chapter also discusses the pros and cons of the different formulations and their specific applications. In addition, the chapter covers the methods and mathematical equations used to calculate the correct amounts and rates of application for various types of equipment.

Chapter 5 discusses the different pesticide classification schemes, the different methods used to estimate their toxicity, and the specific mode of action of the different chemical groups of pesticides. This chapter also discusses the three main functional groups of pesticides (insecticides, herbicides, and fungicides) and the characteristics of various products in each group.

Pesticide handling activities and the recommended personal protective equipment (PPE) are addressed in Chapter 6. Pesticide storage, disposal, and spill management are also covered in this chapter.

Chapter 7 describes the harmful effects of pesticides and the recommended first aid treatments. In addition, heat stress and ways to manage work activities to minimize heat illness are also discussed.

The different types of equipment used to apply pesticides and their advantages and limitations are addressed in Chapter 8. This chapter also discusses the different parts of the application equipment and their selection and care.

The last chapter, Chapter 9, discusses the environmental effects of pesticides. The topics covered include; sources of contamination, pesticide fate and transport, and endangered species.

Part Two:

Pesticide and Chemical Tables

There are three important tables in Part Two. The first of these, Table 1, lists the trade names or brands of pesticides of current importance alphabetically and assigns a pesticide guide to each product.

Table 2 provides the user with the common name, and the family or group of chemicals that the chemical belongs to or is derived from.

Table 3 lists alphabetically the chemicals that are important to users of pesticides and assigns a chemical guide to each item in the table.

Pesticide and Chemical Guides:

The pesticides listed under trade name in Tables 1 and 2 have each been assigned to a specific pesticide guide. These guides assist the user in handling various tasks from mixing and applying pesticides to cleaning up spilled pesticide and disposing of the waste properly.

Over 500 formulated pesticides are listed in Tables 1 and 2. In order to provide usable information, these pesticides are organized into groups of pesticides with similar attributes and hazards. Each group was defined by these attributes and hazards and a guide was developed to provide pertinent information for each group. As a result there are 61 unique pesticide guides.

The agrochemicals (over 100) were treated in a similar fashion and 18 unique chemical guides were developed to provide similar information.

Finally, a glossary has been developed to assist the reader define the various terms and acronyms used throughout the handbook.

The information in this book is focused toward those involved in handling, mixing, and applying pesticides. It should be especially useful to commercial pesticide applicators, formulators, and handlers as well as employees of city, county, state, and federal agencies. Pesticide dealers, salespeople, consultants, and trainers should also find it helpful in their work.

# ACKNOWLEDGMENTS

The writing of this textbook required approximately one year, but during that period much time and effort was devoted to this project. Even though it is part of each faculty member's duties to transfer technology or impart knowledge gained from the University to the communities which it serves, I would like to acknowledge the Department of Engineering Professional Development and the University of Wisconsin-Madison for supporting my efforts and allowing the resources to complete this work.

The information and illustrations for this text were drawn from many sources including the U.S. Environmental Protection Agency, the State of Wisconsin Department of Agriculture, Trade and Consumer Protection, University of Wisconsin-Extension, Michigan State University, Fire Protection Association, Abbott Laboratories, AgrEvo USA Company, BSAF Corporation, Ciba-Geigy Corporation, Cyanamid Company, DowElanco, DuPont, ISK Biotech Corporation, FMC Corporation, Growmark, Micro Flo Corporation, Monsanto Company, Rhone-Poulenc, Sandoz Agro, Inc., Terra International, Inc., Uniroyal Chemical Company, and Zeneca Ag Products.

I have tried to acknowledge all illustrations and materials, but I may have inadvertently missed a source because of the extensive interchange of materials with the agencies, organizations, and companies mentioned above.

And finally, a special debt of gratitude is due to my wife, Barbara, for her patience and understanding during the preparation of this manuscript and for her encouragement that enabled me to see it through to fruition.

# ACKNOWLEDGMENTS

# TABLE OF CONTENTS

## PART ONE: INTRODUCTION AND RESOURCE MATERIALS

# PART ONE

# INTRODUCTION AND RESOURCE MATERIALS

# CHAPTER 1

# INTRODUCTION

Pesticides are chemicals or biological substances used to kill or control pests. They fall into three major classes: insecticides, fungicides, and herbicides (or weed killers). There are also rodenticides (for control of vertebrate pests), nematicides (to kill eelworms, etc.), molluscicides (to kill slugs and snails), and acaricides (to kill mites). These chemicals are typically manmade synthetic organic compounds, but there are exceptions which occur naturally that are plant derivatives or naturally occurring inorganic minerals.

Pesticides may also be divided into two main types: contact or nonsystemic pesticides and systemic pesticides. Contact or surface coating pesticides do not appreciably penetrate plant tissue and are consequently not transported, or translocated, within the plant vascular system. The earlier pesticides were of this type; their disadvantages were that they were susceptible to the effects of the weather and new plant growth was not protected.

In contrast, most of recently developed pesticides are systemically active and therefore they penetrate the plant cuticle and move through the plant vascular system. Examples of systemic fungicides are benomyl and hexaconazole. These systemic agents cannot only protect a plant from attack but also inhibit or cure established infections. They are not affected by weathering and also confer immunity to all new plant growth.

The use of pesticides has been traced by historians to before 1000 B.C. Homer mentioned the use of sulfur as a fumigant to avert disease and control insects. Theophrastus, in 300 B.C., described many plant diseases known today such as scorch, rot, scab, and rust. There are also several references in the Old Testament to the plagues of Egypt for which the locust was chiefly responsible, and even today locusts cause vast food losses in the Near East and Africa. Pliny in 79 A.D. advocated the use of arsenic as an insecticide and by 900 A.D., the Chinese were using arsenic and other inorganic chemicals in their gardens to kill insects.

In the seventeenth century the first naturally occurring insecticide, nicotine from extracts of tobacco leaves, was used to control the plum curculio and the lace bug. Hamberg (1705) proposed mercuric chloride as a wood preservative and a hundred years later Prevost described the inhibition of smut spores by copper sulfate.

It was not until the middle of the nineteenth century that systematic scientific methods began to be applied to the problem of controlling agricultural pests. About 1850 two important natural insecticides were developed: rotenone from the roots of derris plants and pyrethrum from the flower heads of a species of chrysanthemum. These insecticides are still widely used. At about the same time, new inorganic materials were introduced for combating insect pests. For instance, an investigation into the use of new arsenic compounds led in 1867 to the introduction of an impure copper arsenite (Paris Green) for control of the Colorado beetle in the state of Mississippi. In 1892 lead arsenate was used for control of gypsy moth.

The Irish Potato Famine of the 1840s illustrates what can occur when a staple food crop is stricken by a disease against which there is no known defense. The potato crop was virtually destroyed by severe attacks of the fungal disease known as potato late blight, resulting in the deaths of more than a million people.

Millardet, in 1882, accidentally discovered a valuable chemical treatment for the control of pathogenic fungi, like potato blight and vine mildew. This discovery came from a local custom of the farmers in the Bordeaux district of France. They daubed the roadside vines with a mixture of copper sulfate and lime in order to discourage pilfering of the crop. At this time the crops of the French vineyards were being destroyed by the downy mildew disease. Millardet observed that although the vines away from the road were heavily infested with mildew, those alongside the road which had been treated with the mixture were relatively free from the disease. Millardet subsequently carried out further experiments which established the effectiveness of the mixture of copper sulfate, lime, and water against vine mildew. The mixture, called the Bordeaux mixture, was widely applied, the disease was arrested, and Millardet became somewhat of a hero.

In 1897 formaldehyde was introduced for the first time as a fumigant. In 1913 organomercurials were first used as fungicidal seed dressings against cereal smut and bunt diseases.

W. C. Piver in 1912 developed calcium arsenate as a replacement for Paris Green and lead arsenate. This mixture soon became important for controlling the boll weevil on cotton in the United States. By the early 1920s the extensive application of arsenical insecticides caused widespread public outcries because fruits and vegetables treated with arsenates were sometimes shown to contain poisonous residues. This stimulated the search for other less dangerous pesticides and led to the introduction of organic compounds, such as tar, petroleum oils, and dinitro-o-cresol. The latter compound eventually replaced tar oil for control of aphid eggs, and in 1933 was patented as a selective herbicide against weeds in cereal crops. Unfortunately, this is also a very poisonous substance.

The 1930s really represents the beginning of the modern era of synthetic organic pesticides—important examples include the introduction of alkyl thiocyanate insecticides, the first organic fungicides (dithiocarbamate fungicides), and a host of other fungicides and insecticides. In 1939 Müller discovered the powerful insecticidal properties of dichlorodiphenyltrichloroethane or DDT. In 1943, DDT was first manufactured and soon became the most widely used single insecticide in the world.

In the 1940s, many chlorinated hydrocarbon insecticides were developed though they did not come into widespread use until the 1950s. Common examples include aldrin, dieldrin, heptochlor, and endrin. However, in spite of their early promise, these organochlorine insecticides are now much less used because of their environmental pollution impact.

The organophosphosphates represent another extremely important class of organic insecticides. They were developed during World War II as chemical warfare agents. Early examples included the powerful insecticide schradan, a systemic insecticide, and the contact insecticide parathion. Unfortunately, both of these compounds are highly poisonous to mammals and subsequent research in this field has been directed toward the development of more selective and less poisonous insecticides. In 1950, malathion, the first example of a wide-spectrum organophosphorus insecticide combined with very low mammalian toxicity, was developed. And at about the same time the phenoxyacetic acid herbicides were discovered. These systemic compounds are extremely valuable for the selective control of broad-leaved weeds in cereal crops. These compounds have a relatively low toxicity to mammals and are therefore relatively safe to use.

The bipyridinium herbicides were introduced in 1958. These are very quick-acting herbicides which are absorbed by the plants and translocated causing desiccation of the foliage. These herbicides are strongly absorbed to the clay components of the soil and become effectively inactivated.

It was not until the late 1960s that effective systemic fungicides appeared on the market, and their development represents an important breakthrough in the field of plant chemotherapy. The major classes of systemic fungicides developed since 1966 are oxathiins, benzimidazoles, thiophanates, and pyrimidines. Other effective systemic fungicides used currently include antibiotics, morpholines, organophosphorus compounds, and most recently, the sterol biosynthesis inhibitors, e.g., triazoles.

Throughout the history of pesticide usage, the manufacturers of pesticides have faced the same challenge that confronts the makers of pesticides today. That is, the development of chemicals that kill or control unwanted insects, weeds, fungi, rodents, and other pests without harming desired plants, beneficial insects, wildlife, and, most important, humans.

Chemicals that control rats are termed rodenticides. The first effective compound was warfarin. It was developed by the Wisconsin Alumni Research Foundation in 1944. It functions as an anticoagulant in human medicine. However, when used against rats and mice, at high concentrations it is extremely effective, causing death by internal hemorrhaging.

In 1962 *The Silent Spring*, written by Rachel Carson, was published. Carson's book was one of the first that attracted national attention to the problems of toxic chemicals and the effects of these chemicals on the environment. *The Silent Spring* recounted how the residues of the pesticide DDT could be found throughout the food chain. In aquatic birds, high levels of DDT were associated with reduced fertility. DDT affected the deposition of calcium in avian ovaries, leading to egg shells too thin to survive, thus causing a widespread reduction in many bird species.

*The Silent Spring* and other books on the dangers of pesticides have served to illustrate that great efforts must be taken to prevent the misuse of pesticides and other chemicals. It is this misuse, overuse, and improper disposal that causes many of the problems that have been reported.

Recently, man has made great advances in the genetic manipulation of genes. It is now possible to create in the laboratory seeds and thus crops which possess the genetic ability to kill or inhibit disease-causing pests.

The term "agrochemical" is broader and includes chemicals which will enhance the growth and yield of crops, but excludes large-scale inorganic fertilizers.

## I. THE MARKET FOR PESTICIDES

### A. CURRENT STATUS

In 1992, approximately \$8.2 billion, and in 1993, approximately \$8.5 billion worth of pesticides were purchased for use in the United States. There is no question that the productivity of American agriculture is due in large part to the success of modern pesticides. There is also no question that we are still grappling with the problem of balancing the usefulness of pesticides with their safety.

The largest market for pesticides as of 1993 was the United States. It represents 34% of the total world market, which has been estimated at over \$25 billion. The retail value of pesticide sales in the United States for 1993 was well over \$8 billion (see Table 1 and Figures 1.1 and 1.2).

**Table 1.1** U.S. and World Conventional Pesticide Sales at User Level, 1993 Estimates.

| Pesticide Class | U.S. Market Million | % | World Market Million | % | U.S. % of World Market |
|---|---|---|---|---|---|
| User Expenditures in Millions of $ | | | | | |
| Herbicides | $4,756 | 56% | $11,700 | 46% | 41% |
| Insecticides | 2,550 | 30% | 7,900 | 31% | 32% |
| Fungicides | 584 | 7% | 4,139 | 16% | 14% |
| Other | 594 | 7% | 1,550 | 6% | 38% |
| **Total** | **$8,484** | **100%** | **$25,280** | **100%** | **34%** |
| Volume of Active Ingredients in Millions of lbs | | | | | |
| Herbicides | 620 | 57% | 2,110 | 47% | 29% |
| Insecticides | 247 | 23% | 1,625 | 36% | 15% |
| Fungicides | 131 | 12% | 535 | 12% | 24% |
| Other | 83 | 8% | 230 | 5% | 36% |
| **Total** | **1,081** | **100%** | **4,500** | **100%** | **24%** |

Note: Totals may not add due to rounding.
Source: EPA estimates based on National Agricultural Chemicals Association.

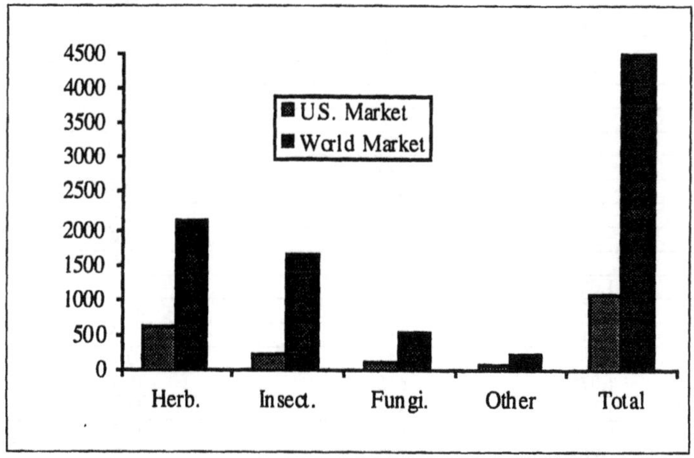


**Figure 1.1** U.S. vs. World Conventional Pesticide Sales: Volume of Active Ingredient, 1993.

Pesticide usage in the U.S. has been relatively stable at about 1.1 billion pounds of active ingredient during recent years. The agricultural share of pesticide usage (see Table 1.3) appears to have stabilized at about three-fourths of the total after increasing steadily throughout the 1960s and 1970s, primarily due to the expanded use of herbicides in crop production. Growth in the use of

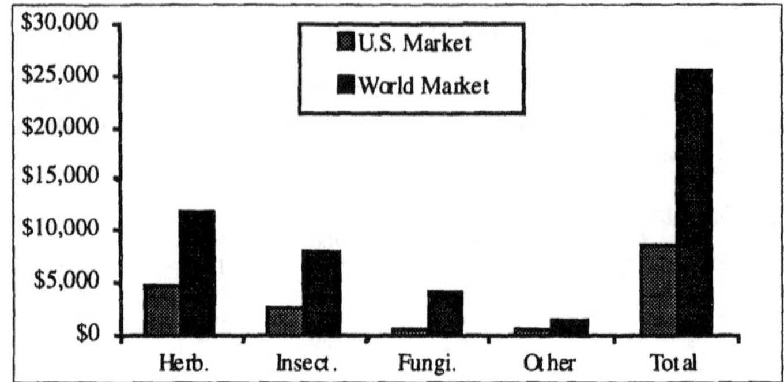


**Figure 1.2** U.S. vs. World Conventional Pesticide Sales: User Expenditures, 1993.

**Table 1.2** United States Conventional Pesticide Usage, Total and Estimated Agricultural Sector, 1964-1993.

| Year | Total U.S. Millions of lbs. | Agricultural Sector Active Ingredient | Percent of Total |
|---|---|---|---|
| 1964 | 540 | 320 | 59% |
| 1965 | 610 | 335 | 55% |
| 1966 | 680 | 350 | 51% |
| 1967 | 735 | 380 | 52% |
| 1968 | 835 | 470 | 56% |
| 1969 | 775 | 430 | 55% |
| 1970 | 740 | 430 | 58% |
| 1971 | 835 | 495 | 59% |
| 1972 | 875 | 525 | 60% |
| 1973 | 910 | 560 | 62% |
| 1974 | 950 | 590 | 62% |
| 1975 | 990 | 625 | 63% |
| 1976 | 1,030 | 660 | 64% |
| 1977 | 1,075 | 720 | 67% |
| 1978 | 1,110 | 780 | 70% |
| 1979 | 1,058 | 840 | 79% |
| 1980 | 1,075 | 846 | 79% |
| 1981 | 1,101 | 860 | 78% |
| 1982 | 1,056 | 815 | 77% |
| 1983 | 963 | 733 | 76% |
| 1984 | 1,080 | 850 | 79% |
| 1985 | 1,112 | 861 | 77% |
| 1986 | 1,096 | 820 | 75% |
| 1987 | 1,087 | 814 | 75% |
| 1988 | 1,130 | 845 | 75% |
| 1989 | 1,070 | 806 | 75% |
| 1990 | 1,086 | 834 | 77% |
| 1991 | 1,077 | 817 | 76% |
| 1992 | 1,103 | 839 | 76% |
| 1993 | 1,081 | 811 | 75% |

* Active ingredient
Note: Excludes wood preservatives and disinfectants.
Source: EPA estimates.

pesticides has been slowed by lower application rates due to the introduction of more potent pesticides, more efficient use of pesticides, and lower farm commodity prices. USDA and EPA are working together with commodity groups to develop plans to reduce use/risk of pesticides as part of a food safety initiative.

The volume of pesticides used for non-agricultural purposes in the U.S. also has been quite stable in recent years at about 275 million pounds of active ingredient (a.i.). This equals about 1.1 pounds per capita in the U.S. (average for 250 million people). Considering all usage, including agricultural, U.S. pesticide usage equals somewhat more than 4 pounds per capita (4.2 pounds in 1993).

Table 1.3 shows that in the United States there are more than 120 manufacturers of pesticides, with only 20 accounting for the bulk of production and sales. The manufacturers supply the pesticidal active ingredients (not including carrier liquids, diluting agents, and inert ingredients found in formulations) to over 2,000 pesticide formulators who mix the active and inactive ingredients to produce over 21,000 registered products. As of 1993, there were about 17,000 distributor-dealers of pesticides, approximately 40,000 pest control firms, almost a million certified private applicators (individual growers), and over 350,000 certified commercial applicators. As of 1994, there were an estimated 17,500 licensed-certified agricultural pest control advisors, of whom over 8,000 were self-employed independent consultants.

**Table 1.3** U.S. Pesticide Production, Marketing and User Sectors; Profile of Number of Units Involved, 1993/1994 Estimates (Approximate Values).

**PRODUCTION AND DISTRIBUTION**

**Basic Production**

| | |
|---|---|
| 1. Major Basic Producers | 20 |
| 2. Other Producers | 100 |
| 3. Active Ingredients Registered | 860 |
| 4. Active Ingredients with Food/Feed Tolerances | 453 |
| 5. Chemical Cases for Re-registration | |
| —Pre-FIFRA 1988 | 612 |
| —Post-FIFRA 1988 | 405 |
| 6. New Active Ingredients Registered | |
| —1992 | 11 |
| —1993 | 20 |
| 7. Total Employment | 6,000-10,000 |
| 8. Producing Establishments | 7,300 |

**Distribution and Marketing**

| | |
|---|---|
| 1. Formulators | |
| —Major national | 150-200 |
| —Other | 2,000 |

**Table 1.3** continued.

| | |
|---|---|
| 2. Distributors and Establishments | |
| —Major national | 250-350 |
| —Other | 16,900 |
| 3. Formulated Products Registered | 21,560 |
| —Federal level | 18,360 |
| —State | 3,200 |

**USER LEVEL**

**Agriculture Sector**

| | |
|---|---|
| 1. Land in Farms | 991M acres |
| 2. Harvested | 289M acres |
| 3. Total No. Farms | 2.1M |
| 4. No. of Farms Using Chemicals for: | |
| —Insect on hay crops | 554,000 |
| —Nematodes | 66.000 |
| —Diseases on crops/orchards | 129,000 |
| —Weed/grass/bush | 913,000 |
| —Defoliation/fruit thinning | 75,000 |
| (Above are 1987 census numbers) | |
| 5. No. Private Pesticide Applicators Registered | 965,700 |

**Industrial/Commercial/Government Sector**

| | |
|---|---|
| 1. No. Commercial Pest Control Firms | 35,000-40,000 |
| 2. No. Certified Commercial Applicators | 351,600 |

**Home and Garden Sector**

| | |
|---|---|
| 1. Total U.S. Households | 94M |
| 2. No. Households Using ('90) | |
| —Insecticides | 52M |
| —Fungicides | 36M |
| —Herbicides | 14M |
| —Repellents | 17M |
| —Disinfectants | 40M |
| —Any pesticides | 69M |

Source: EPA estimates.

In the United States in 1993, 75 percent of the pesticides sold are used in agriculture. Government and industry uses 18 percent, and home and garden consumption accounts for the remaining 7 percent. Industrial and commercial users consist of pest control operators, turf and sod producers, floral and scrub nurseries, railroads, highways, utility rights-of-way, and industrial plant site landscape management (see Table 1.4 and Figure 3).

Pesticides are regulated by the United States Environmental Protection Agency (EPA). The number of active ingredients registered and in production has declined in the last ten years, from over 1,200 active ingredients to 860 in 1993. Of these only 200 are considered major products and manufactured in quantity (see Table 1.5). The table below shows a breakdown of the types of pesticides and numbers in production according to the latest available statistics.

**Table 1.4** Volume of Conventional Pesticide Active Ingredients Used in the U.S. by Class and Sector (Millions of lbs).

| | Herbicides | | Insecticides | | Fungicides | | Other | | Total | |
|---|---|---|---|---|---|---|---|---|---|---|
| Sector | lbs. | % | lbs. | % | lbs. | % | lbs | % | lbs. | % |
| 1993 | | | | | | | | | | |
| Agriculture | 481 | 78 | 171 | 69 | 84 | 64 | 75 | 90 | 811 | 75 |
| Ind./Comm./ Govt. | 112 | 18 | 44 | 18 | 36 | 27 | 5 | 6 | 197 | 18 |
| Home and Garden | 27 | 4 | 32 | 13 | 11 | 8 | 3 | 4 | 73 | 7 |
| **Total** | **620** | **100** | **247** | **100** | **131** | **100** | **83** | **100** | **1,081** | **100** |

Note: Totals may not add due to rounding.
Source: EPA estimates based on National Agricultural Chemicals Association.

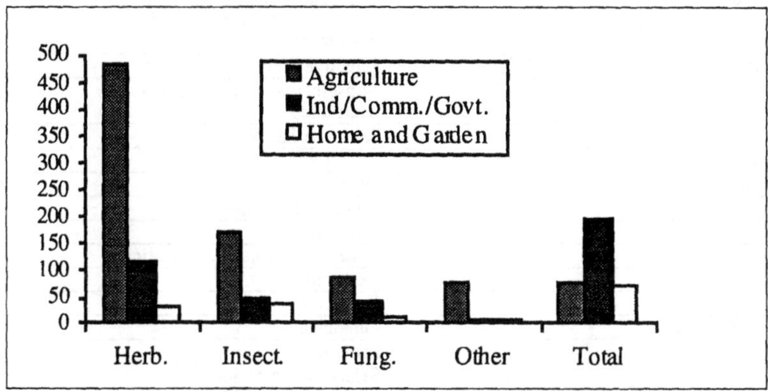


**Figure 1.3** U.S. Volume for Conventional Pesticides, 1993 Estimates.

**Table 1.5** Breakdown of Types of Pesticides in Production.

| Type | Number |
|---|---|
| Disinfectants | 200 |
| Fungicides and Nematicides | 165 |
| Herbicides | 240 |
| Insecticides | 215 |
| Rodenticides | 40 |

The most heavily used pesticides in the agricultural sector in 1993 are listed in Table 1.6. Of these, 17 are herbicides, 3 are insecticides and 5 are fungicides. Table 1.7 lists the most commonly used pesticides in the non-agricultural sectors.

**Table 1.6** Quantities of Pesticides Most Commonly Used in U.S. Agricultural Crop Production (Approximate Quantities, 1993).

| Pesticide | Usage in Millions of lbs Active Ingredient | Intended Use |
|---|---|---|
| Atrazine | 70-75 | Selective Herbicide |
| Metolachlor | 60-65 | Selective Herbicide |
| Sulfur | 45-50 | Fungicide, Acaricide |
| Alachlor | 45-50 | Preemergence Herbicide |
| Methyl Bromide | 30-35 | Fumigant |
| Cyanazine | 30-35 | Selective Herbicide |
| Dichloropropene | 30-35 | Nematicide, Soil Fumigant |
| 2,4-D | 25-30 | Postemergence Herbicide |
| Metam Sodium | 25-30 | Fungicide, Herbicide, Insecticide, Nematicide, Soil Fumigant |
| Trifluralin | 20-25 | Selective Preemergence Herbicide |
| Petroleum Oil | 20-25 | Dormant Spray, Summer Oil Parasiticides, Carrier Fluid, Herbicide, Adjuvants |
| Pendimethalin | 20-25 | Selective Herbicide |
| Glyphosate | 15-20 | Nonselective, Preemergence Herbicide |
| EPTC | 10-15 | Selective Herbicide |
| Chlorpyrifos | 10-15 | Insecticide |
| Chlorothalonil | 10-15 | Fungicide |
| Propanil | 7-12 | Contact Herbicide |
| Dicamba | 6-10 | Herbicide |
| Terbufos | 5-8 | Systemic Herbicide, Nematicide |
| Bentazone | 4-7 | Herbicide |
| Mancozeb | 4-7 | Fungicide |
| Copper Hydroxide | 4-7 | Fungicide |
| Parathion | 4-7 | Insecticide |
| Simazine | 3-6 | Selective Herbicide |
| Butylate | 3-6 | Selective Herbicide |

Source: EPA estimates based on a variety of government sources.

**Table 1.7** Quantities of Pesticides Most Commonly Used in Non-Agricultural Sectors of the U.S. (Approximate Quantities, 1993).

| Pesticide | Usage in Millions of lbs Active Ingredient | Intended Use |
|---|---|---|
| 2,4-D | 12-15 | Postemergence Herbicide |
| Chlorpyrifos | 9-12 | Insecticide |
| Diazinon | 8-10 | Insecticide, Nematicide |
| Glyphosate | 4-6 | Nonselective, Preemergence Herbicide |
| Malathion | 4-6 | Insecticide |
| Dicamba | 3-5 | Herbicide |
| Diuron | 3-5 | Herbicide |
| Naled | 3-5 | Insecticide, Acaricide (non-systemic) |
| MCPP | 3-5 | Herbicide |
| Carbaryl | 2-4 | Broad spectrum insecticide |

## B. FUTURE TRENDS

Rapid advances in the field of biotechnology will lead to novel microbial products and new crop varieties. These new microbial products will become increasingly important crop protection agents and should gradually replace agrochemicals. However, these new methods are not likely to take more than 5% of the crop protection market by the year 2000. In the next 20 years, rapid progress is expected and numerous microbial products and resistant crop varieties are expected to reach the marketplace. The current expectations are that the long-term growth potential for the agrochemicals industry will be approximately 2% per annum in real terms, but higher (5%) in the less developed countries.

Agrochemicals are becoming more potent in terms of the dose required (grams per hectare rather than kilograms per hectare) to control the pest. This efficiency of modern agrochemicals in controlling their target organisms and the resultant increase in crop yields is well illustrated by two examples. The yield of cotton in the United States, after treatment with cypermethrin against cotton bollworm, was 402 kg/ha, whereas the untreated crop yielded 67 kg/ha. The yield of wheat which had been treated with the herbicide diclofop-methyl against infestation by wild oats was 366 g/m$^2$, whereas the yield was 143 g/m$^2$ for untreated wheat.

The greater efficiency of modern agricultural practice liberates land that can be used for recreational purposes; in the United States in 1983, sufficient food was produced from 117 ha, whereas in 1950 the production of the same quantity of food required 243 ha.

In countries like the United States, the development of a pesticide from initial discovery in the laboratory to marketing takes at least eight years. The costs of development have substantially increased; in 1964 the cost was $2.9 million, but by 1987 it had risen to $50 million due to increasingly stringent environmental and toxicological tests required by the EPA.

It is also becoming increasingly difficult to discover a new agrochemical product with significant advantages over existing products. Consequently, the number of compounds which need to be screened to obtain one marketable product has substantially increased, from one in 3,600 in 1964, to one in 16,000 in 1985, and in 1989 was estimated to be one in 20,000. There has been an overall decline in the profitability of the agrochemical industry, from 11.5% (1981) to 7.9% (1986); this is illustrated by the fact that all 10 major agrochemicals used in the United States in 1987 were introduced prior to 1976, namely glyphosate (1972), alachlor (1966), metribuzin (1971), carbaryl (1956), chlorpyrifos (1965), carbofuran (1967), chlorothalonil (1963), trifluralin (1963), bentazone (1975), and dicamba (1965).

A new agrochemical will be developed today only if it is effective in protecting one or more of the following major world crops: corn, rice, soyabeans, cotton, wheat, or oilseed rape. Research and development is now coordinated very closely with marketing to ascertain if this activity will ensure a sufficiently large potential market to justify the development costs.

Agrochemicals today are a very high risk business because the substantial sum of approximately $50 million spent on development during the first eight years must be recovered quickly, since the life of the patent expires after 20 years. Then other companies who did not bear the high development costs can manufacture the product and sell it, often at a lower price and at a higher profit.

The agrochemicals industry today is much more complex. In order to protect the environment and consumers from dangerous agrochemicals, the standards demanded for approval and registration of products have become much more rigorous. To satisfy the criteria may involve the company's expenditure of some $5 million.

Worldwide, even in developed countries, many of the pesticides discovered in the 1950s are still extensively used. There is urgent need for the introduction of more selective agrochemicals, particularly with different modes of action to combat the growing problems presented by resistant fungi and insects. There is a real danger that excessive emphasis on potential environmental hazards, especially in the United States, may result in the elimination of valuable agrochemicals and stifle the development of promising compounds due to overregulation. Such factors have caused a massive increase in the development costs of a new pesticide and for a time caused a reduction in the number of significant new products coming onto the market. The maximum number of new compounds were introduced in the 1950s and 1960s with some 18 per annum, but declined to six in the 1970s.

Random screening has become less successful; consequently there has been more research, with greater resources being concentrated on areas of chemistry of proven biological activity. This approach, coupled with increasing use of computer graphics to provide a three-dimensional model of the active sites, has been quite successful, and the number of new compounds coming onto the market has increased.

This approach has inevitably led to a clustering of new agrochemicals in certain areas, such as the triazole fungicides and synthetic pyrethroids which were launched in 1976 and 1977, respectively. In 1988 there were approximately 14 and 17 members of these groups on the market.

## REFERENCES

Aspelin, A., *Pesticide Industry Sales and Usage: 1992 and 1993 Market Estimates,* U.S. Environmental Protection Agency, 1994.

Carson, Rachel, *The Silent Spring*, Fawcett, Greenwich, 1962.

*Farm Chemicals Handbook'96,* Meister Publishing, Ohio, 1996.

Fest, C. and K. J. Schmidt, *The Chemistry of Organophosphorus Insecticides,* Springer-Verlag, Berlin, 1973.

Green, M. B., Hartley, G. S., and T. F. West, *Chemicals for Crop Improvement and Pest Management,* 3rd ed., Pergamon Press, Oxford, 1987.

Hassall, K. A., *The Biochemistry and Uses of Pesticides,* Macmillan, London, 1990.

McCallan, S. E. A., 'History of fungicides', in *Fungicides, An Advanced Treatise* (Ed., Torgeson, D. C.), Academic Press, New York, 1967.

McMillen, W., *Bugs or People,* Appleton-Century, New York, 1965.

Wain, R. L. and G. A. Carter, 'Historical aspects', in *Systemic Fungicides* (Ed., Marsh, R. W.), 2nd ed., Longman, London, 1977.

# CHAPTER 2

# REGULATIONS

## I. REGULATIONS

### A. INSECTICIDE ACT

The government regulation of pesticides started with the Insecticide Act of 1910, promulgated to prevent consumer fraud and to protect the legitimate manufacture of pesticides. It made it unlawful to manufacture any insecticide or fungicide that was "adulterated or misbranded." The Insecticide Act was essentially a labeling statute; it did not require registration or establish significant safety standards for pesticides. This act remained federal law for 37 years. At that time agricultural chemicals consisted of only insecticides and fungicides. These products contained inorganic compounds as their active ingredients, principally arsenicals such as lead arsenate, Paris green, a derivative of copper acetoarsenite, copper, mercury, and zinc, to name a few.

The Pure Food Law of 1906 was amended in 1938 to include pesticides on foods, primarily the arsenicals, such as lead arsenate and Paris green. It also required the adding of color to white insecticides, including sodium fluoride and lead arsenate, to prevent their accidental use as flour or other look-alike cooking materials. This was the first federal effort toward protecting the consumer from pesticide-contaminated food by providing tolerances for pesticide residues, namely arsenic and lead, in foods where these materials were necessary for the production of a food supply.

By 1950, synthetic organic pesticides had emerged as the effective new products of choice in the marketplace. The organochlorine insecticides such as aldrin, dieldrin, chlordane, heptachlor, lindane, endrin, toxaphene, and DDT were highly effective in controlling insects at very low dosages.

### B. FEDERAL INSECTICIDE, FUNGICIDE, AND RODENTICIDE ACT (FIFRA)

In 1947, Congress passed the Federal Insecticide, Fungicide, and Rodenticide Act (FIFRA). Unlike modern environmental statutes, the impetus for passage came not from environmentalists, but primarily from users and manufacturers. FIFRA required that pesticides be registered by the Secretary of Agriculture before their sale or distribution in interstate or foreign commerce. However, the intrastate sales and distribution were left to the states to the regulate. In addition, FIFRA required specific label and package information to be included with each pesticide sold and also expanded the regulation of the pesticide law to cover rodenticides and herbicides that were not covered under the Insecticide Act of 1910.

Despite the enhanced powers of the U.S. Department Agriculture (USDA), FIFRA (1947) remained fundamentally a labeling statute.

In 1964, FIFRA was amended to authorize the Secretary of Agriculture to deny or cancel a registration. In addition, the regulation also authorized the Secretary to suspend a registration immediately if necessary to prevent an imminent hazard to the public.

In 1954, Congress enacted the Miller Amendment to the Federal Food, Drug, and Cosmetic Act (FFDCA), which required that a maximum acceptable level (tolerance) be established for pesticide residues in foods and animal feed (for example, 10.0 ppm carbaryl in lettuce, or 1.0 ppm ethyl parathion on string beans). In 1958, congress enacted the Food Additives Amendment to FFDCA, which regulated pesticide residues in processed foods.

On December 2, 1970, President Nixon created the EPA. Under the President's reorganization plan the EPA assumed the pesticide regulatory functions of the USDA and the Food and Drug Administration (FDA) under FIFRA and the FFDCA.

In 1972, FIFRA was revised. The Federal Environmental Pesticide Control Act (FEPCA), commonly referred to as FIFRA Amended 1972, was subsequently revised in 1975, 1978, 1980, 1984, and 1988. The new FIFRA regulates the use of pesticides to protect people and the environment and extends federal pesticide regulation to all pesticides, including those distributed or used within a single state. The eight basic provisions of the new FIFRA are as follows:

1. All pesticides must be registered by the EPA.
2. For a product to be registered, the manufacturer is required to provide scientific evidence that the product, when used as directed, (a) will effectively control the pests listed on the label, (b) will not injure humans, crops, livestock, wildlife, or damage the environment, and (c) will not result in illegal residues in food or feed.
3. All pesticides will be classified into general use or restricted use categories.

4. Restricted-use pesticides must be applied by a certified applicator.
5. Pesticide-producing establishments must be registered and inspected by the EPA.
6. Use of any pesticide inconsistent with the label is prohibited.
7. States may register pesticides on a limited basis for local needs.
8. Violations can result in heavy fines and/or imprisonment.

As intended by Congress, FIFRA has four main thrusts:

1. To evaluate the risks posed by pesticides by requiring stringent screening and testing of each pesticide and eventual registration with the EPA before being offered for sale.
2. To classify pesticides for specific uses and to certify pesticide applicators and thus control exposure to humans and the environment.
3. To suspend, cancel, or restrict pesticides that pose a risk to the environment.
4. To enforce these requirements through inspections, labeling notices, and regulation by state agencies.

Most of the current efforts to amend FIFRA concentrate on the tolerance and food safety provisions of FIFRA and the FFDCA. For years the EPA has interpreted the Delaney Clause of the FFDCA to impose a negligible risk standard for pesticide residues on food, whereas others, including influential environmental groups, read the section to allow zero risk. EPA's interpretation was struck down by the Ninth Circuit in Les v. Reilly. The EPA asked the Supreme Court to review that decision. Their petition was denied on March 8, 1993. Legislation has been proposed on several occasions to change the outcome of Les v. Reilly; however, none to date has been adopted.

In addition, EPA has announced an aggressive regulatory agenda for pesticides. It has proposed changes in regulations covering tolerances, registration, data requirements, worker protection standards, recordkeeping requirements, and labeling requirements. In June of 1993, the Clinton administration announced a new policy designed to cut pesticide usage. On July 21, 1993, the EPA followed up with its "Voluntary Reduced Risk Pesticides Incentive."

Whatever changes are made to FIFRA and the regulations issued under it in the next decade, it appears certain that the era of benign neglect in the pesticide industry is over.

### Restricted Use Pesticides (RUPs)

Certain pesticides have been classified as "restricted use" by the EPA. A RUP can be applied only by a certified applicator or under his or her direct supervision.

Some states have their own restricted use list in addition to the EPA list. Some states also have additional requirements that must be satisfied before a permit can be issued for the use of certain pesticides. A pesticide applicator contemplating the use of RUPs must not only be certified to use those pesticides, but must be aware of any state laws that may be in effect regarding the use and application of certain pesticides.

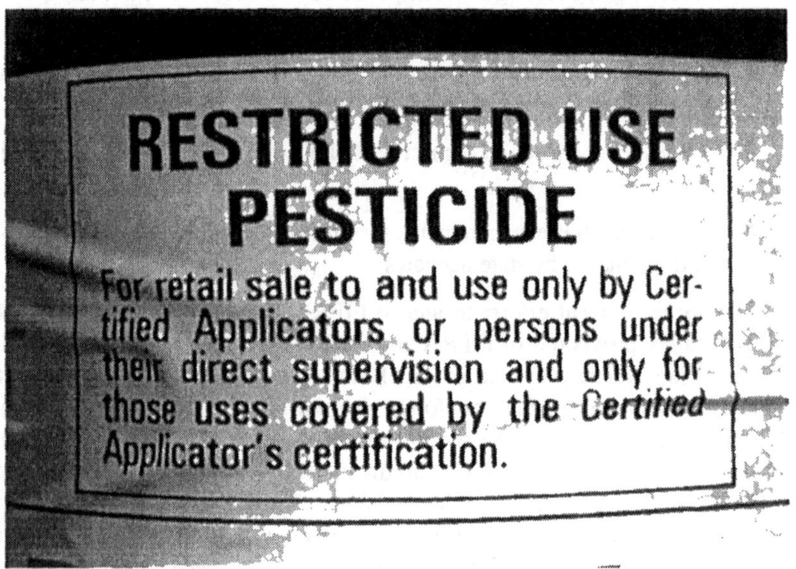


**Figure 2.1** Portion of a labeled container showing the RUP notification.

## C. HAZARD COMMUNICATION STANDARD (HazCom)

The Occupational Safety and Health Administration's (OSHA) Hazard Communication Standard (29 CFR 1910.1200) of 1986 requires that employers at manufacturing facilities, and any other work place where toxic chemicals are handled or processed, be provided Material Safety Data Sheets (MSDS) for all hazardous substances used in their facility, and to make these MSDSs available to all employees potentially exposed to these hazards.

All employers, including those in agriculture, are covered by the Hazard Communication Standard as of May 1988. Every farm, ranch, or business with one or more employees is affected by the standard, which formerly covered only the manufacturing industry. The purpose of this regulation is to ensure that employers evaluate the hazards of all chemicals in the workplace and transmit this information to every employee.

The term "hazardous" refers to any chemical which can affect an employee's health. Effects can range from mildly irritating to potentially highly carcinogenic. If employees may be potentially exposed to hazardous chemicals, the employer is required to develop a written hazard communication program that includes the following:

1. A MSDS for each known hazardous chemical used in the farm or business, along with container labeling or other forms of warning.

2. A written plan which outlines the method that will be used to inform employees, workers, and outside contractors of hazardous chemicals to which they may be exposed while working on the site.

3. A program which explains the dangers of the hazardous chemicals in the work area at initial work assignment and provides information about any new hazard introduced into this work area.

Employers are required to make the written hazard communication program available upon request to employees or their designated representatives and/or officials of OSHA.

The standard requires an employer to identify each hazardous chemical with a label, tag, or other mark clearly identifying the material. All warning labels shall be clearly legible and provided in English, although the employer may in addition voluntarily include the warning in other languages.

An employer may use signs, placards, or other written materials in lieu of individual container labels, as long as the method identifies the containers to which it applies and conveys the necessary labeling information. These sources of information must be readily available to employees.

The MSDS for each chemical in use must include the following information:

- chemical and common name of single hazardous substances
- chemical and common names of any health hazardous ingredient exceeding 1% of a mixed chemical
- chemical and common names of known carcinogens of 0.1% concentration or greater
- chemical and common name of all ingredients which are a physical hazard when present, including vapor pressure, flash point, fire and explosion potential, and reactivity

- exposure signs and symptoms of the health effects of hazardous chemicals, and medical conditions which could be aggravated by exposure to the chemical and potential route(s) of entry into the body
- OSHA permissible exposure limits (PELs)
- all known safe handling precautions
- known work practices, personal protective equipment, and engineering controls for handling the hazardous chemical
- emergency and first aid procedures
- preparation date of the data sheets as well as dates of changes or additions
- name, address, and phone number of the company or persons who prepared the data sheet.

The employee information and training program shall include at least the following topics:

1. Explanation of the Hazard Communication Standard.
2. Location and availability of the written hazard communication program and MSDS.
3. Methods and monitoring equipment used for detecting presence or release of hazardous chemicals in the work area.
4. Measures an employee can use to protect himself from the hazards.

Farmers, ranchers, and businesses are subject to routine general inspections by OSHA. The first time an inspection shows a business is not complying with the Hazard Communication Standard, OSHA can issue a noncompliance citation and a penalty of up to $1,000. Usually, the business is allowed 15 to 30 days to correct the problem. A business that does not correct the noncompliance(s) may be subject to additional penalties of up to $10,000 per violation.

## D. THE ENDANGERED SPECIES ACT

The Federal Endangered Species Act (ESA) was first passed in 1973, amended in 1978, 1979, and 1982, and reauthorized in 1985. The act requires that all federal agencies insure that their actions will not jeopardize endangered or threatened animal and plant species and their habitats.

Under FIFRA, the EPA must take steps to prevent such harm to these species through pesticide use. In this effort, EPA should consult with the

U.S. Fish and Wildlife Service to determine whether use of a certain pesticide will endanger these species.

After proper notification the EPA can require that certain pesticide products be relabeled to prohibit their use in the range of the endangered species.

As presently envisioned, the new provisions would require pesticide labels to list counties where users would need to consult special county bulletins concerning endangered species before applying that product. Users would be obliged to comply with these bulletins for each of the counties in which they operate.

These bulletins would be available from county extension offices, farm supply dealers, and other groups. Each would include:

1. A list of affected species.
2. A county map identifying species habitats, using commonly recognized boundaries such as roads, power lines, and bodies of water.
3. A list of affected pesticides, identified by active ingredient. The list of endangered species will be determined by the Fish and Wildlife Service of the Department of Interior, which will also determine the habitats of these species and which pesticide uses place them in jeopardy.

The EPA anticipates using a "clustered approach" to pesticide selection, so that all products with common use patterns would be classified together. In this way EPA hopes to reduce the unfair economic impact of selective labeling decisions and to avoid case-by-case consultation with the Fish and Wildlife's Office of Endangered Species.

The compliance date for use on range and/or pastureland, corn, wheat, soybeans, sorghum, oats, barley, rye, and/or cotton was originally set for February 1, 1988. After EPA delayed partial implementation of the program in 1989 due to dissatisfaction of the methods of implementation by most states, the American Farm Bureau Federation, and even members of Congress, compliance was finally initiated in 1992. More than 60% of all agricultural pesticides and 1,000 U.S. counties will be affected by the program's provisions.

Unlike the FEPCA, the Endangered Species Act does not require the EPA to weigh risks and benefits in prohibiting a product's use. The economic cost of protecting a species is not a consideration under the act.

## E. SUPERFUND AMENDMENTS AND REAUTHORIZATION ACT (1986)

The Superfund Amendments and Reauthorization Act (SARA), Title III, the Community Right-to-Know Act, has four major reporting requirements for facilities that handle, store, transport, and/or dispose of hazardous materials: emergency planning notification, emergency release notification, community right-to-know, and toxic chemical release forms. Each reporting provision has different requirements for the chemicals and facilities covered.

1. Emergency Planning Notification

This law was designed to identify all facilities (including farms) that have in inventory any of more than 400 extremely hazardous substances present in excess of its threshold planning quantity (TPQ). The TPQ is based on the amount of any one of the hazardous substances that could, upon release, present human health hazards that warrant emergency planning. The TPQ emergency planning requirement is based on these public health concerns rather than the type of facility where the chemical might be located. The type of facility and degree of hazard presented at any particular site are relevant factors considered by the local emergency planning committees (LEPCs).

If one or more of the extremely hazardous substances is present in excess of its TPQ, the facility must notify its state emergency response commission, which will in turn notify the appropriate local planning committee and other relevant local agencies and departments.

2. Emergency Release Notification

Releases (spills) of any of the extremely hazardous chemicals including pesticides in excess of their reportable quantity (RQ) must be reported to the National Response Center, state and local emergency planning committees, and other agencies requiring notification. The reportable quantity of a substance can be found in 40 CFR 355.40 Appendix A or B. (The CERCLA list of hazardous substances is found in 40 CFR 302.4 and has similar reporting requirements.)

The release (other than spills or other accidents) of a pesticide registered under FIFRA when used generally in accordance with its intended purpose (during routine agricultural applications according to approved product label instructions) is exempted from this reporting requirement.

3. Community Right-to-Know

Community right-to-know reporting pertains to the material safety data sheets (MSDS) for hazardous chemicals from facilities (mainly manufacturers and importers) that are required to report and make MSDSs available to wholesale and retail outlets.

4. Toxic Chemical Release Forms

The chemical release reporting is limited to facilities that are manufacturing or importing hazardous chemicals. A complete listing of hazardous pesticides with their TPQs is available from the county offices of the USDA Natural Resources Conservation Service. For information on any new changes to hazardous chemicals lists, the EPA maintains a chemical emergency preparedness program hotline, which is (800) 535-0202.

## F. AGRICULTURE CHEMICALS IN GROUNDWATER: PESTICIDE STRATEGY

In October 1991, the EPA issued its final version of the Agriculture Chemicals in Groundwater: Pesticide Strategy. The strategy provides a framework for EPA's approach to generic issues posed by pesticides in groundwater. The strategy addresses three major issue areas:

1. Environmental Goals

The EPA goal is to manage pesticides to protect the groundwater resource. This would include the use of maximum contamination levels (MCLs), the enforceable drinking water standards, established under the Safe Drinking Water Act (SDWA), as reference points to determine the unacceptable levels of pesticides in groundwater sources.

2. Prevention

The EPA's strategy includes a national registration of pesticides and state-directed management by monitoring and pesticide use restrictions based on local groundwater conditions.

3. Response Actions

The response actions would be coordinated jointly by the federal and state governments. Under the specific state management plans, the state would take the necessary action to prevent further contamination when unacceptable levels of contamination in groundwater are reported. Under this strategy the states and the EPA would coordinate closely on the enforcement of FIFRA, SDWA, and CERCLA to identify parties responsible for groundwater contamination as a result of the misuse, spills, leaks, and/or illegal disposal of pesticides.

## G. SAFE DRINKING WATER ACT (1974)

The Safe Drinking Water Act (SDWA) is designed to ensure that public water systems provide water meeting the minimum national standards for protection of public health. The act mandates establishment of uniform federal standards for drinking water quality, and sets up a system to regulate underground injection of wastes and other substances that could contaminate groundwater sources. (Surface water is protected under the Clean Water Act.)

The protection afforded under this act is based on a system of drinking water standards set by the EPA. Each standard limits the amount of a specific contaminant that may be in drinking water. Primary standards—maximum contamination levels (MCLs), which apply to substances that may have an adverse effect on health, are enforced by the states and must be complied with. Secondary standards, which provide guidelines on substances that affect color, taste, smell, and other physical characteristics of water, are advisory. Both types of standards are established by the EPA but enforced by the states.

The primary standards (MCLs) established included fluoride, arsenic, a variety of pesticides, mercury, lead, nitrates, and several additional inorganic and organic chemicals.

The intent of the SDWA was to fill the gap left by the enactment of the Clean Water Act (CWA) passed two years earlier. While the CWA was passed to clean up and protect streams and other surface waters, SDWA was intended to protect undergroundwater sources.

The Safe Drinking Water Act Amendments of 1996 (PL 104-182) established a new charter for the nation's public water system, states, and the EPA in protecting the safety of drinking water. The amendments include, among other things, new prevention approaches, improved consumer information, changes to improve the regulatory program, and funding for states and local water systems.

## H. CLEAN WATER ACT (1972)

The Clean Water Act (CWA) provides the legislative vehicle for regulating the discharge of non-toxic and toxic pollutants into surface waters by municipal sources, industrial sources, and other specific and non-specific sources.

The act's ultimate goal is to eliminate all discharges of pollutants into surface waters. Its interim goal is to make all waters in the United States usable for fishing and swimming.

In 1987, the CWA was extensively amended. The amendments also set up programs to reduce polluted runoff from "non-point" sources, such as city streets, farm land, and mining sites.

## I. RESOURCE CONSERVATION AND RECOVERY ACT (1976)

The Resource Conservation and Recovery Act (RCRA) regulates the generation, treatment, storage, transportation, and disposal of hazardous wastes, including pesticide wastes that may create a hazard. Pesticide-containing wastes that are considered hazardous wastes under RCRA are subject to extensive regulatory requirements governing storage, transportation, treatment, and disposal.

Solid wastes are defined as hazardous under RCRA when they are included in one of the following lists (commonly referred to as listed wastes):

- F-List - Hazardous wastes from non-specific sources (40 CFR 261.31). Five wastes from pesticide manufacturing or use and three from wood preserving processes are included.
- K-List - Hazardous wastes from specific sources (40 CFR 261.32). One waste from wood preservation and 22 from pesticide manufacture are listed.
- P-List - Acutely toxic commercial chemical products (40 CFR 261.33[e]).
- U-List - Toxic and other commercial products (40 CFR 261.33[f]).

Both the P-List and the U-List contain several commercial pesticides.

Solid wastes are also hazardous when they meet one of the following defined characteristics (40 CFR 261.21 through 261.24):

- Ignitability (waste code D001) - Based on the waste's low flash point or other physical characteristics, this material poses a fire hazard.
- Corrosivity (waste code D002) - Liquid wastes that have a pH $\leq 2$ or $\geq 12.5$, or corrodes steel at a rate greater than 0.25 inches per year. These wastes, therefore, have the ability to dissolve steel drums, tanks, and other containers and care should be taken when containerizing or shipping these wastes.
- Reactivity (waste code D003) - Wastes that are unstable and can readily undergo violent change. These wastes can react violently with water to generate toxic gases, vapors, or fumes.
- Toxicity (waste code D004-D043) - Liquid wastes or extracts from waste solids that fail the Toxicity Characteristics Leaching Procedure (TCLP) analytical test because they contain certain designated metals, pesticides, or organic chemicals at concentrations equal to, or in excess of, specified regulatory limits.

The following are examples of pesticide wastes which can be regulated under RCRA:

- Discarded, unused pesticides that are listed or meet one or more of the characteristics of hazardous waste.
- Discarded residue or rinsate from drums, tanks, or other containers, depending on the RCRA classification of the pesticide/rinsate.
- Non-empty pesticide containers which held a listed pesticide or held a pesticide exhibiting a hazardous waste characteristic.

- Pesticide residue consisting of contaminated soil, water, or other debris resulting from the cleanup of a spilled pesticide.

## J. COMPREHENSIVE ENVIRONMENTAL RESPONSE, COMPENSATION, AND RECOVERY ACT (1980)

The Comprehensive Environmental Response, Compensation, and Recovery Act (CERCLA), commonly called the Superfund Law, requires the cleanup of releases of hazardous substances in air, water, groundwater, and on land. Both new spills and leaking or abandoned waste sites are covered.

The act establishes a trust fund (Superfund) to finance government responses to releases or threats of releases of hazardous substances. However, if groundwater contamination results from normal application of pesticides, the law does not allow the agency to recover costs from pesticide applicators or private users.

## K. TOXIC SUBSTANCES CONTROL ACT (TSCA)

The Toxic Substances Control Act (TSCA) was enacted in 1976 to:

- Ensure that adequate data are developed on the effects of chemical substances on health and the environment.
- Regulate the production, distribution, use, and disposal of those chemicals deemed to present an unreasonable risk of injury to health or the environment.
- Provide authority to EPA to take action on those chemicals which present an "imminent hazard."

The regulations promulgated to implement TSCA are found in 40 CFR 700-799, and are administered by the Office of Toxic Substances (OTS).

Pesticides are specifically excluded from TSCA's definition of a chemical substance, when manufactured, processed, or distributed as a pesticide. EPA generally considers a product to be a pesticide once it requires a FIFRA (Section 5) Experimental Use Permit. It is important to note that inert and raw materials are not considered pesticides until they become part of a pesticide product. Intermediates used in the manufacture of a pesticide are not excluded from regulation under TSCA, unless they are pesticides themselves and are being used in a mixture for their pesticidal properties, or are not isolated during the manufacturing process.

While pesticides are exempted from TSCA, fertilizers and other substances used in the growing of crops are not exempt. Thus, any chemical to be used for such purposes must be listed on the TSCA Inventory.

As mentioned above, a product becomes a pesticide once it has received an Experimental Use Permit. This interpretation means that chemicals that are being tested in the laboratory or in small field trials are regulated by TSCA. However, current OTS policy is to exempt such activities from submission of risk information, notification, and recordkeeping, provided the requirements in 40 CFR 720.36 are met.

Under Section 8(b) of TSCA, EPA is responsible for compiling and publishing an Inventory of Existing Chemical Substances. No one may manufacture or import a "new" chemical unless a Premanufacture Notification (PMN) is submitted to EPA at least 90 days before manufacture or import commences. During this 90-day period EPA will review data on the chemical to determine whether specific requirements must be established for its handling, use, or disposal. If EPA does not take any specific action with regard to the chemical by the end of the review period, manufacture or import may begin. A Notice of Commencement of Manufactufacture or Import (NOC) must be filed with EPA no later than 30 days after either occurs. When the NOC is received, the chemical is added to the inventory.

As part of its premanufacture review, EPA can use Section 5(e) to require the submitter to develop data and to obtain hazards which the "new" chemical may pose, and can impose controls or restrictions on its use. These requirements apply only to the company which submits the PMN. Once a chemical is added to the inventory, one may manufacture or use it without notifying EPA, without the same restrictions or controls. However, EPA can use its authority under Section 5 of TSCA to issue Significant New Use Rules (SNURS) which extend the limitations in Section 5(e) orders to other manufacturers, importers, and processors. This ensures that everyone is treated in essentially the same manner, and that the original PMN submitter is not put at a disadvantage compared to subsequent manufacturers, importers, and processors.

## L. THE FOOD, AGRICULTURE, CONSERVATION AND TRADE ACT (Commonly referred to as the 1990 Farm Bill)

Since the Agricultural Adjustment Acts of the 1930s were originally adopted, they have gradually become more of environmental protection statutes than price supports for producers. With these environmental initiatives, many of the Farm Bill's provisions directly impact the use of pesticides and fertilizers on the farm. Title XIV, XVI, and XXI covering Conservation, Research, and Organic Certification, respectively, impact agrochemical use directly.

For the first time, farm legislation requires direct action by dealer-applicators. The 1990 Farm Bill requires certified applicators, including farmers, to maintain records on the use of restricted pesticides. These records, which must be maintained for two years, can be requested by federal and state agencies, as well as health care personnel. Fines of $500 to $1000 for each violation are authorized. Records are to be kept confidential.

## 1. Title XIV, Subtitle H - Pesticides

### a. *Farm Bill Pesticide Recordkeeping*

The pesticide recordkeeping program of the 1990 Farm Bill became final on April 9, 1993. The rule affects both private and commercial applicators who use pesticides classified for restricted use under the Federal Insecticide, Fungicide, and Rodenticide Act (FIFRA). The regulations became effective on May 10, 1993.

Unless records are currently prescribed by the state (retail dealers who commercially apply restricted use pesticides (RUPS) may use the records they already keep under FIFRA to comply with the Farm Bill recordkeeping requirements and distribute these to their customers), the applicator shall maintain the following data elements for each RUP application:

1. The brand or product name and the EPA registration number of the RUP that was applied.
2. The total amount of the RUP applied. The total amount refers to the total quantity of pesticide product used, with each RUP being listed separately. This amount does not refer to the percent of active ingredient, nor does it include the amount of water used as a carrying agent.
3. The location of the application, size of area treated, and the crop, commodity, stored product, or site to which an RUP was applied.
   - Location of the pesticide application shall be recorded as actual location where the application of RUP was made.
   - Size of area treated should reflect label language, which is provided in the label's direction of use section. Examples of "size of area treated" are acres for field crops, linear feet for fence-rows, square feet for greenhouses and nurseries, or other applicable designations. When recording size of area treated for livestock and poultry, enter the number of animals treated.
   - Crop, commodity, stored product, or site shall include general references such as corn, cotton, or wheat, and not specific scientific or variety names.
4. The month, day, and year when the RUP application occurred.
5. The name and certification number of the certified applicator who applied or who supervised the application of the RUP.

The law stipulates that violations of federal pesticide recordkeeping requirements will require the imposition of a fine not to exceed $500 for the first violation and a fine of not less than $1000 for the second violation unless it is determined that a good faith effort to comply was made.

*b. Reduction of Waiver of Fees for Pesticides Registered for Minor Agricultural Uses*

The 1990 Farm Bill also amended FIFRA to allow the administrator of the Environmental Protection Agency (EPA) to reduce or waive the payment of fees for a pesticide that is registered for a minor agricultural use if it is determined that the fee would significantly reduce the availability of the pesticide for the use.

*c. Voluntary Cancellation*

The Farm Bill also amended FIFRA to allow a registrant to request that a pesticide registration be canceled or amended to terminate one or more uses. In the case of a minor use pesticide, if the administrator determines that the cancellation or termination of uses would adversely affect the availability of the pesticide for use, 90 days must be given for affected parties to comment on the proposal. During the 90-day comment period, the registrant may enter into a transfer of registration agreement with a third party that would allow the continued registration of the minor use.

*d. Pest Control*

This policy establishes a study to allow USDA and EPA to identify available methods of pest control and pest control problems. It also directs the secretary to develop Integrated Pest Management (IPM) methods.

*e. Inter-Regional Research Project Number 4 (IR-4) Program*

The IR-4 program was established to assist in the collection of residue and efficacy data in support of the registration and reregistration of minor use pesticides and tolerances for residues under FIFRA.

*f. Water Policy With Respect to Agrochemicals*

This policy gives the USDA the principal responsibility and accountability for the development and delivery of educational programs, technical assistance, and research programs for the users and dealers of agrochemicals. It also gives USDA the responsibility to ensure that the use, storage, and disposal of agrochemicals by users is prudent, economical, and environmentally sound, and that users, dealers, and the general public understand the implications of their actions and potential effects on water. However, the USDA responsibility does not affect the EPA's authority under FIFRA.

### 2. Title XVI - Research

The 1990 Farm Bill establishes several research programs aimed at sustainable agricultural research and education, integrated management systems, sustainable agricultural technology, development, and transfer, along with alternative agricultural research and commercialization.

### 3. Title XVI - Organic Certification

Because of the moderate increase in organic farming (using nonman-made pesticides and fertilizers), Congress established national standards governing the marketing of organically produced farm products.

## M. IMPORTANT DEFINITIONS

### 1. Tolerances

A pesticide tolerance is the maximum amount of a pesticide residue that can legally be present on a food or feed. The tolerance is expressed in ppm, or parts of the pesticide per million parts of the food or feed by weight, and usually applies to the raw agricultural commodity. Pesticide tolerances are set by the EPA and enforced by the FDA or, in the case of meat, poultry, and eggs, by USDA agencies.

### 2. Acceptable Daily Intake

The tolerance on each food is set sufficiently low that daily consumption of the particular food or of all foods treated with the particular pesticide will not result in an exposure that exceeds the acceptable daily intake (ADI) for the pesticide. The tolerance is set still lower if the effective use of the pesticide results in lower residues.

The acceptable daily intake (ADI) is the level of a residue to which daily exposure over the course of the average human life span appears to be without appreciable risk on the basis of all facts known at the time.

### 3. No Observable Effect Level

The ADI is usually set one hundred times lower than the no observable effect level (NOEL). A much greater safety factor is required if there is evidence that the pesticide causes cancer in test animals. Although the Delaney Amendment to the Federal Food, Drug, and Cosmetic Act prevents the addition of an animal carcinogen to foods, it does not apply to pesticide residues that occur inadvertently in the production of the crop.

The no observable effect level (NOEL) is the dosage of a pesticide that results in no distinguishable harm to experimental animals in chronic toxicity studies that include the minute examination of all body organs for abnormality.

It is extremely unlikely that any American will ever be exposed to anything near the ADI in his or her food. Numerous, detailed, continuing, nationwide studies show the actual pesticide residues in food to be far below the ADI and the established tolerances.

The pesticide residues on crops at the time of harvest are usually less than the tolerances. Residues decrease during storage and transit. They are reduced further by operations such as washing, peeling, and cooking when the food is prepared for eating. Every acre of a food crop will not have been treated with the same pesticide, if any. And no one eats every food for which there is a tolerance for a particular pesticide every day. Further, the ADI refers to lifetime exposure. A minor excess of the ADI for a short period should be inconsequential.

## II. WORKER SAFETY REGULATIONS

## WORKER PROTECTION STANDARD FOR AGRICULTURAL PESTICIDES (40 CFR Part 170, 1992)

The Worker Protection Standard (WPS) is a federal regulation designed to protect agricultural workers (people involved in the production of agricultural plants) and pesticide handlers (people mixing, loading, or applying pesticides or doing other tasks involving direct contact with pesticides). Employers must provide the following required information, training, or equipment:

### A. PESTICIDE APPLICATION INFORMATION AT A CENTRAL LOCATION

1. In an easily seen central location on each agricultural establishment, the following information must be displayed:

- EPA WPS safety poster
- name, address and telephone number of the nearest emergency medical facility
- these following facts about each pesticide application:

  —product name, EPA registration number, and active ingredi ent(s)

  —location and description of treated area

  —time and date of application, and REI.

2. Inform workers and handlers where the information is posted and permit them access.

3. Display emergency information, which must include the name, telephone number, and address of the nearest emergency medical facility.

4. Keep the posted information legible.

## B. PESTICIDE SAFETY TRAINING

Unless workers and handlers possess a valid EPA-approved training card, training must occur before they begin work and at least once each 5 years (see Figure 2.2). This training should:

- use written and/or audiovisual materials
- use EPA WPS training materials
- use only certified pesticide applicators to conduct training.

## C. DECONTAMINATION SITES

1. Establish a decontamination site within 1/4 mile of all workers and handlers. At this site the following should be supplied:

- enough water for routine and emergency whole-body washing and for eyeflushing
- plenty of soap and single-use towels
- clean coveralls.

2. Provide water that is safe and cool enough for washing, eyeflushing, and drinking. Do not use tank-stored water that is also used for mixing pesticides.

3. Make at least one pint of eyeflushing water immediately accessible to each handler.

4. Do not put worker decontamination sites in areas being treated or under an REI.

**Figure 2.2** On-site training using the EPA Worker Protection Standard as a training guide (EPA, Protect Yourself from Pesticides poster, EPA 735-H-93-001).

5. If a decontamination site must be located in an area being treated, put decontamination supplies in enclosed containers. Figure 2.3 shows an emergency field decontamination station.

## D. EMPLOYER INFORMATION EXCHANGE

1. Before any application, commercial handler employers must make sure the operator of the agricultural establishment where a pesticide will be applied is aware of:

- location and description of area to be treated
- time and date of application
- product name, EPA registration number, active ingredient(s), and REI
- whether the product label requires both oral warnings and treated area posting
- all other safety requirements on labeling for workers or other people.

**Figure 2.3** Emergency decontamination site (Protect Yourself from Pesticides poster, EPA 735-H-93-001).

2. Operators of agricultural establishments must make sure any commercial pesticide establishment operator they hire is aware of

- specific location and description of all areas on the agricultural establishment where pesticides will be applied or where an REI will be in effect while the commercial handler is on the establishment
- restrictions on entering those areas.

## E. EMERGENCY ASSISTANCE

When an accident occurs resulting in an employee poisoning or injury involving pesticides:

1. Promptly make transportation available to an appropriate medical facility.

2. Promptly provide to the victim and to medical personnel:

   - product name, EPA registration number, and active ingredient(s)
   - all first aid and medical information from label
   - description of how the pesticide was used
   - information about victim's exposure.

## F. RESTRICTIONS DURING APPLICATIONS

1. In areas being treated with pesticides, allow entry only to appropriately trained and equipped handlers.

2. Keep nursery workers at least 100 feet away from nursery areas being treated.

3. Allow handlers to be in a greenhouse only

   - during a pesticide application
   - until labeling-listed air concentration level is met or, if there is no such level, until after two hours of ventilation with fans.

## G. RESTRICTED-ENTRY INTERVALS (REIS)

A pesticide's restricted-entry interval (REI) is the time immediately after application when entry into the treated area is limited. Some pesticides have one REI, such as 12 hours, for all crops and uses. Other products have different REIs depending on the crop or method of application. When two or more pesticides are applied at the same time and have different REIs, the longer interval must be followed.

The REI is listed on the pesticide labeling under the heading "Agricultural Use Requirements" in the "Directions for Use" section of the pesticide labeling.

In 1995, EPA reduced the REI for certain "low-risk" pesticide active ingredients from 12 to 4 hours. Registrants may apply to reduce the REI for products that contain these active ingredients and meet certain other criteria. Check the product label for Class III and IV products to see if a four-hour REI has been obtained.

Some pesticide labeling requires a different REI for arid areas. Average rainfalls can be obtained from any nearby weather bureau.

**Figure 2.4** Worker being notified of treated field entry restrictions.

In general, the REI is:

- 48 hours for products in Toxicity Category I (signal word Danger). This is extended to 72 hours if applied outdoors in areas with less than 25 inches of rain per year.
- 24 hours for products in Toxicity Category II (signal word Warning).
- 12 hours for categories III and IV (signal word Caution).

Signal words can be found on the product label. See Chapter 5, Table 5.1, for toxicity categories and their respective signal words.

Table 5.2 shows how signal words can be used to select the appropriate personal protective equipment (PPE) for different toxicity categories. In all cases, follow label instructions.

During any REI, do not allow workers to enter a treated area and contact anything treated with the pesticide to which the REI applies.

## H. NOTICE ABOUT APPLICATION

- Warn workers and post treated areas if required by the pesticide labeling.
- Post all greenhouse applications.

*Posted Warning Signs:*

1. Post legible 14" X 16" WPS-designed signs just before application; keep posted during the REI; remove signs before workers enter and within three days after the end of the REI.

**Figures 2.5A** and **2.5B** Posted warning signs to be used at entrances to treated areas. (Figure 2.5A, Protect Yourself from Pesticides poster, EPA 735-H-93-001. Figure 2.5B, The Worker Protection Standard for Agricultural Pesticides-How to Comply, EPA, 1993.)

2. Post signs so they can be seen at all entrances to treated areas.

*Oral Warnings:*

1. Before each application, tell workers who are on the establishment:

- location and description of treated area
- REI, and not to enter during REI.

2. Workers who enter the establishment after application starts must receive the same warning at the start of their work period.

## I. APPLICATION RESTRICTIONS AND MONITORING

1. Do not allow handlers to apply a pesticide so that it contacts, directly or through drift, anyone other than trained and PPE-equipped handlers.

2. Make sight or voice contact at least every two hours with anyone handling pesticides labeled with a skull and crossbones (poison).

3. Make sure a trained handler equipped with labeling-specified PPE maintains constant voice or visual contact with any handler in a greenhouse who is doing fumigant-related tasks, such as application or air monitoring.

## J. SPECIFIC INSTRUCTIONS FOR HANDLERS

1. Before handlers do any handling task, inform them of all pesticide labeling instructions for safe use.

2. Keep pesticide labeling accessible to each handler during entire handling task.

3. Before handlers use any assigned handling equipment, tell them how to use it safely.

4. When commercial handlers will be on an agricultural establishment, inform them beforehand of:

   - areas on the establishment where pesticides will be applied or where an REI will be in effect
   - restrictions on entering those areas.

## K. EQUIPMENT SAFETY

1. Inspect pesticide handling equipment before each use, and repair or replace as needed.

2. Allow only appropriately trained and equipped handlers to repair, clean, or adjust pesticide equipment that contains pesticides or residues.

## L. PERSONAL PROTECTIVE EQUIPMENT (PPE)

*Duties Related to PPE:*

1. Provide handlers with the PPE the pesticide labeling requires for the task, and be sure it is

   - clean and in operating condition
   - worn and used correctly
   - inspected before each day of use
   - repaired or replaced as needed.

2. Be sure respirators fit correctly.

3. Take steps to avoid heat illness.

4. Provide handlers a pesticide-free area for

   - storing personal clothing not in use
   - putting on PPE at start of task
   - taking off PPE at end of task.

5. Do not allow used PPE to be worn home or taken home.

*Care of PPE:*

1. Store and wash used PPE separately from other clothing and laundry.

2. If PPE will be reused, clean it before each day of reuse, according to the instructions from the PPE manufacturer unless the pesticide labeling specifies other requirements. If there are no other instructions, wash in detergent and hot water.

3. Dry the clean PPE before storing.

4. Store clean PPE away from other clothing and away from pesticide areas.

*Replacing Respirator Purifying Elements:*

1. Replace dust/mist filters:

   - when breathing becomes difficult
   - when filter is damaged or torn
   - when respirator label or pesticide label requires (whichever is shorter)
   - at the end of day's work period, in the absence of any other instructions or indications.

2. Replace vapor-removing cartridges/canisters:

   - when odor/taste/irritation is noticed
   - when respirator label or pesticide label requires (whichever is shorter)
   - at the end of day's work period, in the absence of any other instructions or indications.

*Disposal of PPE:*

1. Discard coveralls and other absorbent materials that are heavily contaminated with undiluted pesticide having a DANGER or WARNING signal word.

2. Follow federal, state, and local laws when disposing of PPE that cannot be cleaned correctly.

*Instructions for People Who Clean PPE:*

1. Inform people who clean or launder PPE:

- that PPE may be contaminated with pesticides
- the potentially harmful effects of exposure to pesticides
- how to protect themselves when handling PPE
- how to clean PPE correctly.

## III. THE PESTICIDE LABEL

The label contains the key information about any pesticide you are planning to use. The label on the package of a pesticide tells the most important facts you must know about that particular pesticide for safe and effective use. The information on the label gives not only the directions on how to mix and apply the pesticide, but also offers guidelines for safe handling, first aid, storage, and protection of the environment.

Information on pesticide labels has been called the most expensive literature in the world. The research and development that is responsible for the wording on a label frequently costs millions of dollars. The combined knowledge of laboratory and field scientists and others in industry, universities, and government is used to develop the information found on the label.

Before we discuss the information found on pesticide labels, we must distinguish *label* from *labeling*. The *label* is the information printed on or attached to the pesticide container or wrapper. *Labeling* refers to both the actual label and to all additional product information, such as brochures and handouts that you receive from the manufacturer or dealer when you buy the product.

The EPA must approve all language that the manufacturer proposes to include in the product labeling. Thus, information in handouts and brochures cannot differ in meaning from the information on the label. The label and supplemental labeling are legally binding documents that must be followed explicitly.

Although we focus on the pesticide label in this chapter, remember that for many products, much of the information presented may appear in supplemental labeling rather than on the label itself.

Some of the information on labels is required by law; manufacturers may choose to include additional information. While some information must appear on a certain part of the label, the rest may be placed wherever the manufacturer chooses. Information is often grouped under headings to help you find what you need.

In this section, we will describe the principal types of information found on labels. Product labels vary and do not always present information under specific headings. However, you can find in the labeling all the information needed to safely and effectively use the product. Figures 2.6A and 2.6B show workers reading the label prior to use.

## A. THE INFORMATION ON THE LABEL

The pesticide label contains basic information that clearly identifies the product. Some of this information must appear on the front panel. Figure 2.7 shows an example of a label.

### 1. Trade Name

The trade name is the producer's or formulator's proprietary name of the pesticide. Most companies register each brand name as a trademark and will not allow any other company to use that name. The brand name is usually the largest print on the front panel of the label. Beware of choosing a product by trade name alone. Companies often use the same basic name with slight variations for different products; the products may contain different amounts or types of active ingredients and will probably be registered for different uses. Always check the label for the active ingredient(s) in the product and for registered uses.

### 2. Use Classification

All pesticides are evaluated and if additional care over and above simple label compliance is required for their safe use, they receive a restricted use classification. Pesticides classified for restricted use will bear the following statement on the label:

**RESTRICTED USE PESTICIDE**

For retail sale and use only by certified applicators or persons under their direct supervision and only for those uses covered by the certified applicator's certification.

Some labels state why the product is classified as restricted use. Examples include:

- Very high toxicity to humans and birds
- Oncogenicity (cancer causing)

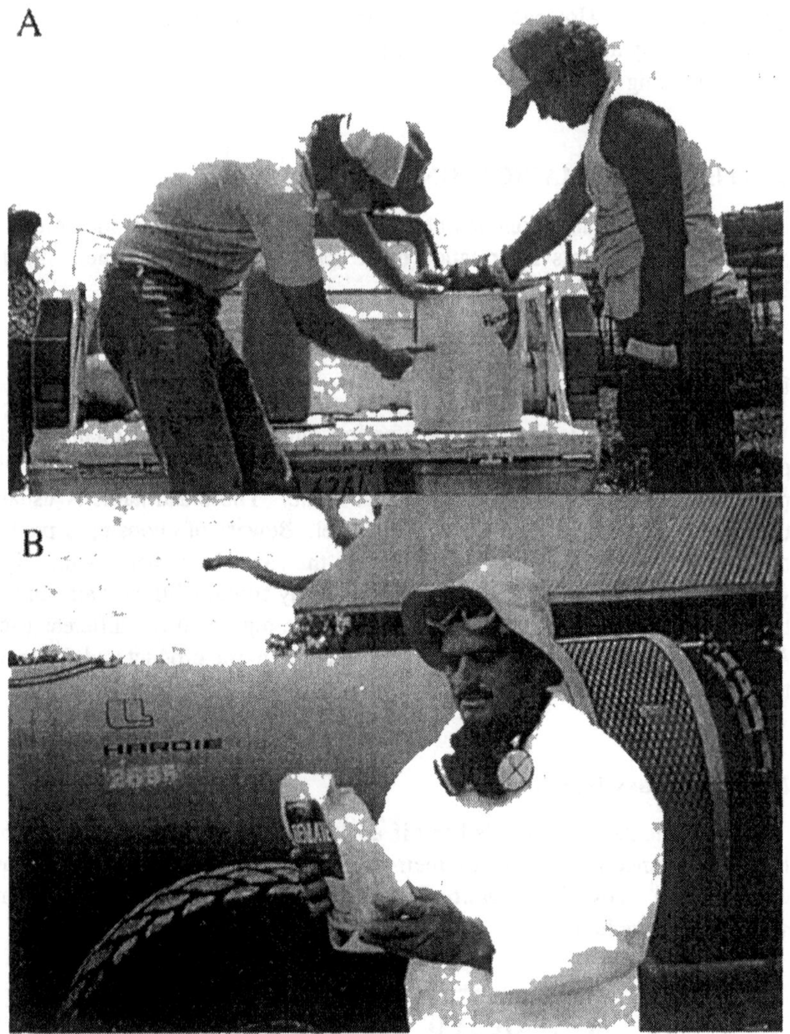


**Figures 2.6A** and **B**. In Figure A the supervisor points out important information on the label. Worker in Figure B reads the label for product use instruction prior to applying product with the air-blast sprayer.

- Groundwater concern.

Sometimes, different companies sell similar products under different trade names. Again, in many cases, similar products are registered for different uses.

### 3. Type of Pesticide

The type of pesticide usually is listed on the front panel of the pesticide label. This short statement indicates in general terms what the product will control. For example:

- Insecticide-ovicide for control of certain insects on cotton
- Herbicide for nonselective weed control on industrial sites in noncrop areas
- Fungicide for the control of certain diseases in broccoli, cabbage, cauliflower, cucurbit vegetables, onions, potatoes, and tomatoes.

### 4. Formulation

Sometimes, the front panel states how the product is formulated. The formulation may be abbreviated as part of the trade name (e.g., Aaterra EC). EC is the abbreviation for emulsifiable concentrate. If the label does not state the formulation, you can often figure out what it is by reading the directions for use.

### 5. Ingredient Statement

The ingredient statement tells you what is in the container. Ingredients are usually listed as active or inert (inactive). The active ingredient is the chemical that does the job that the product is intended to do. The ingredient statement must list the official chemical names and/or common names for the active ingredients. Inert ingredients need not be named, but the label must show what percent of the total content they comprise. Inert ingredients may be the wetting agent, the solvent, the carrier, or the filler which give the product its desirable qualities for application. The *common name is* the generic name accepted for the active ingredient of the pesticide product regardless of the brand name. Only common names that are officially accepted by the EPA may be used in the ingredient statement on the pesticide label. The *chemical name* of the pesticide is the one that is long and difficult to pronounce unless you are a chemist. It identifies the chemical components and structure of the pesticide. For example, AAtrex® is a brand name, and the common name for this fungicide is Atrazine. The chemical name of the active ingredient in Aatrex® is 6-chloro-N-ethyl-N'-(1-methylethyl)-1,3,5-trazine-2,4-diamine. Always check the common or chemical names of pesticides when you make a purchase to be certain of getting the right active ingredient.

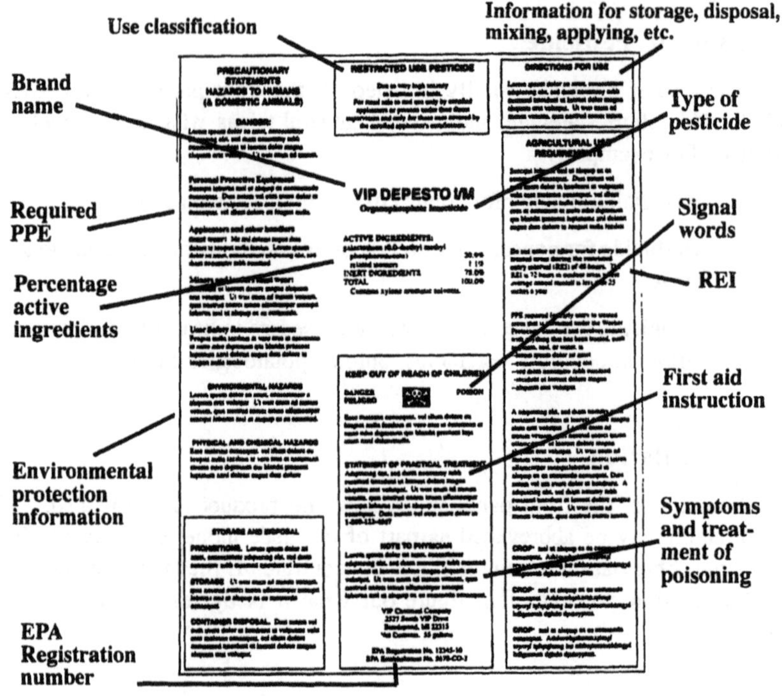


**Figure 2.7** Sample label with major sections delineated.

The ingredient statement provides the amount of each active ingredient as a percentage of the product's total weight. The amount may also be given in another manner (e.g., pounds per gallon) just below the ingredient statement.

Sometimes it is not the listed active ingredient, but rather one of its breakdown products, that kills pests. In such cases, the "equivalent" amount of the toxic chemical is usually listed as well, again just below the ingredient statement.

## 6. Contents

The bottom of the front panel of the pesticide label tells you how much product is in the container. This is expressed as ounces, pounds, grams, or kilograms for dry formulation, and as gallons, quarts, pints, milliliters, or liters for liquids. Also listed is the amount of active ingredient per unit of product. This information is usually found in or near the ingredient statement.

### 7. Name and Address of Manufacturer

The law requires that the name and address of the manufacturer or distributor of the product be put on a label. This is to inform the purchaser or end-user of the product as to the manufacturer or seller of the product in case he or she wish to contact them with specific questions.

### 8. EPA Registration Number and Establishment Number

A registration number appears on every pesticide label. It shows that the product has been registered with the federal government. It is the responsibility of the EPA to assign a registration number to every product that it registers. This number usually consists of two sets of numbers; for example, EPA Reg. No. 55947-136. The first set, 55947, identifies the manufacturer and the second set, 136, identifies the product.

The establishment number is a special number assigned to the plant that manufactured the pesticide for the company. Some companies manufacture pesticides in more than one manufacturing plant, and the establishment number identifies the place of manufacture. In case something goes wrong, the facility that made the product can be traced.

### 9. Health Hazards (Signal Words and Symbols)

These precautionary statements appear on the label's front panel. They tell the relative toxicity of the product and how you should respond if you or a coworker is exposed to the product.

Usually, the label will not give specific information on the pesticide's toxicity to humans. It must, however, indicate the pesticide's relative toxicity. It does so by using *signal words* that are established by law that identify the toxicity category to which the product is assigned. Signal words must appear in large letters on the label's front panel.

The signal word **DANGER** must appear on labels of all highly toxic products; this word alone is sufficient for products for which the principal concerns are eye and/or skin hazards. If a product is highly toxic due to oral, dermal, or inhalation exposure, the label must also bear the word POISON and the skull-and-crossbones symbol. A taste to a teaspoonful taken by mouth could kill an average-sized adult. Any product that is highly toxic orally, dermally, or through inhalation will be labeled DANGER-POISON.

The word **WARNING** signals that the product is moderately toxic. A teaspoonful to a tablespoonful by mouth could kill the average-sized adult. Any product that is moderately toxic orally, dermally, or through inhalation or that causes moderate eye and skin irritation will be so labeled.

**CAUTION**—This word signals that the product is slightly toxic. An ounce to more than a pint taken by mouth could kill the average adult. Any product that is slightly toxic orally, dermally, or through inhalation or that causes slight eye or skin irritation must be labeled **CAUTION** and **CAUTION** appears on labels of pesticides that are slightly toxic (Toxicity Category III) and relatively nontoxic (Toxicity Category IV), respectively.

The statements that immediately follow the signal word, either on the front or side of the pesticide label, indicate which route or routes of entry into the body (mouth, skin, lungs, eyes) must be particularly protected. Many pesticide products are hazardous by more than one route, so study these statements carefully. DANGER-POISON followed by "May be fatal if swallowed or inhaled" gives a far different warning than "DANGER: CORROSIVE—causes eye damage and severe skin burns."

All pesticide labels contain additional statements to help you decide the proper precautions to take to protect yourself, your workers, and other persons (or domestic animals) that may be exposed. Sometimes these statements are listed under the heading "Hazardous to Humans and Domestic Animals." The statements usually follow immediately after the route-of-entry statements. They recommend specific action you should take to prevent poisoning accidents. These statements are directly related to the toxicity of the pesticide product (signal word) and the route or routes of entry that must be protected. These statements contain such items as "Do not breathe vapors or spray mist" or "Avoid contact with skin or clothing" or "Do not get in eyes." There is a distinct difference between "May be harmful if swallowed" and "Harmful if swallowed," as well as the difference between "Do not get in eyes" and "Avoid getting in eyes." The variation and the strength of the statement depends on the signal word for the product. These specific-action statements help prevent pesticide poisoning by taking the necessary precautions and wearing the correct protective clothing and equipment.

## 10. Statement of Practical Treatment

Most pesticide labels tell how to respond to an exposure. The instructions usually provide first aid measures, describe exposures that warrant medical attention, and contain specific instructions for the attending physician. Most labels also provide a phone number to use in case of poisoning or some other accident involving the product.

Typical statements include:

- In case of contact with the skin, wash immediately with plenty of soap and water.

- In case of contact with eyes, flush with water for 15 minutes and get medical attention.
- In case of inhalation exposure, move from contaminated area and give artificial respiration if necessary.
- If swallowed, drink large quantities of milk, egg white, or water. Do not induce vomiting.
- If swallowed, induce vomiting.

All DANGER-POISON labels and some WARNING and CAUTION labels will contain a note to physicians describing the appropriate medical procedures for poisoning emergencies and may identify an antidote.

If the statement of practical treatment is not on the front panel, there must be a statement on the front panel that tells you where the information is in the labeling.

### 11. Acute Effects

The label will state which routes of exposure may lead to acutely toxic effects, and will also describe those effects. For example, the label may state that the product is "injurious to eyes" or "may be fatal if swallowed." The label may also tell how to avoid poisoning, including what protective clothing to use. Figure 2.8 shows the routes of exposure that may lead to acute effects.

### 12. Delayed or Chronic Effects

The label will tell if the pesticide may cause chronic or other delayed effects. For example, any pesticide that causes tumors in laboratory animals must bear a "Cancer/ Tumor Statement" on its label. Similarly, products that cause birth defects in test animals must have a "Birth Defects Statement." The statements describe these risks and tell how to avoid them.

### 13. Allergic Effects

If data indicate that the pesticide may cause allergic effects, such as skin irritation or asthma, the label must say so. Labels sometimes refer to allergic effects as "sensitization."

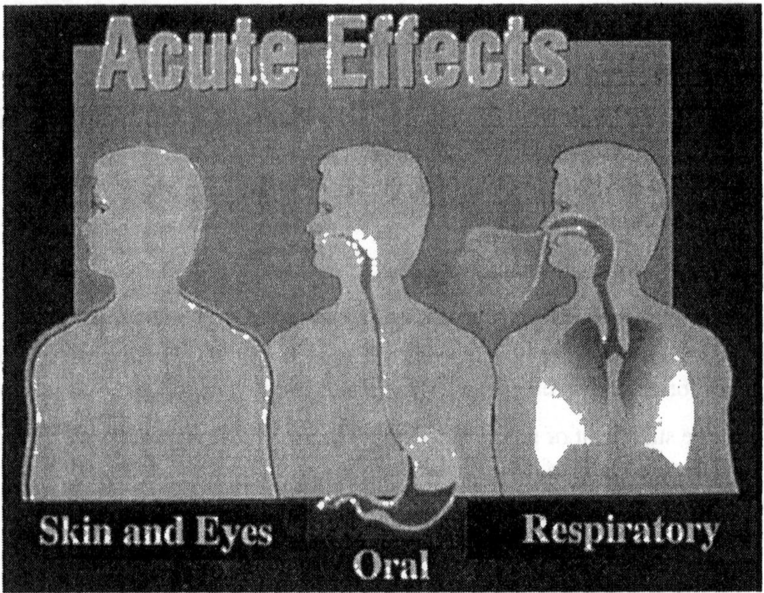


**Figure 2.8** Shows the routes of exposure that may lead to acute effects.

## 14. Personal Protective Equipment

The Worker Protection Standard requires that each pesticide label give information on the personal protective equipment (PPE) that applicators and other handlers must wear. A good way to double check for the correct type of clothing and equipment is to use the signal word, the route-of-entry statement, and the specific-action statement, along with the basic guidelines listed in Chapter 6. Figure 2.9 lists the precautionary statements that may be present on the label and suggests with the help of illustrations the proper attire to address each precautionary statement.

Some pesticide labels fully describe appropriate protective clothing and equipment. A few list the kind of respirator that should be worn when handling and applying the product. Others require the use of a respirator but do not specify the type or model to be used. You should follow all advice for protective clothing or equipment that appears on the label. However, the lack of any statement or the mention of only one piece of equipment does not rule out the need for additional protection. Even though the label may not specifically require them, you should wear a long-sleeved shirt, long-legged trousers, and gloves whenever handling pesticides. You should consider wearing rubberized or waterproof clothing if you will be in prolonged contact with the pesticide or may be wet by an overhead spray application.

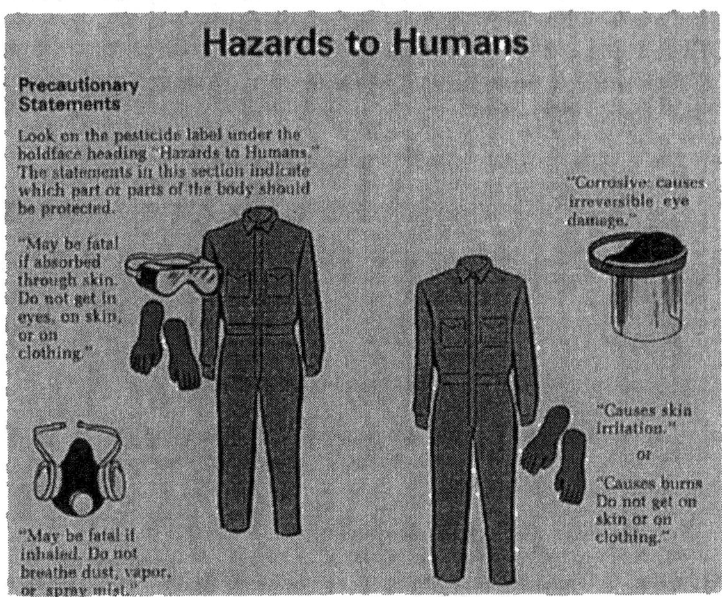


**Figure 2.9** Precautionary statements under "Hazards to Humans" section on the label (EPA, Protective Clothing for Pesticide Users poster).

## 15. Environmental Hazard Statement

Pesticides are useful tools. But wrong or careless use could cause undesirable effects. To help avoid this the label provides precautions for protecting nontarget organisms and the environment. Some general statements appear on the label of almost all pesticides; for example, most labels warn not to contaminate water when you apply or dispose of pesticides. The label will contain specific precautionary statements if the pesticide poses a specific hazard to the environment. Some examples are:

- "This product is toxic to fish, birds, and other wildlife. Treated granules exposed on soil surface may be hazardous to birds and other wildlife."
- "This product is highly toxic to bees exposed to direct treatment or residues on blooming crops or weeds. Do not apply this product or allow it to drift to blooming crops or weeds if bees are visiting the treatment area."

These statements alert you to the special hazards that the use of the products may pose. They should help you choose the safest product for a particular job and remind you to take extra precautions. Other general environmental statements appear on nearly every pesticide label. They are reminders of common sense actions to follow to avoid contaminating the environment. The absence of any or all of these statements does not indicate that you do not have to take adequate precautions. Examples of general environmental statements include the following:

- Do not apply when runoff is likely to occur.
- Do not apply when weather conditions favor drift from treated areas.
- Do not contaminate water by cleaning of equipment or disposal of waste.
- Keep out of any body of water.
- Do not allow drift on desirable plants or trees.
- Do not apply when bees are likely to be in the area.

## 16. Physical or Chemical Hazards

The label will specify any physical or chemical hazard. For example, the product may be flammable, explosive if exposed to heat, or corrosive to metals, rubber or skin. For example:

- Flammable: Do not use, pour, or store near heat or open flame.
- Corrosive: Store only in a corrosion-resistant tank.

Some products pose no such hazards and so there will be no statement on the label.

## 17. Directions for Use

The instructions on how to use the pesticide are an important part of the label. This is the best way to find out the correct manner in which to apply the product. The instructions include the following:

- The sites for which the product is registered.
- The pests that the product will control.
- The crop, animal, or site the product is intended to protect.
- How much to use.
- Mixing directions.
- Compatibility with other often-used products.
- Phytotoxicity and other possible injury or staining problems.

- Where the material should be applied.
- Application rates and methods to use and the proper equipment to be used.
- The correct form in which the product should be applied.
- The length of the reentry interval.
- How to store and dispose of the product.
- When the material should be applied.

Directly under the heading "Directions for Use" on every label is the following statement:

> It is a violation of Federal law to use this pesticide in a manner inconsistent with this labeling.

**Figure 2.10** Warning statment on each pesticide label. Misuse is a violation of FIFRA.

### 18. Reentry Statement

The label will specify the restricted entry interval (REI) for the product. The restricted entry statement includes the amount of time that must elapse after a pesticide has been applied before it is safe to enter the treated area without wearing full protective clothing and equipment. If this interval applies to only certain uses or sites, the label will say so. The label may also list posting requirements associated with the reentry interval. Refer back to Section II, part G, for information on restricted entry intervals (REIs) in connection with the Worker Protection Act.

Sometimes the use of one pesticide soon after the use of another can cause phytotoxicity. The label of at least one of the products will warn of the potential problem. Likewise, in addition to providing instructions on mixing pesticides, the label will tell which mixtures to avoid (usually because of physical or chemical incompatibility).

### 19. Storage and Disposal Statements

Each label tells how to store the pesticide. The instructions may include general statements as well as specific precautions such as "Do not store at temperatures below 32°F."

While the label also tells how to dispose of excess pesticide and its container, remember that it will not address individual state laws regarding disposal. You are responsible for knowing and complying with these laws.

Often, this section of a label tells how to respond if the product is spilled. Again, comply with local state regulations.

## REFERENCES

Code of Federal Regulations, 29 CFR Parts 1900-191O, Department of Labor, Occupational Safety and Health Administration, Washington, D.C.

Code of Federal Regulations, 29 CFR Part 191O.1200, Hazard communication Standard, Department of Labor, Occupational Safety and Health Administration, Washington, D.C. (Effective 1986).

Code of Federal Regulations, 29 CFR Part 191O.120, Hazardous Waste Operations and Emergency Response, Department of Labor, Occupational Safety and Health Administration, Washington, D.C. (Effective 1990).

Code of Federal Regulations, 29 CFR Part 191O.1200, Department of Labor, Occupational Safety and Health Administration, Washington, D.C.

Code of Federal Regulations, 40 CFR Federal Insecticide, Fungicide, and Rodenticide Act of 1947, U.S. Environmental Protection Agency, Washington, D.C. (Amended 1972).

Code of Federal Regulations, 40 CFR Parts 1-99, Clean Air Act of 1963, U.S. Environmental Protection Agency, Washington, D.C. (Amended 1970, 1977, and 1990).

Code of Federal Regulations, 40 CFR Parts 122-125,131, and 401-471, Clean Water Act (formerly the Federal Water Pollution Control Act of 1972), U.S. Environmental Protection Agency, Washington, D.C. (1972, Amended 1977 and 1987).

Code of Federal Regulations, 40 CFR Part 170, Worker Protection Standard for Agricultural Pesticides, U.S. Environmental Protection Agency, Washington, D.C.

Code of Federal Regulations, 40 CFR Parts 191-192, U.S. Environmental Protection Agency, Washington, D.C.

Code of Federal Regulations, 40 CFR Parts 259-270, Resource Conservation and Recovery Act, U.S. Environmental Protection Agency, Washington, D.C. (1976).

Code of Federal Regulations, 40 CFR Parts 300, Comprehensive Environmental Response, Compensation Liability Act (Superfund), U.S. Environmental Protection Agency, Washington, D.C. (1980).

Code of Federal Regulations, 40 CFR Parts 355, 370, and 372, Superfund Amendments and Reauthorization Act (SARA also referred to as Emergency Planning and Community Right-to-Know Act), U.S. Environmental Protection Agency, Washington, D.C. (1986).

Code of Federal Regulations, 40 CFR Parts 700-799, Toxic Substance Control Act (TSCA) of 1976, U.S. Environmental Protection Agency, Washington, D.C.

*Environmental Law Deskbook*, Environmental Law Institute, Washington, D.C., 1989.

*Environmental Statutes*, Government Institutes, Inc., Rockville, MD, 1992.

U.S. Environmental Protection Agency, *Applying Pesticides Correctly: A Guide for Private and Commercial Applicators*, Revised 1991.

U.S. Environmental Protection Agency, *Protect Yourself from Pesticides: Safety Training for Agricultural Workers*, November 1993.

U.S. Environmental Protection Agency, *Protect Yourself from Pesticides: Safety Training for Pesticide Handlers*, December 1993.

U.S. Environmental Protection Agency, *The Worker Protection Standard for Agricultural Pesticides—How to Comply: What Employers Need to Know*, July 1993.

U.S. Environmental Protection Agency, *Protective Clothing for Pesticide Users*—Poster, 1993.

# CHAPTER 3

# PESTS AND PEST CONTROL

A pest is anything that:

- competes with humans, domestic animals, or crops for food, feed, or water
- injures humans, animals, crops, structures, or possessions
- spreads disease to humans, domestic animals, or crops
- annoys humans or domestic animals.

Pests can be placed into five main categories:

- insects (and related animals)
- plant disease agents
- weeds
- mollusks
- vertebrates.

As a person that uses pesticides, you must be familiar with the pests you will be likely to encounter on the job, at home, or at recreation. To be able to identify and control the pests, you need to know about some aspects of:

- the common features of pest organisms
- characteristics of the damage they cause
- pest development and biology.

You can get identification aids, publications, and pictures from your Cooperative Extension Service agent or ask university, regulatory agency personnel, or other experts for advice.

To solve pest problems, the applicator must:

- identify the pest
- know what control methods are available

- evaluate the benefits and risks of each method or combination of methods
- choose the methods that are most effective and will cause the least harm to people and the environment
- use each method correctly
- observe local, state, and federal regulations that apply to the situation.

The most important principle of pest control is this:

*Use a pest control method only when that method will prevent the pest from causing more damage than is reasonable to accept.*

Even though a pest is present, it may not do very much harm. It could cost more to control the pest than would be lost because of the pest's damage.

The three main objectives of pest control are:

- prevention—keeping a pest from becoming a problem
- suppression—reducing pest numbers or damage to an acceptable level
- eradication—destroying an entire pest population.

## I. PEST CONTROL METHODS

The use of a combination of methods to control pests is basic to all pest control. Successful pest control is based on the ability to:

- keep pest damage to a minimum by choosing an appropriate combination of control methods
- recognize when direct action, such as a pesticide application, is necessary
- cause as little danger as possible to the environment.

The combination of methods chosen will depend on the kind and amount of control needed.

### A. NATURAL FORCES

Some natural forces act on pests, causing the populations to rise and fall. These natural forces act independently of humans and may either help or hinder pest control. You usually cannot alter the action of natural forces on a pest population, but you should be aware of their influence and take advantage of them whenever possible. Some forces which affect the pest population include climate, natural enemies, topography, and food and water supply.

### 1. Climate

Weather conditions, especially temperature, day length, and humidity, affect pests' activity and their rate of reproduction. Pests may be killed or suppressed by rain, frost, freezing temperatures, drought, or other adverse weather.

Climate also affects pests indirectly by influencing the growth and development of their hosts. The population of plant-eating pests is related to the growth of host plants. Unusual weather conditions can change normal patterns so that increased or decreased damage results.

### 2. Natural Enemies

Birds, reptiles, amphibians, fish, mammals, and predatory and parasitic insects feed on some pests and help control their numbers. More than half of all insect and insect-like species feed on other insects, some of which are pests. Disease organisms often suppress pest populations.

### 3. Topography

Features such as mountains and large bodies of water restrict the spread of many pests. Other features of the landscape can have similar effects. Soil type is a prime factor affecting wireworms, grubs, nematodes, and other soil organisms. Some pests live in heavy, poorly drained soil, others in light, sandy soils. Soil type also affects the distribution of plants (including weeds), which in turn affects the population of insects and other plant pests.

### 4. Food and Water Supply

Pest populations can thrive only as long as their food and water supply lasts. Once the food source, plant or animal, is exhausted, the pests die or become inactive. The life cycle of many pests depends on the availability of water.

## B. OTHER METHODS

Unfortunately, natural controls often do not control pests quickly enough to prevent unacceptable injury or damage. In these cases, other pest control methods must be initiated. Those available include:

- host resistance
- biological control
- cultural control
- mechanical control
- sanitation
- chemical control.

### 1. Host Resistance

Some crops, animals, and structures resist pests better than others. Some varieties of crops, wood, and animals are immune to certain pests. Use of resistant types helps keep pest populations below harmful levels by making the environment less favorable for the pests. Host resistance works in two main ways:

- Chemicals in the host prevent the pest from completing its life cycle.
- The host is more vigorous or tolerant than other varieties and thus less likely to be seriously damaged by pest attacks.

### 2. Biological Control

Biological control involves the use of naturally occurring enemies—parasites, predators, and disease agents (pathogens). It also includes methods by which the pest is biologically altered, as in the production of sterile males and the use of pheromones or juvenile hormones. Most kinds of biological control agents occur naturally. Releasing more of a pest's enemies or predators into the target area can supplement this natural control.

Biological control is never complete. The degree of control fluctuates. There is always a time lag between pest population increase and the corresponding increase in natural controls. But, under proper conditions, sufficient control can be achieved to eliminate the threat to the crop or animal needing protection. Biological control can be a low-cost control method, particularly suited to low-value crops (pastureland, clover, and hay crops) or in areas where some injury can be tolerated (golf course fairways or forest areas).

### 3. Cultural Control

Cultural practices are agricultural practices used to alter the environment, the condition of the host, or the behavior of the pest to prevent or suppress an infestation. Planting, growing, harvesting, and tillage practices sometimes can be manipulated to reduce pest populations. Other practices such as crop rotation, pasture rotation, varying the time of planting, and use of trap crops also affect pests.

### 4. Mechanical Control

Devices and machines used to control pests or alter their environment are called mechanical controls. Traps, screens, barriers, radiation, and electricity can sometimes be used to prevent the spread of pests or reduce an infestation. Lights, heat, and refrigeration can alter the environment sufficiently to suppress or eradicate some pest populations.

### 5. Sanitation

Sanitation practices help to suppress some pests by removing sources of food and shelter. Other forms of sanitation which help prevent pest spread include using pest-free seeds or using decontaminating equipment or methods on plants, livestock, and other possible carriers before allowing them to enter a pest-free area.

### 6. Chemical Control

Pesticides are chemicals used to destroy pests, control their activity, or prevent them from causing damage. Some pesticides either attract or repel pests. Chemicals which regulate plant growth or remove foliage may also be classified as pesticides.

Pesticides are generally the fastest way to control pests. In many instances, they are the only weapon available. Choosing the best chemical for the job is important.

#### a. *Pest Resistance to Pesticides*

The ability of pests to resist poisoning is called pesticide resistance. Consider this when planning pest control programs that rely on the use of pesticides.

Rarely does any pesticide kill all the target pests. Each time a pesticide is used, it selectively kills the most susceptible pests. Some pests avoid the pesticide. Others are able to withstand its effects. Pests that are not destroyed may pass along to their offspring the trait that allowed them to survive.

When we use one pesticide repeatedly in the same place, against the same pest, the surviving pest population may show greater resistance to the pesticide than did the original population. Some pests have become partially resistant to the action of the pesticide.

Not every pesticide failure is caused by pest resistance. Make sure that you have used the correct pesticide and the correct dosage, and have applied the pesticide correctly. Also, remember that the pests that are present may be part of a new infestation that occurred after the chemical was applied.

*b. Factors Affecting Pesticide Use Outdoors*

**Soil Factors**—Organic matter in soils may "tie up" pesticides, limiting their activity. Soils with high organic matter content may need higher application rates of some pesticides for best control.

Soil texture also affects the way pesticides work. Soils with fine particles (silts and clays) have the most surface area. They may need higher rates for total coverage. Coarser soils (sands) have less surface area and, therefore, they may require lower rates.

**Surface Moisture**—Pesticides work best with moderate surface moisture. Wetness may keep the pesticide from adequately contacting the protected surface. Dryness may prevent the pesticide from spreading evenly over the surface and contacting the target pest.

Rain may interfere with pest control by causing pesticides to run off or to leach down through the soil. Rain during or soon after over-the-top or foliar applications may wash pesticides off the plant. However, some protectant fungicides are sometimes purposely applied just before periods of expected high humidity and light rain. When preemergence pesticides are applied to the surface, moderate rainfall aids in carrying them down through the soil to the pests. Rain may also release pesticide action after some granular applications.

**Humidity and Temperature**—Humidity also affects the way pesticides work. Herbicides often work best when weeds are growing fast—usually in high humidity and optimum temperature. However, these same conditions may make the protected plant more susceptible to pesticide injuries.

High temperature and sunlight will cause some pesticides to break down when they are left exposed on top of the soil or on other surfaces. Low temperatures may slow down or stop the activity of some pesticides.

**Wind**—Wind speed and direction can greatly alter the effectiveness of a pesticide application. Excessive wind can blow the pesticide off target and result in inadequate control. Even moderate winds can greatly alter the coverage of ULV and mist blower applications. Sometimes the applicator can compensate for minor winds by applying the pesticides at an angle where the winds blow the chemical towards the area to be protected.

## II. INSECTS

There are more kinds of insects on earth than all other living animals combined. They are found in soil, hot springs, water, snow, air, and inside plants and animals. They eat the choicest foods from our table. They can even eat the table.

The large number of insects can be divided into three categories according to their importance to man:

- species of minor importance—About 99 percent of all species are in this category. They are food for birds, fish, mammals, reptiles, amphibians, and other insects. Some have aesthetic value.
- beneficial insects—In this small but important group are the predators and parasites that feed on destructive insects, mites, and weeds. Examples are ladybird beetles, some bugs, ground beetles, tachinid flies, praying mantids, many tiny parasitic wasps, and predaceous mites. Also in this category are the pollinating insects, such as bumblebees and honeybees, some moths, butterflies, and beetles. Without pollinators, many kinds of plants could not grow. Honey from honeybees is food for humans. Secretions from some insects are made into dyes and paints. Silk comes from the cocoons of silkworms.
- destructive insects—Although this is the category which usually comes to mind when insects are mentioned, it includes the fewest number of species. These are the insects that feed on, cause injury to, or transmit disease to humans, animals, plants, food, fiber, and structures. In this category are, for example, aphids, beetles, fleas, mosquitoes, caterpillars, and termites.

### A. PHYSICAL CHARACTERISTICS

All insects in the adult stage have two physical characteristics in common. They have three pairs of jointed legs, and they have three body regions

—the head, thorax, and abdomen. Figure 3.1 shows the adult ant and its body regions.

## 1. Head

The head contains antennae, eyes, and mouthparts. The antennae vary in size and shape and can be a help in identifying some pest insects. Insects have compound eyes made up of many individual eyes. These compound eyes enable insects to discern motion, but probably not clear images.

The four general types of mouthparts are:

- chewing
- piercing-sucking
- sponging
- siphoning.

**Chewing mouthparts** contain toothed jaws that bite and tear the food. Cockroaches, ants, beetles, caterpillars, and grasshoppers are in this group. **Piercing-sucking mouthparts** consist of a long slender tube which is forced into plant or animal tissue to suck out fluids or blood. Insects with these mouthparts are stable flies, sucking lice, bed bugs, mosquitoes, true bugs, and aphids. **Sponging mouthparts** have a tubular tonguelike structure with a spongy tip to suck up liquids or soluble food. This type of mouthpart is found in the flesh flies, blow flies, and house flies. **Siphoning mouthparts** are formed into a long tube for sucking nectar. Butterflies and moths have this type.

## 2. Thorax

The thorax contains the three pairs of legs and (if present) the wings. The various sizes, shapes, and textures of wings and the pattern of the veins can be used to identify insect species.

The forewings take many forms. In beetles, they are hard and shell-like; in grasshoppers, they are leathery. The forewings of flies are membranous; those of true bugs are part membranous and part hardened. Most insects have membranous hindwings. The wings of moths and butterflies are membranous but are covered with scales.

### 3. Abdomen

The abdomen is usually composed of 11 segments. Along each side of most of the segments are openings (called spiracles) through which the insect breathes. In some insects, the tip end of the abdomen carries tail-like appendages.

## B. INSECT DEVELOPMENT

Most insect reproduction results from the males fertilizing the females. The females of some aphids and parasitic wasps produce eggs without mating. In some of these insect species, males are unknown. A few insects give birth to living young; however, life for most insects begins as an egg. Temperature, humidity, and light are some of the major factors influencing the time of hatching. Eggs come in various sizes and shapes—elongate, round, oval, and flat. Eggs of cockroaches, grasshoppers, and praying mantids are laid in capsules. Eggs may be deposited singly or in masses on or near the host—in soil or on plants, animals, or structures.

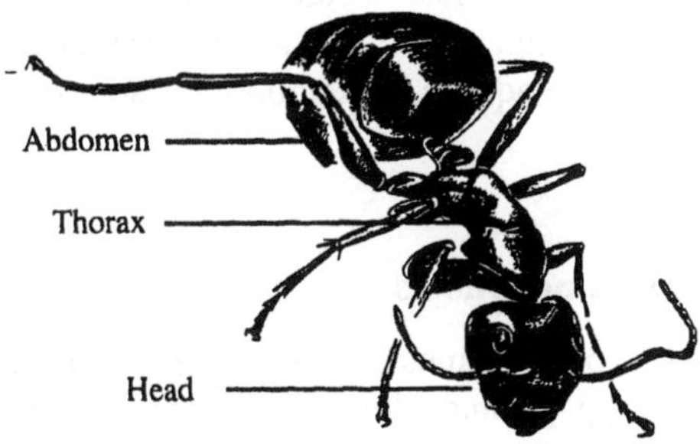


**Figure 3.1** The adult ant (EPA, *Urban Integrated Pest Management: A Guide for Commercial Applicators*).

### 1. Metamorphosis

The series of changes through which an insect passes in its growth from egg to adult is called metamorphosis.

When the young first hatch from an egg, it is either a larva, nymph, or naiad. After feeding for a time, the young grows to a point where the skin cannot stretch further; the young molts, and new skin formed. The number of these stages (called instars) varies with different insect species and, in some cases, may vary with the temperature, humidity, and availability and kinds of food. The heaviest feeding generally occurs during the final two instars.

## 2. No Metamorphosis

Some insects do not change except in size between hatching and reaching the adult stage. The insect grows larger with each successive instar until it reaches maturity. Examples are silverfish, firebrats, and springtails. The food and habitats of the young (called nymphs) are similar to those of the adult.

## 3. Gradual Metamorphosis

Insects in this group pass through three quite different stages of development before reaching maturity: egg, nymph, and adult. The nymphs resemble the adult in form, eat the same food, and live in the same environment. The change of the body is gradual and the wings become fully developed only in the adult stage. Examples are cockroaches, lice, termites, aphids, and scales.

## 4. Incomplete Metamorphosis

The insects with incomplete metamorphosis also pass through three stages of development: egg, naiad, and adult. The adult is similar to the young, but the naiads are aquatic. Examples include: dragonflies, mayflies, and stoneflies.

## 5. Complete Metamorphosis

The insects with complete metamorphosis pass through four stages of development: egg, larva, pupa, and adult. The young, which may be called larvae, caterpillars, maggots, or grubs, are entirely different from the adults. They usually live in different situations and in many cases feed on different food than adults. Examples are the beetles, butterflies, flies, mosquitoes, fleas, bees, and ants.

Larvae hatch from the egg. They grow larger by molting and passing through one to several instar stages. Moth and butterfly larvae are called caterpillars; some beetle larvae are called grubs; most fly larvae are called maggots. Caterpillars often have legs; maggots are legless. Weevil grubs are legless; other kinds of beetle larvae usually have three pairs of legs.

The pupa is a resting stage during which the larva changes into an adult with legs, wings, antennae, and functional reproductive organs.

## C. INSECT-LIKE PESTS

Mites, ticks, spiders, sowbugs, pillbugs, centipedes, and millipedes resemble insects in size, shape, life cycle, and habits. Pest species usually can be controlled with the same techniques and materials used to control insects.

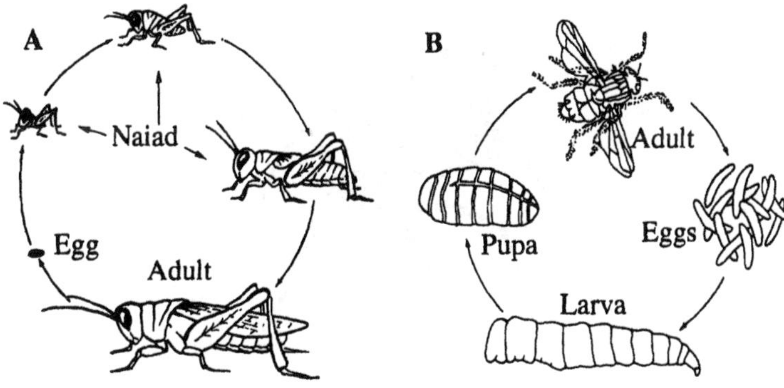


**Figure 3.2A** Incomplete metamorphosis. Grasshopper passes through three stages of development: egg, naiad, and adult.

**Figure 3.2B** Complete metamorphosis. Fly passes through four stages of development: egg, larva, pupa, and adult.

### 1. Arachnids

Ticks, scorpions, spiders, and mites have eight legs and only two body regions. They are wingless and lack antennae. The metamorphosis is gradual and includes both larval and nymphal stages. Eggs hatch into larvae (six legs) which become nymphs (eight legs) and then adults. Ticks and mites have modified piercing-sucking mouthparts; spiders and scorpions have chewing mouthparts. Figure 3.3 is an illustration of a brown recluse spider.

### 2. Crustaceans

Sowbugs, also known as pillbugs, water fleas, and wood lice have 14 legs. They are wingless and contain only one segmented body region. They have two pairs of antennae and chewing mouthparts. Sowbugs and pillbugs have a hard, protective shell-like covering and are related to the aquatic lobsters, crabs, and crayfish. The metamorphosis is gradual, and there may be up to 20 instars before adulthood is reached. Figures 3.4 and 3.5 picture the pillbug, sometimes called the sowbug.

### 3. Centipedes and Millipedes

Centipedes are made up of 30 segments, each containing one pair of legs. They have chewing mouthparts. Some species can inflict painful bites on humans.

**Figure 3.3** The adult brown recluse spider (EPA, *Urban Integrated Pest Management: A Guide for Commercial Applicators*).

**Figure 3.4** A sowbug, an example of the crestacea (EPA, *Urban Integrated Pest Management: A Guide for Commercial Applicators*).

Millipedes contain 30 segments and are cylindrical like an earthworm. The body is wingless, and each segment bears two pairs of legs. The antennae are short, and mouthparts are comblike. Millipedes feed on decaying organic matter, seeds, bulbs, and roots.

There is no metamorphosis; centipedes and millipedes do not change except in size between hatching and reaching the adult stage. Figure 3.6 illustrates a centipede, and Figure 3.7 illustrates a millipede.

## D. CONTROLLING INSECTS

Control of insects and their relatives may involve any of the three basic pest control objectives. Control is usually aimed at **suppression** of pests to a point where the presence or damage level is acceptable. **Prevention and eradication** are useful only in relatively small, confined areas such as indoors or in programs designed to keep foreign pests out of a new area.

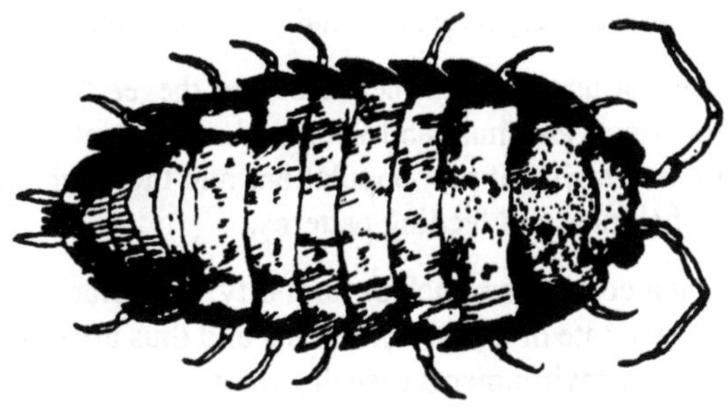

**Figure 3.5** Diagram of a pillbug (sowbug) showing the seven pairs of legs (EPA, *Urban Integrated Pest Management: A Guide for Commercial Applicators*).

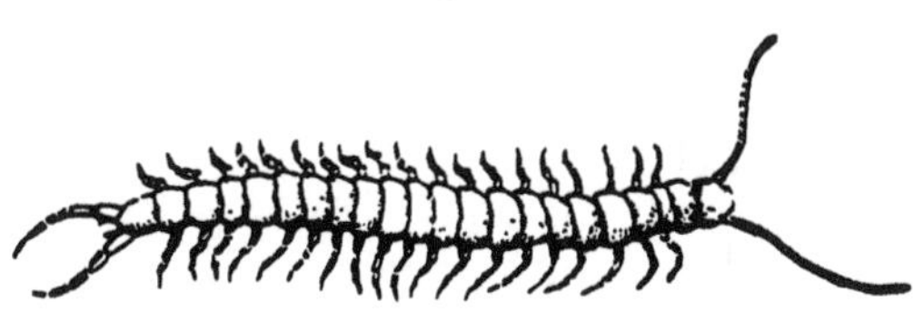

**Figure 3.6** The centipede, a many-segmented arthrodod with one pair of legs attached to each segment (EPA, *Urban Integrated Pest Management: A Guide for Commercial Applicators*).

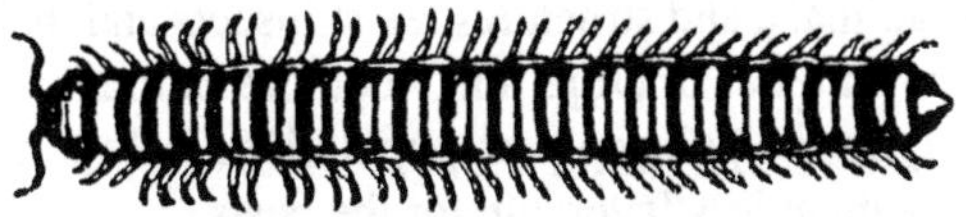

**Figure 3.7** The millipede, a many-segmented arthropod with two pair of legs attached to each segment (EPA, *Urban Integrated Pest Management: A Guide for Commercial Applicators*).

The key to successful control of insect and insect-like pests is knowledge of the stage(s) of their life cycle in which they are most vulnerable. It is generally difficult to control insects in either the egg or pupal stage, because these stages are inactive: not feeding, immobile, and often in inaccessible areas such as underground, in cocoons or cases, and in cracks or crevices.

Controlling insects in the late instar and adult stages is moderately successful. The insects, because of their size, are most visible in these stages and usually are causing the most destruction. Therefore, control attempts are often begun at these times. However, the larger insects are often more resistant to pesticides, and adults already may have laid eggs for another generation.

The best control usually is achieved during the early larval or nymphal stages when the insects are small and vulnerable. Control during these stages requires careful monitoring of pest populations and thorough knowledge of the pest's life cycle, habitats, and feeding patterns.

Environmental conditions, such as humidity, temperature, and availability of food, can alter the rate of growth of insects and thus affect the length of the life cycle. Optimum environments (usually warm and humid) can decrease the time of development from egg to adult.

## 1. Insect Control Strategy

Control methods used for insects include:

- host resistance
- biological control
- cultural control
- mechanical control
- sanitation
- chemical control.

### *a. Host Resistance*

Some crops, animals, and structures resist insects and their relatives better than others. Some varieties of crops and wood are immune to certain insects. Use of resistant types helps keep pest populations below harmful levels by making the environment less favorable for the pests.

*b. Biological Control*

Biological control of insects includes:

- predators and parasites
- pathogens
- sterile males
- pheromones
- juvenile hormones.

**Predators and Parasites**—Organisms known to attack insect (and insect-like) pests in their native environment can be imported or reared in laboratories and released in infested areas. This is done only after the parasites or predators are determined to be harmless to man, animals, plants, and other beneficial organisms. For example, several kinds of parasites and predators of the alfalfa weevil have been imported from Europe and Asia and released in the infested areas in this country. Several species have become established and are helping to reduce pest numbers. However, they do not always prevent serious outbreaks and the resultant damage.

**Pathogens**—Parasitic bacteria, viruses, and fungi may be introduced into an infested area to control insects by subjecting them to disease. These disease agents, like predators, are often found in the pest's native environment. They can be imported or they can be reared in laboratories.

For example, the use of pathogens is an important part of the pest control program for Japanese beetles. Japanese beetles are subject to attack by two naturally occurring species of bacteria which cause the fatal milky disease. Preparations containing spores of the contagious bacteria are produced commercially and released in infested areas.

**Sterile Males**—Males of some pest insect species may be reared and sterilized in laboratories and released in large numbers into infested areas to mate with native females. These matings produce infertile eggs or sterile offspring and help reduce the pest population. This technique has been used successfully in only a few species and is still being developed. The screw worm, which attacks cattle, is one insect on which this technique has been effective.

**Pheromones**—Some insects (and insect-like organisms) produce natural chemicals, called pheromones, which cause responses in other insects of the same or very closely related species. Once a particular insect pheromone is identified and the chemical is synthetically produced, it can be used to disrupt the behavior of that insect species. Synthetic pheromones may be used to disrupt normal reproduction, or they may be used to attract the pests into a trap.

Because each pheromone affects only one specific group of insects, their use poses no risk of harm to other organisms, including man. Unfortunately, only a few have been discovered and produced synthetically, and the use of pheromones is still in the experimental stages. It is very costly to discover, produce, and market a chemical which will be useful in controlling only one pest species.

**Juvenile Hormones**—Another type of species-specific chemical is also being developed. Juvenile hormones interrupt the metamorphosis of insects (and insect-like organisms). These chemicals prevent reproduction by keeping immature insects from maturing into adults. Each chemical acts against a single pest species and has the same advantages and disadvantages as pheromones. The few juvenile hormones available are usually applied as a broadcast spray to reach as many target pests as possible.

*c. Cultural Control*

Cultural control methods for insects include:

- crop rotation
- trap crops
- delay of planting
- harvest timing.

**Crop Rotation**—Taking infested fields out of production and leaving them fallow or planting an alternate crop may deprive pests of host plants on which to feed and reproduce. Rotations are most effective against insects which have long life cycles and infest the crop during all stages of growth. Many of the traditional rotational schemes were developed to reduce pest problems.

**Trap Crops**—Other crops attractive to the pests may be planted early or nearby to draw pests away from the main crop. Destruction of such crops at the proper time breaks the reproductive cycle of the pest before the desired crops are infested. To control the pickle worm in cucumbers, for example, the grower might also plant yellow squash, to which the pest is more attracted. The squash crop can be sprayed or destroyed before the pest can complete its development.

**Delay of Planting**—Delaying the date of planting may reduce the population of certain pests by eliminating the host plant needed for food and reproduction when the pest population is at its peak. For example, prevention of Hessian fly damage in wheat can be avoided by delaying planting until fly reproduction has ended for the year.

**Harvest Timing**—Crops should not be left in the field after maturity if they are susceptible to pest attack. For example, wireworm damage to mature potatoes causes a serious quality reduction. Damage increases if the crop is left in the ground even for a very short time after maturity.

*d. Mechanical Control*

Mechanical controls used on insects are:

- screens and other barriers
- traps
- light
- heat and cold
- radiation and electrocution.

**Screens and Other Barriers**—A major aspect of insect control indoors is the use of screens and other barriers to keep insects out. Flying insects, such as mosquitoes, wasps, and flies, are kept outside by blocking any openings with screening. The effective mesh size depends on the size of the smallest flying insect pests in that environment. Crawling insects are also kept outside by screens or by other barriers such as tightly sealed doors and windows. Barriers made of sticky substances sometimes can be used to stop crawling insects from entering an area.

**Traps**—Traps are sometimes used to control the target pest. More often, however, they are used to survey for the presence of insect pests and to determine when the pest population has increased to the point where control is needed.

**Light**—Many insect pests may be attracted to artificial light at night. However, since not all the pests are killed, the light attractant may actually help create infestations.

**Heat and Cold**—It is sometimes possible to expose insect pests to the killing effects of the heat of summer or cold of winter. Insects that feed on stored grain and flour, for example, can sometimes be controlled by ventilating grain elevators in winter.

**Radiation and Electrocution**—Radiation and electrocution are sometimes used to kill pests in a limited area. The electric screens in such places as outdoor restaurants and amusement parks are used to attract and electrocute a variety of nocturnal insect pests. Ionizing radiation is used to sterilize pests by destroying reproductive tissues, and ultrasonic radiation is used to kill pests in some products.

*e. Sanitation*

Cultivation, moldboard plowing, and burning of crop residues soon after harvest greatly aid in the control of some insect pests on agricultural crops. Pink bollworm infestations in cotton, for example, can be greatly reduced by plowing the field immediately after harvest.

Removing litter from around buildings helps control pests which use it for breeding or shelter. Ants, termites, and some other indoor pests may be suppressed by using this technique.

Sanitation is important in the control of animal parasites and filth flies. Fly control in and around barns and livestock pens, for example, is greatly aided by proper manure management. A major aspect of fly control in residential areas and cities is weekly or biweekly garbage removal. This scheduling prevents fly eggs and maggots in the garbage from reaching adult fly stage, since the fly's life cycle is 10 to 14 days even in very warm weather.

Indoors, sanitation is a major method of preventing insect pest problems. Keeping surfaces in restrooms and food preparation areas immaculately clean and dry is an important factor in suppressing or eliminating ant, fly, and cockroach infestations.

*f. Chemical Control*

Chemicals used to control insects and insect-like pests include insecticides, miticides, and acaricides. Most chemicals used to control insects act in one of two ways:

- repellents—These products keep pests away from an area or from a specific host. Products designed to keep mosquitoes, chiggers, and ticks off humans are an example.
- direct poisons—Common insecticides include chemicals that poison one or more life systems in the pest. Some will poison an insect if they are eaten (stomach poisons); others require only contact with the insect's body (contact poisons).

A few insecticides interfere mechanically with the insect's functions. For example, mineral oils suffocate insects; silica dusts destroy their body water balance by damaging their protective wax covering.

**Outdoors**—With few exceptions, insecticides labeled for outdoor use are designed to be used for full coverage of an area. The objective is to cover the entire surface to be protected with a residue of active insecticide. Insects which then eat or otherwise contact the treated surface are killed.

Thorough knowledge of the target insects helps determine the frequency of application and the choice of chemicals. One well-timed application of an effective pesticide may provide the desired control. Sometimes repeated insecticide applications will be necessary as the insect infestation continues and pesticide residues break down.

The pesticide label, Cooperative Extension Service recommendations, and other sources usually indicate a range of treatment intervals and dosages. By carefully observing the pest problem and applying chemicals when the pests are most vulnerable, you often will be able to use lower doses of pesticides and apply them less often. Over a long growing period, this can mean considerable savings in time, money, and total pesticide chemicals applied.

Most control strategies take advantage of the natural controls provided by the pest's natural enemies. When choosing a pesticide, consider what effect it will have on these beneficial organisms. Ask your pesticide dealer, agricultural extension agent, or other experts for advice.

**Indoors**—Most indoor insect control is aimed at prevention or eradication of the pest problem while minimizing the exposure of humans and animals to chemicals. The most common application techniques are crack and crevice treatments, spot treatments, and fumigation of entire structures, commodities, or individual pieces of equipment.

## III. PLANT DISEASE AGENTS

A plant disease is any harmful condition that makes a plant different from a normal plant in its appearance or function. Plant diseases caused by biological agents (pathogens) are of primary interest to pesticide applicators because they often can be controlled with pesticides. Pathogens include:

- fungi
- bacteria
- viruses and mycoplasmas
- nematodes.

Parasitic seed plants, discussed in the section on weeds, are sometimes considered plant disease agents because of the type of injury they cause to the host plant.

## A. PATHOGENIC PLANT DISEASES

Pathogens which cause plant disease are parasites which live and feed on or in host plants. They can be passed from one plant to another. Three factors are required before a pathogenic disease can develop—a susceptible host plant, a pathogenic agent, and an environment favorable for development of the pathogen.

A pathogenic disease depends on the life cycle of the parasite. The environment greatly affects this cycle. Temperature and moisture are especially important. They affect the activity of the parasite, the ease with which a plant becomes diseased, and the way the disease develops.

The disease process starts when the parasite arrives at a part of a plant where infection can occur. If environmental conditions are favorable, the parasite will begin to develop. If the parasite can get into the plant, the infection starts. The plant is diseased when it responds to the parasite.

The three main ways a plant responds are:

- overdevelopment of tissue, such as galls, swellings, and leaf curls
- underdevelopment of tissue, such as stunting, lack of chlorophyll, and incomplete development of organs
- death of tissue, such as blights, leaf spots, wilting, and cankers.

The parasites which cause plant diseases may be spread by wind, rain, insects, birds, snails, slugs, and earthworms, transplant soil, nursery grafts, vegetative propagation (especially in strawberries, potatoes, and many flowers and ornamentals), contaminated equipment and tools, infected seed stock, pollen, dust storms, irrigation water, and people. Figure 3.8 illustrates the effects of microbial infection of various fruit trees.

### 1. Fungi

Fungi are plants that lack chlorophyll and cannot make their own food. They get food by living on other organisms. Some fungi live on dead or decaying organic matter. Most fungi are beneficial because they help release nutrients from dead plants and animals and thus contribute to soil fertility. These fungi are a pest problem when they rot or discolor wood. They can do considerable damage to buildings and lumber which are improperly ventilated or in contact with water or high humidity.

Most fungi which cause plant diseases are parasites on living plants. They may attack plants and plant products both above and below the soil surface. Some fungus pathogens attack many plant species, but others are restricted to only one host species. See Figures 3.8 and 3.9.

**Figure 3.8** Different types of microbial infestation of fruit trees.

Most fungi reproduce by spores, which function about the same way as seeds. Fungal spores are often microscopic in size and are produced in tremendous numbers. Most of them die because they do not find a host plant to attack. Some can survive for weeks or months without a host plant. Water or high humidity (above 90 percent) are nearly always essential for spore germination and active fungal growth. Spores can spread from plant to plant and crop to crop. Mildew and smut are examples of fungus diseases.

## 2. Bacteria

Bacteria are microscopic, one-celled organisms. They usually reproduce by single cell division. Each new cell possesses all the characteristics of the parent cell. Bacteria can multiply fast under warm, humid weather conditions. Some can divide every 30 minutes. Bacteria may attack any part of a plant, either above or below the soil surface. Many leaf spots and rots are caused by bacteria. Table 3.8 shows examples of leaf spots and fruit rot.

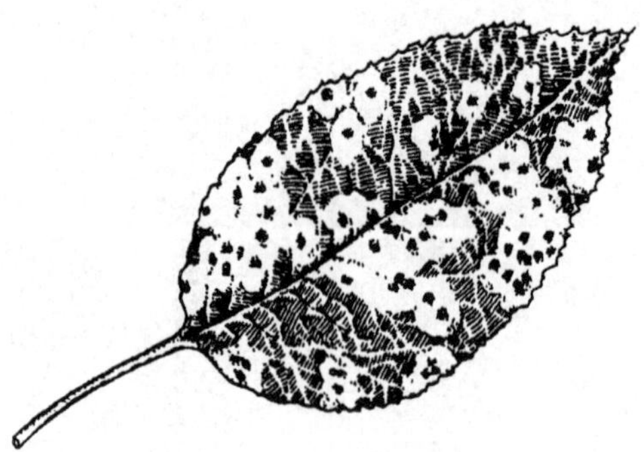

**Figure 3.9** Fungal infection of fruit tree (EPA, *Applying Pesticides Correctly: A Guide for Private and Commercial Applicators*, 1983).

## 3. Viruses and Mycoplasmas

Viruses and mycoplasmas are so small that they cannot be seen with an ordinary microscope. They are generally recognized by their effects on plants. Often it is difficult to distinguish between diseases caused by viruses or mycoplasmas and those caused by other plant disease agents such as fungi and bacteria.

Viruses depend on other living organisms for food and to reproduce. They cannot exist separately from the host for very long and technically are not considered to be living organisms. They can induce a wide variety of responses in the host plants. A few can kill the plant. More commonly the response is lowered product quality and reduced yields. Mosaic diseases, for example, are usually caused by viruses.

Mycoplasmas are the smallest known independently living organisms. They can reproduce and exist apart from other living organisms. They obtain their food from plants. Yellow diseases and some stunts are caused by mycoplasmas.

### 4. Nematodes

Nematodes are small, usually microscopic, eel-like roundworms. Many nematodes are harmless. Others attack and feed on plants grown for food, feed, ornamentals, turf, or forests. Some species attack above-ground plant parts, such as leaves, stems, and seeds, but most pest nematode species feed on or in the roots. They may feed in one location, or they may constantly move throughout the roots. The root-feeding nematodes directly interfere with water and nutrient uptake. Typical host plant symptoms include stunting, yellowing, loss of vigor, and general decline. Nematode damage may go unrecognized or be blamed on something else, such as nutrient deficiencies.

All nematodes that are parasites on plants have a hollow style which they use to puncture plant cells and feed on the cell contents. Nematodes may develop and feed either inside or outside of a plant. They move with an eel-like motion in water, even water as thin as the film of moisture around plant cells or soil particles. Their life cycle includes egg, several larval stages, and adult. Most larvae look like adults, but are smaller.

In adverse conditions, the females of some species, such as root knot and cyst nematodes, form an inactive, resistant form called a cyst. The cyst is the hard, leathery, egg-filled body of the dead female. It is difficult to penetrate with pesticides. Cysts may provide protection for several hundred eggs for as long as 10 years. Figure 3.10 shows the life cycle of a typical nematode.

## B. DIAGNOSIS OF PLANT DISEASE

Attempting to control plant diseases without sufficient information usually results in failure. For maximum effectiveness, the first step is to diagnose the disease correctly.

Diseased plants may be recognized by comparing them with healthy plants. Knowledge of normal growth habits is necessary for recognition of a diseased condition. To identify the cause of plant disease, you must observe:

- symptoms—reaction of the host plant to the disease agent
- signs—actual evidence of the presence of the disease agent

Many plant diseases cause similar symptoms in the host plants. Such things as leaf spots, wilts, galls on roots, or stunted growth may be caused by many different agents, including many that are not pathogens. For example, the symptoms may be a result of mechanical injury, improperly applied fertilizers and pesticides, or frost. Often the only way to pinpoint the cause is by finding the signs of the particular disease agent—such as fungal spores and mycelium or bacterial ooze. Many pathogenic disease agents, including some fungi, bacteria, and nematodes, may have to be positively identified by an expert with access to sophisticated laboratory procedures. However, other pathogenic diseases occur regularly on specific agricultural, ornamental, and forestry plantings, and the appearance of specific symptoms is enough to correctly identify the cause.

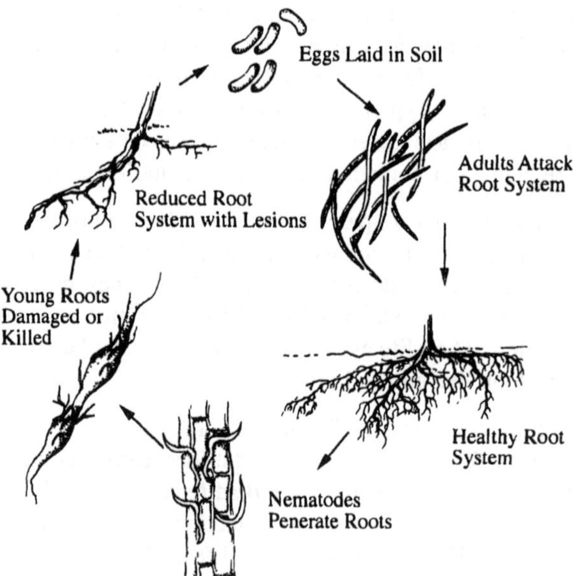


**Figure 3.10** Life cycle of a disease-causing nematode.

## C. CONTROLLING PLANT DISEASE

At present, plant disease control measures are mainly preventive. Once a plant or plant product is infected and symptoms appear, few control methods, including pesticides, are effective.

### 1. In Greenhouses

In greenhouse production, the keys to prevention of plant disease buildup are:

- host resistance—use of disease-resistant varieties

- mechanical control—use of temperature and humidity control, and soil sterilization by heat
- chemical control—use of chemicals for specific pest outbreaks and chemical soil sterilization.

The major elements of the environment can be manipulated in a greenhouse to prevent pathogens from building up rapidly enough to damage crops.

## 2. On Stored Food and Feed

As in greenhouses, the environment in storage facilities can be manipulated. To achieve disease control in storage, you must have good sanitation in the storage facility before storage, make sure that the crop is relatively pathogen-free at time of storage, and provide adequate ventilation to control the buildup of temperature and humidity.

Ideal storage conditions vary from crop to crop. Favorable environmental conditions must be maintained to ensure the quality of the product being stored. However, if these same conditions favor a rapid development of plant diseases, the two factors must be weighed carefully.

## 3. Outdoors

The main methods for control of plant disease outdoors include:

- host resistance
- cultural control
- mechanical control
- sanitation
- chemical control.

### a. *Host Resistance*

The use of disease-resistant varieties is usually one of the most effective, long-lasting, and economical ways to control plant disease, if the resistant varieties are otherwise acceptable. Resistant varieties have long been one of the major factors in maintaining high levels of crop productivity in the United States. For certain crops, 95 to 98 percent of the acreage grown is planted with varieties that are resistant to specific diseases.

In some cases, resistant varieties are the only way to ensure continued production. For many diseases in low-value forage and field crops, for example, chemical controls are too costly. For other diseases, such as many soil-borne pathogens, no economical or effective chemical control method is available.

### *b. Cultural Control*

A pathogen and its host must be brought together under specific environmental conditions for a plant disease to develop. Cultural practices are used to alter the environment, the condition of the host, or the behavior of the pathogen to prevent an infection.

**Crop Rotation**—Pathogenic organisms nearly always can be carried over from one growing season to the next in the soil or in plant debris. Continual production of the same or closely related crops on the same piece of land leads to a disease buildup. Crop rotation reduces the buildup of pathogens, but seldom provides complete disease control. Obviously, crop rotation is not always possible, practical, or desirable. Perennial crops such as trees, woody ornamentals, and turfgrass must remain in one location for many years. Some crops, such as corn, cotton, or wheat, often are more practical to grow on the same land year after year despite the potential for a buildup of plant disease pathogens.

**Planting Time**—Cool-weather crops, such as spinach, peas, and some turfgrass, are subject to attack by certain diseases if planted when the temperatures are warmer. They often emerge and establish poorly under such conditions. Conversely, beans, melons, and many flowers should be planted under warm conditions to avoid disease.

**Seed Aging**—Some seed pathogens can be deactivated by holding the seed in storage. Proper storage conditions are essential to ensure that seed viability is not lowered.

### *c. Mechanical Control*

Hot water treatments are effective in producing clean seed and planting materials. Seed and vegetative propagation materials (such as roots, bulbs, corms, and tubers) may be treated before planting to eliminate some fungal, bacterial, and viral diseases.

### *d. Sanitation*

**Pathogen-Free Seed Stock**—Production of clean seed stock is important in reducing plant disease spread. Often, seeds are grown in arid areas where the amount of moisture is controlled by an irrigation system. This eliminates infection by diseases which require high moisture and humidity levels.

**Pathogen-Free Propagation**—Plant disease pathogens are frequently carried in or on vegetative propagation materials (such as roots, bulbs, tubers, corms, and cuttings). Production of clean planting stock is especially important in the culture of certain high-value agricultural and ornamental crops. These plants must be grown in pathogen-free greenhouses or in sites isolated from growing areas for these crops. When planning for isolation, consider how far the pathogen may spread, how the pathogen is spread, and the distance between potential growing sites.

**Clean Planting Sites**—In some crops, certain plant disease pathogens can be controlled or reduced by eliminating other nearby plants which are hosts for the same disease organisms. These may be:

- plants which harbor the pathogens, such as weeds around field borders, ditch banks, and hedgerows
- plants which the organism requires for one stage of its life cycle. An apple grower, for example, can control cedar apple rust by eliminating nearby cedar (juniper) trees.

**Removing Infected Plants**—Diseases often can be controlled by systematically removing infected plants or plant parts before the disease pathogen spreads to other "clean" plants. This method can be especially important for the control of some viral and mycoplasma pathogens for which no other controls are available.

**Crop Residue Management**—Infected crop residues often provide an ideal environment for carryover of many pathogens. In some cases the pathogens increase greatly in the residues. Three basic techniques are used in crop residue management:

- deep plowing buries pathogen-infested residues and surface soil and replaces them with soil which is relatively free from pathogens
- fallowing reduces pathogen carry-over because their food source decays and is no longer available
- burning kills some pathogens and removes the residue on which they live. This practice may not be legal in some areas.

**Disinfecting Equipment and Tools**—Some plant diseases can be spread from plant to plant, field to field, and crop to crop by workers and their equipment. Disinfecting equipment, tools, and clothing before moving from an infected area to a disease-free area can prevent or delay disease spread. This method of disease spread is especially important in high humidity and wet field conditions, because the pathogens are transported in the droplets of water which form on the equipment, tools, and skin.

*e. Chemical Control*

Chemicals used to control plant disease pathogens include fungicides, bactericides (disinfectants), and nematicides. The general term "fungicide" is often used to describe chemicals which combat fungi and bacteria. Fungicides may be classified as protectants, eradicants, and systemics.

**Protectants** must be applied before or during infection by the pathogen. In order to be effective, they must either persist or be applied repeatedly. Most chemicals now available to combat plant diseases are protectants.

**Eradicants** are less common and are applied after infection has occurred. They act on contact by killing the organism or by preventing its further growth and reproduction.

**Systemics** are used to kill disease organisms on living plants. Systemic chemicals are transported in the sap stream from the application site to other plant parts. This type of chemical may act as both a protectant and an eradicant.

Successful chemical control of plant diseases requires proper timing. Plant disease control on some crops must begin **before** infection occurs. The protectant chemical must be applied when environmental conditions are expected to be ideal for the development of plant pathogens. If the protectant is not applied in time, major crop damage may result, or the application of the more expensive eradicant sprays may be needed. Label directions often call for routine protectant applications every 7 to 10 days during periods of prime infection risk. Almost all plant disease control chemicals are applied as cover sprays. The purpose is to reach and protect all potential sites of infection.

## IV. WEEDS

Any plant can be considered a weed when it is growing where it is not wanted. Weeds are a problem because they reduce crop yields, they increase costs of production, and they reduce the quality of crop and livestock products. In addition, some cause skin irritation and hay fever, and some are poisonous to man and livestock. Weeds also can spoil the beauty of turf and landscape plants.

Weeds harm desirable plants by:

- competing for water, nutrients, light, and space

- contaminating the product at harvest
- harboring pest insects, mites, vertebrates, or plant disease agents
- releasing toxins in the soil which inhibit growth of desirable plants.

Weeds may become pests in water by:

- hindering fish growth and reproduction
- promoting mosquito production
- hindering boating, fishing, and swimming
- clogging irrigation ditches, drainage ditches, and channels.

Weeds can harm grazing animals by:

- poisoning
- causing an "off-flavor" in milk and meat.

Weeds are undesirable in rights-of-way because they:

- obscure vision, signs, guideposts, crossroads, etc.
- increase mowing costs
- hinder travel
- provide cover for rodents and other pest animals
- clog drainage areas.

In cultivated crops, the weeds that are favored by the crop production practices do best. The size and kind of weed problem often depends more on the crop production method, especially the use or nonuse of cultivation, than on the crop species involved. In noncrop areas, the weed problem may be affected by factors such as:

- weed control programs used in the past
- frequency of mowing or other traffic in the area
- susceptibility to herbicides.

## A. DEVELOPMENT STAGES

All plants have four stages of development:

- seedling—small, vulnerable plantlets
- vegetative—rapid growth: production of stems, roots, and foliage. Uptake and the movement of water and nutrients are rapid and thorough.

- seed production—energy directed toward production of seed. Uptake of water and nutrients is slow and is directed mainly to flower, fruit, and seed structures.
- maturity—little or no energy production or movement of water and nutrients.

## B. LIFE CYCLES OF PLANTS

### 1. Annuals

Plants with a one-year life cycle are annuals. They grow from seed, mature, and produce seed for the next generation in one year or less. They are grasslike (crabgrass and foxtail) or broad-leaved (pigweed and cocklebur). There are two types:

**Summer annuals** are plants that grow from seeds which sprout in the spring. They grow, mature, produce seed, and die before winter. Examples include: crabgrass, foxtail, cocklebur, pigweed, and lambsquarters.

**Winter annuals** are plants that grow from seeds which sprout in the fall. They grow, mature, produce seed, and die before summer. Examples include: cheat, henbit, and annual bluegrass.

### 2. Biennials

Plants with a two-year life cycle are biennials. They grow from seed and develop a heavy root and compact cluster of leaves (called a rosette) the first year. In the second year they mature, produce seed, and die. Examples include: mullein, burdock, and bull thistle. Figures 3.11 and 3.12 show several biennials.

### 3. Perennials

Plants which live more than two years and may live indefinitely are perennials. Perennial plants may mature and reproduce in the first year and then repeat the vegetative, seed production, and maturity stages for several following years. In other perennials, the seed maturity and production stages may be delayed for several years. Some perennial plants die back each winter; others, such as trees, may lose their leaves, but do not die back to the ground. Most perennials grow from seed; many also produce tubers, bulbs, rhizomes (belowground rootlike stems), or stolons (above-ground stems that produce roots). Examples of perennials are johnsongrass, field bindweed, dandelion, and plantain. Figure 3.13 shows several pictures of the parts of the dandelion, and Figure 3.14 shows the Trumpet-weed or Joepyeweed which is also a perennial.

**Simple perennials** normally reproduce by seeds. However, root pieces which may be left by cultivation can produce new plants. Examples are: dandelions, plantain, trees, and shrubs.

**Bulbous perennials** may reproduce by seed, bulblets, or bulbs. Wild garlic, for example, produces seed and bulblets above ground and bulbs below ground.

**Creeping perennials** produce seeds but also produce rhizomes (below-ground stems), or stolons (above-ground stems that produce roots). Examples include johnsongrass, field bindweed, and Bermudagrass.

## C. WEED CLASSIFICATION

### 1. Land Plants

Most pest plants on land are grasses, sedges, or broadleaves.

#### *a. Grasses*

Grasses have fibrous root systems. The growing point on seedling grasses is sheathed and located below the soil surface. Some grass species are annuals; others are perennials. Grass seedlings have only one leaf as they emerge from the seed. Their leaves are generally narrow and upright with parallel veins.

#### *b. Sedges*

Sedges are similar to grasses except that they have triangular stems and three rows of leaves. They are often listed under grasses on the pesticide label. Most sedges are found in wet places, but principal pest species are found in fertile, well-drained soils. Yellow and purple nutsedge are perennial weed species which produce rhizomes and tubers.

#### *c. Broadleaves*

Broadleaf seedlings have two leaves as they emerge from the seed. Their leaves are generally broad with netlike veins. Broadleaves usually have a taproot and a relatively coarse root system. All actively growing broadleaf plants have exposed growing points at the end of each stem and in each leaf axil. Perennial broadleaf plants may also have growing points on roots and stems above and below the surface of the soil. Broadleaves contain species with annual, biennial, and perennial life cycles.

**Figure 3.11** The Yellow Daisy or Coneflower is a plant with a two-year life cycle (biennial).

## 2. Aquatic Plants

### *a. Vascular Plants*

Many aquatic plants are similar to land plants and have stems, leaves, flowers, and roots. Most act as perennial plants—dying back and becoming dormant in the fall and beginning new growth in the spring. They are classified as:

- emergent (emersed)—most of the plant extends above the water surface. Examples are cattails, bulrushes, arrowheads, and reeds (Figure 3.15 shows the common cattail).
- floating—all or part of the plant floats on the surface. Examples are waterlilies, duckweeds, waterlettuce, and waterhyacinth.
- submergent (submersed)—all of the plant grows beneath the water surface. Examples are watermilfoil, elodea, naiads, pond weeds (Potamogeton), and coontails.

**Figure 3.12** Bull thistle is also a biennial.

Emergent and floating plants, like some land plants, have a thick outer layer on their leaves and stems which hinders herbicide absorption. Submergent plants have a very thin outer layer on their leaves and stems and are very susceptible to herbicide injury.

**Figure 3.13** Dandelion is a simple perennial that can reproduce by seed or by root cutting.

*b. Algae*

Algae are aquatic plants without true stems, leaves, or vascular systems. For control purposes, they may be classified as:

- plankton algae—microscopic plants floating in the water. They sometimes multiply very rapidly and cause "blooms" in which the surface water appears soupy green, brown, or reddish brown, depending on the algal type. Figure 3.16 shows a floating algal mass.
- filamentous algae—long, thin strands of plant growth which form floating mats or long strings extending from rocks, bottom sediment, or other underwater surfaces. Examples are cladophora and spirogyra.
- macroscopic freshwater algae—these larger algae look like vascular aquatic plants. The two should not be confused because their control is different. Many are attached to the bottom and grow to 2 feet tall. However, they have no true roots, stems, or leaves. Examples are chara and nitella.

**Figure 3.14** The Trumpet-weed or Joepyeweed is a native perennial with leaves in whorls and pink to purple flowers.

### 3. Parasitic Seed Plants

Dodders, broomrape, witchweed, and some mosses are important weeds on some agricultural, ornamental, and forest plants. They live on and get their food from the host plants. They can severely stunt and even kill the host plants by using the host plant's water, food, and minerals. These plants reproduce by seeds. Some can also spread from plant to plant in close stands by vining and twining.

**Figure 3.15** The common cattail is an emergent perennial weed that extends above the water surface.

## D. CONTROLLING WEEDS

Weed control is nearly always designed to suppress a weed infestation. Prevention and eradication are usually only attempted in regulatory weed programs.

To control weeds which are growing among or close to desirable plants, you must take advantage of the differences between the weeds and the desired species. Be sure that the plants you are trying to protect are not susceptible to the chosen weed control method. Generally, the more similar the desirable plant and the weed species are to one another, the more difficult weed control becomes. For example, broadleaf weeds are most difficult to control in broadleaved crops, and grass weeds are often difficult to control in grass crops.

**Figure 3.16** Floating algae mass.

## 1. Weed Control Strategy

A plan to control weeds may include:

- biological control
- cultural control
- sanitation
- chemical control.

### a. *Biological Control*

Biological weed control usually involves the use of insects and disease-causing agents which attack certain weed species. An example is the control of St. Johnswort by the Chrysolina beetle in the western United States. To be effective, biological control requires two things:

- the insect or disease must be specific to the weed to be controlled; otherwise, it may spread to other species—such as crops and ornamentals—and become a pest itself.
- the insects must have no natural enemies that interfere with their activity.

Grazing is another form of biological control sometimes used to control plant growth along ditches, fence rows, and roadsides. Sheep and goats are used most often, but geese are used for weeding some crops.

*b. Cultural Control*

**Tillage**—This is an effective and often-used method to kill or control weeds in row crops, nurseries, and forest plantings. However, tillage may bring buried seeds to the surface where they can either germinate and compete with the newly planted crop or be spread to nearby fields. Tillage also may increase soil erosion and may help to spread established plant diseases to uninfected areas of the field.

**Time of Planting**—Crops and turfgrass planted in the spring compete well against winter annual weeds. Sometimes the planting date can be delayed until after weeds have sprouted and have been removed by cultivation or by herbicides.

**Nurse Crops**—Plant species (usually annuals) which germinate quickly and grow rapidly are sometimes planted with a perennial crop to provide competition with weeds and allow the crop to become established. The nurse crop is then harvested or removed to allow the perennial crop to take over. For example, oats are sometimes used as a nurse crop to help establish alfalfa or clover. Annual ryegrass is sometimes used in mixtures to provide a nurse crop for perennial rye, fescue, or bluegrass.

**Burning**—Fire may be used to control limited infestations of annual or biennial weeds. Fire destroys only the above-ground parts of plants and is usually not effective against many herbaceous perennial weeds.

**Mulching**—Mulching is used to prevent light from reaching weed seeds, thus preventing weed growth between rows, around trees and shrubs, or in other areas where no plants are desired.

**Mowing**—Mowing may be used to reduce competition between weeds and crops and to prevent flowering and seeding of annual or biennial weeds. Mowing is often used in orchards to control weeds and prevent soil erosion. To be most effective, mowing height must be adequate to ensure control of weed plants and encourage desired vegetation. Mowing is an important aspect of turfgrass weed control.

Mowing and harvesting is good for both short-term and long-term control of aquatic weeds. It depletes the nutrients, removes seeds, and reduces vegetative spread.

**Flooding**—Flooding has long been used for weed control in rice. The water covers the entire weed, killing it by suffocation.

**Reduced Tillage**—This method has been used successfully to reduce weed growth and to reduce soil erosion. With limited tillage, weed seeds are not turned up and those that do germinate do not have as much light or space to get started. However, the remaining debris may harbor insects and plant disease agents.

**Shading**—Aquatic weeds are sometimes controlled by shading them with floats of black plastic, adding dye to the water, or using similar methods for shading out the sunlight. Land weeds can be shaded by planting crops so closely together that they block the light from emerging weeds.

### c. *Sanitation*

The use of seeds with few weed-seed contaminants is important in reducing weed problems.

### d. *Chemical Control*

Chemicals used to control weeds are called herbicides. They kill plants by contact or systemic action. **Contact herbicides** kill only the plant parts which the chemical touches. **Systemic herbicides** are absorbed by roots or foliage and carried throughout the plant. Systemic herbicides are particularly effective against perennial weeds because the chemical reaches all parts of the plant—even deep roots and woody stems, which are relatively inaccessible. Contact herbicides are usually used to control annuals and biennials and are characterized by the quick die-back they cause. Systemics may take a longer time to provide the desired results—up to 2 or 3 weeks, or even longer for woody perennials.

Herbicide activity is either selective or nonselective. Selective herbicides are used to kill weeds without significant damage to nearby plants. They are used to reduce weed competition in crops, lawns, and ornamental plantings. Nonselective herbicides are chemicals that kill all plants present if applied at an adequate rate. They are used where no plant growth is wanted, such as fence rows, ditch banks, driveways, roadsides, parking lots, and recreation areas.

Herbicide selectivity may vary according to the application rate. High rates of selective herbicides usually will injure all plants at the application site. Some nonselective herbicides can be used selectively by applying them at a lower rate. Other factors that affect selectivity include the time and method of application, environmental conditions, and the stage of plant growth.

Several factors affect a plant's susceptibility to herbicides:

**Growing Points**—Those that are sheathed or located below the soil surface are not reached by contact herbicide sprays.

**Leaf Shape**—Herbicides tend to bounce or run off narrow, upright leaves. Broad, flat leaves tend to hold the herbicide longer.

**Wax and Cuticle**—Foliar sprays may be prevented from entering the leaf by a thick wax and cuticle layer. The waxy surface also tends to cause a spray solution to form droplets and run off the leaves.

**Leaf Hairs**—A dense layer of leaf hairs holds the herbicide droplets away from the leaf surface, allowing less chemical to be absorbed into the plant. A thin layer of leaf hairs causes the chemical to stay on the leaf surface longer than normal, allowing more chemical to be absorbed into the plant.

**Size and Age**—Young, rapidly growing plants are more susceptible to herbicides than are larger, more mature plants.

**Deactivation**—Certain plants can deactivate herbicides and are less susceptible to injury from these chemicals. Such plants may become dominant over a period of time if similar herbicides are used repeatedly.

**Stage in life Cycle**—Seedlings are very susceptible to herbicides and to most other weed control practices. Plants in the vegetative and early bud stages are very susceptible to translocated herbicides. Plants with seeds or in the maturity stage are the least susceptible to weed control practices.

**Timing of Stages in the Life Cycle**—Plants that germinate and develop at different times than the crop species may be susceptible to carefully timed herbicide applications.

## 2. Chemicals Which Change Plant Processes

Plant growth regulators, defoliants, and desiccants are classified as pesticides in federal laws. These chemicals are used on plants to alter normal plant processes in some way. They must be measured carefully, because they usually are effective in very small amounts. Overdosing will kill or seriously damage the plants.

A plant growth regulator will speed up, stop, retard, prolong, promote, start, or in some other way influence vegetative or reproductive growth of a plant. These chemicals are sometimes called growth regulators or plant regulators. They are used, for example, to thin apples, control suckers on tobacco, control the height of some floral potted plants, promote dense growth of ornamentals, and stimulate rooting.

A defoliant causes the leaves to drop from plants without killing the plants. A desiccant speeds up the drying of plant leaves, stems, or vines. Desiccants and defoliants are often called "harvest-aid" chemicals. They usually are used to make harvesting of a crop easier or to advance the time of harvest. They are often used on cotton, soybeans, tomatoes, and potatoes.

## V. MOLLUSKS

Mollusks are a large group of land and water animals including slugs, oysters, clams, barnacles, and snails. They have soft, unsegmented bodies and often are protected by a hard shell.

### A. SNAILS AND SLUGS

Land snails and slugs are soft-bodied and have two pairs of antennae-like structures. Their bodies are smooth and elongated. Snails have a spiralshaped shell into which they can completely withdraw for protection when disturbed or when weather conditions are unfavorable. Slugs do not have a shell and must seek protection in damp places.

Snails and slugs feed on plants at night. They tear holes in foliage, fruits, and soft stems, using a rasp-like tongue. They may eat entire seedlings. As they move, snails and slugs leave a slime-like mucous trail which dries into silvery streaks. These streaks are undesirable on floral and ornamental crops and on those portions of crops to be sold for human food.

Snails and slugs deposit eggs in moist, dark places. The young mature in a year or more, depending on the species. Adults may live for several years. They overwinter in sheltered areas. They are active all year in warm regions and in greenhouses. Figure 3.17 shows a land snail and Figure 3.18 pictures a soft-bodied slug.

### B. SHIPWORMS

Shipworms are marine mollusks which cause extensive damage by boring into wood which is in contact with salt or brackish water. Of all the marine borers, shipworms are the most rapid destroyers of untreated wood. Adults lay eggs which hatch into free-swimming larvae. The larvae bore into submerged wood (marine pilings, wharf timbers, wooden boats) using a pair of boring shells located on their heads. The tail remains at the entrance, but the body grows in length and diameter as the animal extends its tunnel up to several feet in length. Shipworms use the wood for food and shelter. The internal structure of the wood may be honeycombed with tunnels, but only a tiny (1/16 to 1/8 inch) entrance hole will be visible on the surface.

**Figure 3.17** Land snails are soft-bodied and have two pairs of antennae-like structures. They have a spiral shaped shell which they can use for protection.

**Figure 3.18** Soft-bodied slug. Slugs do not have shells.

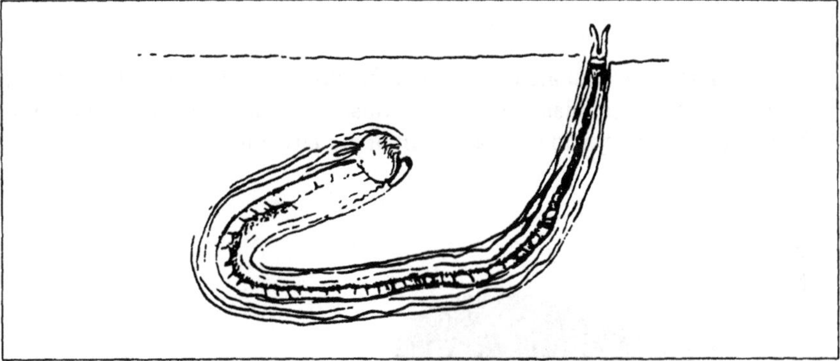

**Figure 3.19** Shipworms are marine mollusks which cause extensive damage by boring into wood (EPA, *Applying Pesticides Correctly: A Guide for Private and Commercial Applicators*, 1983).

## C. PHOLADS

Pholads are another species of marine mollusks which attack and destroy submerged wood. Pholads look like small clams and remain enclosed in a shell even as adults. Like shipworms, they burrow within the wood. After boring a small (1/8-inch) entrance hole, they enlarge their burrow to accommodate the growth of their bodies. Pholads are of economic importance mainly along the Gulf Coast and in Hawaiian waters.

## D. CONTROLLING MOLLUSKS

Mollusk pests on land (usually snails and slugs) can be controlled by many of the same techniques that are used to control insects outdoors. Effective techniques include:

- cultural practices—especially cultivation and trap crops
- mechanical controls—especially traps and barriers
- sanitation—especially eliminating crop debris and other sources of moisture
- chemicals—many insecticide formulations also control mollusks. In addition, specific molluscicides are available, usually as baits.

Marine mollusk control usually depends on impregnating wood with preservatives or applying repellent coatings to prevent infestations.

## VI. VERTEBRATE PESTS

All vertebrate animals have jointed backbones. They include mammals, birds, reptiles, amphibians, and fish. Most vertebrate animals are not pests. They are a necessary and enjoyable part of our environment.

**Figure 3.20** Typical rodent (vertebrate) has a widespread habitat. They cause damage to dwellings, stored products, and cultivated crops.

A few vertebrate animals can be pests in some situations. Some, such as birds, rodents, raccoons, or deer, may eat or injure agricultural and ornamental crops. Birds and mammals may eat newly planted seed. Birds and rodents consume stored food and often contaminate and ruin even more than they eat. Birds and mammals which prey on livestock and poultry cause costly losses to ranchers each year. Large numbers of roosting birds can soil populated areas.

Rodents, other mammals, and some birds are potential reservoirs of serious diseases of humans and domestic animals such as rabies, plague, and tularemia. Rodents are an annoyance and a health hazard when they inhabit homes, restaurants, offices, and warehouses.

Burrowing and gnawing mammals may damage dams, drainage and irrigation tunnels, turf, and outdoor wood products such as furniture and building foundations. Beavers may cause flooding in low-lying land by building dams.

Undesirable fish species may crowd out desirable food and sport species. The few poisonous species of snakes and lizards become a problem when humans, livestock, or pets are threatened. Water snakes and turtles may cause disruption or harm in fish hatcheries or waterfowl nesting reserves.

## A. CONTROLLING VERTEBRATES

As in insect pest control, techniques for control of vertebrate pests depend on whether the pest problem is indoors or outdoors.

Indoor vertebrate pest control usually is aimed at preventing pest entrance and eradicating pest infestations. Nearly all indoor vertebrate pests are rodents, but others, such as bats, birds, and raccoons, may also require control.

Outdoors, the strategy usually is to suppress the pest population to a level where the damage or injury is economically acceptable.

Local and state laws may prohibit the killing or trapping of some animals such as birds, coyotes, muskrats, and beavers without special permits. Always check with local authorities before beginning a control program.

### 1. Vertebrate Control Strategy

Methods of vertebrate pest control include:

- mechanical control
- sanitation
- chemical control.

#### a. *Mechanical Control*

Mechanical control methods for vertebrate pests include traps, barriers, gunning, attractants, and repellents.

**Traps**—Traps are sometimes desirable in vertebrate pest control. Leg-hold traps have been used traditionally, but such traps are nonselective and may injure nontarget animals. Traps which quickly kill only target pests are more desirable. Traps should be checked daily to maintain their effectiveness.

**Barriers**—Barriers are designed to prevent pests from passing. These include fences, screens, and other barriers which cover openings, stop tunneling, and prevent gnawing. Materials used include sheet metal, hardware, cloth, concrete, asbestos board, and similar materials. This kind of approach is especially effective in control of rodents, bats, and birds in structures.

**Gunning**—Gunning, though highly selective, is expensive and time consuming. It works best in combination with other methods. It will often affect larger predators not controlled by traps or toxic devices.

**Attractants**—Many techniques such as light and sound are used to attract pests to a trap. Predator calling can increase the efficiency of gunning efforts on larger predators.

**Repellents**—Repellents include a variety of devices aimed at keeping pests from doing damage. Automatic exploders, noisemakers, recordings of scare calls, moving objects, and lights are some of the repellents used. The efficiency of some of these devices is variable and may be highly dependent on placement.

**Figure 3.21** Typical leg-hold trap (EPA, *Applying Pesticides Correctly: A Guide for Private and Commercial Applicators,* 1983).

### b. *Sanitation*

Removing sources of food and shelter helps to suppress some vertebrate pests. Sanitation techniques are used widely to control rodents in and around homes, institutions, restaurants, food-processing facilities, and other related areas.

### c. *Chemical Control*

Pesticides for rodent pest control usually are formulated in baits. The chemicals may be highly toxic to humans, livestock, and other animals. Therefore, correct bait placement is important in order to control the pest while protecting non-pest species. Thorough knowledge of the pest's habits is necessary.

Few pesticides are available for non-rodent vertebrate pest control, and most require special local permits for use. The chemicals which are registered are usually bait applications. A few chemicals designed for aquatic pests or massive populations of pest birds are used as broadcast applications.

The chemicals used to control vertebrate pests include rodenticides, piscicides (fish), avicides (birds), and predacides (predators).

## REFERENCES

Bohmont, B. L., *The Standard Pesticide User's Guide,* 4th ed., Prentice Hall, NJ, 1997.

Flint, M. L., *Pests of the Garden and Small Farm—A Grower's Guide to Using Less Pesticide,* Publication 3332, University of California, Div. of Agriculture and Natural Resources, Davis, CA, 1990.

Johnson, J. M. and G. W. Ware, *Pesticide Litigation Manual,* Clark Boardman Callaghan, New York, NY, 1996.

Johnson, W. T. and H. H. Lyon, *Insects that Feed on Trees and Shrubs,* Cornell University Press, Ithaca, NY, 1987.

Meister R. T., (Ed.) *Farm Chemicals Handbook '96,* Meister Publishing Co., Willoughby, OH, 1996.

Philbrick, H. and J. Philbrick, *The Bug Book— Harmless Insect Controls,* Garden Way Publishing, Charlotte, VT, 1974.

Tashiro, H., *Turfgrass Insects of the United States and Canada,* Cornell University Press, Ithaca, NY, 407 pp., 1987.

U.S. Environmental Protection Agency, *Applying Pesticides Correctly: A Guide for Private and Commercial Applicators,* revised 1991.

U.S. Environmental Protection Agency, *Applying Pesticides Correctly: A Guide for Private and Commercial Applicators,* 1983.

Westcott, C., *Plant Disease Handbook*, 3rd ed., Van Nostrand Reinhold, New York, NY, 1971.

Westcott, C., *The Gardener's Bug Book*, 4th ed., Doubleday, Garden City, NJ, 1973.

# CHAPTER 4

# PESTICIDE FORMULATION AND DILUTION

## I. PESTICIDE FORMULATIONS

The components of pesticide products which have pesticidial activity are called active ingredients. A particular product may contain more than one active ingredient. Active ingredients are rarely applied in their pure form. Instead, they are normally mixed with inert (inactive) ingredients so that you can handle them more conveniently and safely, and apply them more easily and efficiently. This mixture of active and inert ingredients is called a pesticide formulation. The final pesticide formulation is ready for use, either as packaged or after dilution with solvents (e.g., water or petroleum) or other carriers (e.g., silica or silicates).

Pesticides, then, are formulated into many usable forms for satisfactory storage, for effective application, for safety to the applicator and the environment, for ease of application with readily available equipment, and for economy. These goals are not always easily accomplished, due to the chemical and physical characteristics of the technical grade pesticide. For example, some materials in their "raw" or technical condition are liquids, others solids; some are stable to air and sunlight, whereas others are not; some are volatile, others are not; some are water soluble, some oil soluble, and others may be insoluble in either water or oil. These characteristics pose problems to the formulator, since the final formulated product must meet the standards of acceptability by the user.

There are many kinds of pesticide formulations available in the marketplace. There are two main reasons for this. First, the chemistry of the active ingredient itself dictates what formulations are possible. For example, some active ingredients are water soluble and others are not; the latter, then, are often formulated in organic solvents (e.g., petroleum, hexane, or xylene) instead of water. Second, different formulations offer different advantages; thus, one formulation may be better than another for a given application. For example, some active ingredients that are commonly formulated as granules may also be available as liquids so that they can be applied through irrigation systems.

Pesticides are more extensively applied as liquids than as solids. Even some dry formulations, such as wettable powders, are diluted or suspended in a liquid before being applied. Different formulations present different hazards to the applicator. Figure 4.2 describes the hazards of various formulations.

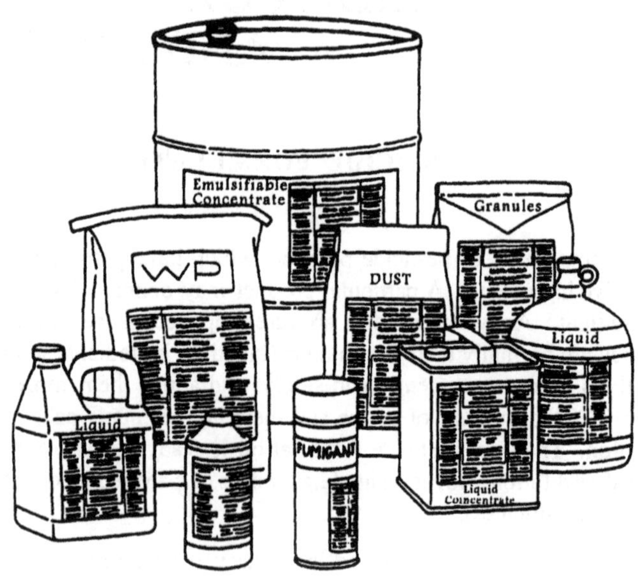


**Figure 4.1** Illustration of the different pesticide formulations and packages available (EPA, *Protect Yourself from Pesticides*, 1993).

## A. TYPES OF FORMULATIONS

### 1. Emulsifiable Concentrates

Emulsifiable concentrates (EC or E are commonly used abbreviations following the trade name on the package) are the most commonly used formulations. The active ingredients in these formulations are insoluble in water. Thus, they are dissolved in an organic or petroleum-based solvent (which gives ECs their strong odor); emulsifiers are then added to the solution. The emulsifiers have properties similar to household detergents and soaps and allow the pesticide to be effectively mixed with water. In water, ECs form "milky" suspensions, called emulsions. Only slight agitation is required to maintain the suspesion.

On the product label, the active ingredient in emulsifiable concentrates is given as a percentage or as pounds per gallon; concentrations normally range from 2 to 8 pounds per gallon. Because of the relatively high percentage of active ingredient, you usually do not need to handle a large amount of product for a particular job. At the same time, it is relatively easy to apply too little or too much of the chemical.

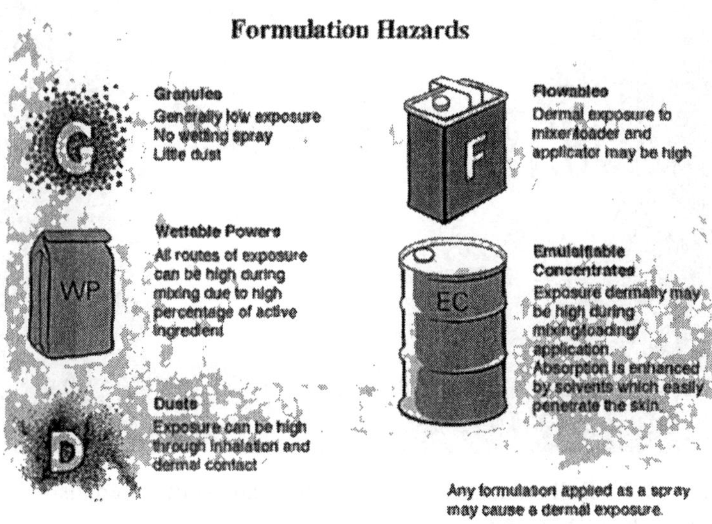


**Figure 4.2** The hazards of various formulations (EPA, *Protective Clothing for Pesticide Users,* poster).

Because of their high concentrations and liquid form, emulsifiable concentrates may be hazardous to the applicator because the skin readily absorbs the solvent carrier. Care must be taken when handling this type of formulation. Figure 4.3 shows a bulk container of an emulsifiable concentrate with a DOT label.

Emulsifiable concentrates leave little visible residue on plants. However, some plants are sensitive to the solvents and additives and damage (sometimes termed phytotoxicity) may occur. For this reason, an emulsifiable concentrate formulation may not be registered for a particular plant, even though wettable powder and dust formulations of the same active ingredient can be used.

Because little agitation of the spray suspension is needed, emulsifiable concentrates are especially suitable for low-pressure, low-volume sprayers and for mist blowers. You can also use them with many other types of application equipment, including dilute hydraulic sprayers, low-volume ground sprayers, mist blowers, and low-volume aerial sprayers.

**Figure 4.3** Bulk container showing labels indicating an emulsifiable concentrate with a flammable carrier solution.

Emulsifiable concentrates are not abrasive and won't separate when the sprayer isn't running. The solvents in emulsifiable concentrates may cause rubber hoses, gaskets, and pump parts to deteriorate rapidly unless they are made of neoprene rubber or more resistant materials.

## 2. Wettable Powders

Wettable powders (WP or W) are finely divided, relatively insoluble powders. The active ingredients are combined with a dry diluent (inert carrier) such as clay, talc, or silica and with wetting agents (surfactants) and/or dispersing agents. Without the wetting agent, the powder would float when added to water, and the two would be almost impossible to mix. The formulated product normally contains 15% to 95% active ingredient; usually 25% or more. Wettable powders form a suspension rather than a true solution when added to water or liquid fertilizer. Therefore, constant and vigorous agitation is needed in the spray tank to maintain the suspension because the wettable powder separates quickly when the sprayer is turned off.

Wettable powders are easy to store, transport, and handle, but are hazardous to the applicator, particularly in mixing where the concentrated dust may be inhaled; however, they are usually not absorbed through the skin as readily as emulsifiable concentrates.

Many of the insecticides sold for garden use are in the form of wettable powders because there is very little chance that this formulation will burn foliage, even at high concentrations. In contrast, the original carrier in emulsifiable concentrates is usually an aromatic solvent, which in relatively moderate concentrations can cause foliage burning at temperatures above 32.5°C (90°F).

Wettable powders can be used with most types of application equipment. As long as the suspension is adequately maintained, they will not clog nozzles, provided strainers and nozzle screens are sufficiently large. WPs are abrasive, however, and may cause both pumps and nozzles to wear more quickly.

### 3. Water-Soluble Powders or Soluble Powders

In water-soluble powders or soluble powders (WSP or SP), the technical grade material is a finely ground, water-soluble solid and may contain a small amount of wetting agent to assist its solution in water. It is simply added to the spray tank, where it dissolves immediately. Unlike the wettable powders and flowables, these formulations do not require constant agitation. They form true solutions and form no precipitate. Because of their sometimes dusty quality, soluble powders may be packaged in convenient, water-soluble bags which are simply dropped into the spray tank. Examples of these are the nicosulfuron herbicide Accent® SP and the linuron herbicide Lorox® SP.

### 4. Water-Soluble Bags

Some wettable and soluble powders are sold in water-soluble bags or packs (WSB or WSP). These products are formulated in a soluble package so that the inner bag is placed directly in the spray tank and subsequently dissolves. This type of packaging eliminates the need for measuring and minimizes your risk of exposure. Examples of this packaging are Pounce® WSB, Ammo® WSB, and Brigade® WSB, all of which are products of the FMC Corporation.

### 5. Water-Soluble Concentrates, Liquids or Solutions

The active ingredient in water-soluble liquids or concentrates (S, WS, WSC or WSL) is soluble in water and is formulated either with water or with a solvent, such as alcohol, which mixes readily with water. When added to water in the spray tank, WSCs form a true solution and require no further agitation after they are mixed. Water-soluble concentrates are often liquid, salt, or amine solutions (e.g., Touchdown® 38-SL, a water-soluble liquid).

### 6. Oil Solutions

The active ingredient in oil solutions is formulated either with oil or with some other organic solvent. The formulation may be highly concentrated so that further dilution with oil is required before application or it may be sold in a dilute form ready for application. In either case, they are true solutions and agitation is not necessary to maintain them.

In their commonest form, oil solutions are the ready-to-use household and garden insecticide sprays sold in a variety of bottles, cans, and plastic containers, all usually equipped with a handy spray atomizer. Not to be confused with aerosols, these sprays are intended to be used directly on pests or places they frequent.

Oil solutions may be used as roadside weed sprays, for marshes and standing pools to control mosquito larvae, in fogging machines for mosquito and fly abatement programs, or for household insect sprays purchased in supermarkets. Commercially they may be sold as oil concentrates of the pesticide to be diluted with kerosene or diesel fuel before application or in the dilute, ready-to-use form. In either case, the compound is dissolved in oil and is applied as an oil spray; it contains no emulsifier or wetting agent.

As with the emulsifiable concentrates, oil solutions may cause damage to vegetation because of the high concentration of solvent; they also may cause significant deterioration of rubber sprayer components. The high-concentrate solutions must be handled with particular caution.

Examples of oil solutions are the dormant oils which are used to control scale insects in trees (see Figure 4.4).

### 7. Flowables or Sprayable Suspensions

Flowables (F, FL), sometimes called water-dispersible suspensions, are particularly useful formulations for active ingredients that are not soluble in either water or the more commonly available organic solvents. The active ingredient is impregnated on a diluent such as clay which is then milled to an extremely fine powder. The powder is then suspended in a small amount of liquid so that the resulting formulation is thick, like paste or cream. Flowables are essentially WPs in suspension that can be measured by volume. They combine the benefits of both WPs and ECs and are becoming increasingly popular. They are mixed with water or liquid fertilizer to form suspensions which require moderate agitation in the spray tank. Because they are more finely divided than WPs, flowables generally require less agitation and can be applied in less total liquid per acre. They seldom clog spray nozzles and usually handle as well as ECs, although they do require larger strainers and nozzle screens than ECs. There are fewer phytotoxicity problems with flowables than with ECs.

**Figure 4.4** Dormant Oil Spray® is used to control scale and mite eggs on fruit, shade trees, roses, and shrubs. It contains paraffinic oil as the active ingredient.

### 8. Dry Flowables

Dry flowables (DF) are finely divided powders that are formulated into concentrated, dustless granules. Dry flowables form a suspension in water and require some agitation to maintain a uniform spray mixture. The principal advantage of this formulation is that although it is sold in the dry form, it is not a dust and can be handled easily. Dry flowables are not meant to be applied directly through a granular applicator; unlike granules designed for application in the dry form, these formulations contain a high percentage of active ingredients, often as much as 75% to 90%.

## 9. Ultralow-Volume Concentrates

An ultralow-volume concentrate (ULV) application is defined as a spray application of undiluted formulation at a rate less than or equal to 1/2 of a gallon per acre. The formulation may contain only the active ingredient or the active ingredient in a small amount of solvent. Always apply ULV formulations without further dilution. These formulations often require specialized application equipment and are usually, though not always, applied by air.

The principal advantage of a ULV application is that many acres can be sprayed with a small volume of liquid. There is some concern that there may be a greater likelihood of significant drift with ULV sprays than with dilute solutions, but the evidence is not conclusive. ULV applications are now limited principally to a small number of insecticides on a few sites.

ULV application is prohibited unless it is specifically designated on the label or is based on an official written or published recommendation of the EPA and the state agency with the regulatory authority for pesticides.

## 10. Encapsulated or Microencapsulated Formulations

The active ingredient (liquid or dry) is incorporated by a special process in small, permeable spheres of polymer or plastic, 15 to 50 μm (1 μm = $10^{-6}$ m) in diameter. These spheres are then mixed with wetting agents, thickeners, and water to give the desired concentration of pesticide in a flowable form, usually 2 pounds per gallon. After the required dilution, the pesticide spray mixture can be applied with conventional sprayers. The pesticide is released gradually over a period of time.

Encapsulated materials can be handled with relative ease and safety. They are effective longer than other formulations containing the same active ingredient. Insecticides formulated this way may, however, pose a significant hazard for bees because bees may take capsules with pollen back to the hive.

## 11. Dusts

Dusts (D) have been the simplest formulations of pesticides to manufacture and the easiest to apply. Dusts are formulated for application in the dry form and with few exceptions, they should be used as purchased.

A prepared dust is a finely ground, dry mixture consisting of a low concentration of active ingredient (usually 1% to 10% by weight) combined with an inert carrier. Because dust particles are finely ground, they may drift long distances from the treated area even when wind velocities are low. Herbicides are not formulated as dusts principally because of this potential for drift. Dusts present a significant inhalation hazard to the applicator and leave a visible residue on plants. They are of limited importance in ornamental and turf pest control.

Despite their ease in handling, formulation, and application, dusts are the least effective and, ultimately, the least economical of the pesticide formulations. The reason is that dusts have a very poor rate of deposit on foliage, unless it is wet from dew or rain. In agriculture, for instance, an aerial application of a standard dust formulation of pesticide will result in 10 percent to 40 percent of the material reaching the crop. The remainder drifts upward and downwind.

## 12. Granules

Granules are very much like dusts except that the inert particles are much larger. Granules are normally made by applying a liquid formulation of the active ingredient (ranging from 2% to 40% by weight) to particles of clay or other porous materials such as corn cobs or walnut shells.

Following application, the active ingredient is released gradually from the inert material. Granules are generally less susceptible to degradation and leaching than other soil-applied formulations. Because granules are relatively large, they drift less than most other formulations and there is little inhalation hazard. However, it is often difficult to accurately calibrate granular spreaders and to obtain uniform distribution of the granules.

Granular formulations are used almost exclusively as soil treatments. They may be applied either directly to the soil or over the plants; they do not cling to plant foliage. They may be used to control pests living at or below ground level or they may be absorbed by roots and translocated throughout the plant. The latter are, for the most part, either systemic insecticides or systemic herbicides.

Only insecticides and a few herbicides are formulated as granules. They range from 2 to 25 percent active ingredient and are used almost exclusively in agriculture, although systemic insecticides as granules can be purchased for lawn and ornamentals. Granular materials may be applied at virtually any time of day, since they can be applied aerially in winds up to 20 mph without problems of drift, an impossible task with sprays or dusts. They also lend themselves to soil application in the drill at planting time to protect roots from insects or to introduce a systemic to roots for transport to above-ground parts in lawns and ornamentals.

## 13. Poisonous Baits

A poisonous bait is a pesticide mixed with an edible material that is attractive to a particular pest. The pests are killed by consuming a lethal dosage of the poison either in a single feeding or over a period of time. Baits are sometimes used to control rodents. Although you may not need to cover the whole area, the bait must be placed where it is likely to be consumed.

The percentage of active ingredient in bait formulations is quite low, usually less than 5%; in addition, only small amounts of pesticide are used in relation to the effective area of treatment. Because the "method of application" also limits the amount of pesticide introduced into the environment, environmental contamination can be minimized. Baits may, however, be attractive to non-target organisms. If a bait is not adequately selective, extra caution should be used to prevent non-target organisms from reaching it, either by placing it properly or by using screens, boxes, or other types of physical barriers.

## 14. Fumigants

Fumigants are a rather loosely defined group of formulations. They are a substance or mixture of substances which produce gas, vapor, fume, or smoke intended to destroy insects, bacteria, or rodents. Fumigants may be volatile liquids and solids as well as gases. They are used to disinfest the interiors of buildings, objects, and materials that can be enclosed so as to retain the fumigant. Soil fumigants are also used in horticultural nurseries, greenhouses, and on high-value cropland, such as tobacco, to control nematodes, insect larvae and adults, and sometimes to control diseases and weed seeds. Depending on the fumigant, the treated soils may require covering with plastic sheets for several days to retain the volatile chemical, allowing it to exert its maximum effect.

## 15. Adjuvants (Additives)

An adjuvant or additive is a chemical added to a pesticide to increase its effectiveness or to reduce its phytotoxicity or drift. Adjuvants are used most extensively in products designed for foliar applications. Most pesticide formulations already contain adjuvants. Several different types of adjuvants are:

**Wetting agents** and **emulsifiers** are often added so that the pesticide will mix with water and/or coat treated surfaces more effectively.

**Spreaders** are substances that increases the area that a given volume of liquid will cover on a solid or another liquid. They also allow pesticides to spread evenly over treated surfaces.

**Stickers** increase the adherence of the chemical to the treated surface, thus increasing its persistence, particularly under adverse weather conditions. Many spreaders also possess wetting and spreading characteristics.

**Penetrants** are wetting agents, oils, or oil concentrates that enhance the absorption of a systemic pesticide by the plant. Examples are Agri-Dex®, Induce®, and Penetrator®.

**Dispersing agents** are materials that reduce the cohesiveness of like particles, either solid or liquid. Dispersants and suspending agents are added during the formulation of emulsifiable concentrates and wettable powders to aid in the dispersion and suspension of the ingredients. Examples are Adherex®, Tamol®, and Lomar®.

**Foaming aids** and **suppressants** are surface-active substances that form a fast draining foam to provide maximum contact of the spray to the plant surface. Foaming aids insulate the surface and reduce the rate of evaporation. Foaming suppressants reduce the ability of a pesticide or formulation to foam during mixing and application.

Formulations as manufactured often contain all necessary adjuvants in appropriate amounts for all or most uses. Sometimes, however, it may be desirable for the applicator to add specific adjuvants prior to application. Wetting agents and spreader-stickers are probably the adjuvants added most frequently by the applicator. These materials are often referred to as surfactants. Surfactants are "surface active agents" which reduce the surface tension of water and therefore enhance spreading of the spray solution on the treated surface. Compatibility agents are being added with increasing frequency to allow the effective mixing of two or more pesticides or a pesticide with a fertilizer. Thickening agents, also called drift reduction agents, are being used more extensively as drift continues to be of increasing concern.

Adjuvants are added only if recommended on the product label; otherwise, you will do so at your own risk. Some labels expressly prohibit the use of adjuvants. Always bear in mind that while increasing the effectiveness of a particular pesticide, you also may be increasing the potential for phytotoxicity and perhaps harm to nontarget organisms and the environment.

## II. DILUTING PESTICIDES CORRECTLY

Unless you have the correct amount of pesticide in your tank mix, even a correctly calibrated sprayer can apply the wrong dose of pesticide to the target.

Formulations such as wettable and soluble powders, emulsifiable concentrates, and flowables are sold as concentrates and must be diluted in the spray tank with an appropriate carrier. Water is the most common carrier, but kerosene, oil, and other liquids are sometimes used. The label or other recommendations will tell:

- how much to dilute the formulation
- how much of the dilute pesticide to apply per unit of area.

### A. MIXING SOLUBLE AND WETTABLE POWDERS

#### 1. Pounds Per 100 Gallons

Directions for wettable or soluble powders may be given in pounds of pesticide formulation per 100 gallons of carrier. You must know the capacity in gallons of your sprayer tank (or the number of gallons you will be adding to the spray tank if the job requires only a partial tank load). Then use the following formula:

$$\frac{\text{Gallons in tank} \times \text{lbs. per 100 gal. Recommended}}{\text{100 gallons}} = \text{Pounds needed in tank}$$

**Example 1:**

Your spray tank holds 500 gallons. The label calls for 2 pounds of formulation per 100 gallons of water. How many pounds of formulation should be added to the tank?

$$\frac{\text{500 gallons x lbs. per 100 gallons (2)}}{\text{100 gallons}} = \text{Pounds needed in tank (10)}$$

$500 \times 2 \div 100 = 10$ You should add 10 pounds to the tank

**Example 2:**

You need to spray only one acre and your equipment is calibrated to spray 60 gallons per acre. The label calls for 2 pounds of formulation per 100 gallons of water. How many pounds of formulation should be added to the tank to make 60 gallons of finished spray?

$$\frac{\text{Gallons in tank (60) x lbs. per 100 gallons (2)}}{\text{100 gallons}} = \text{1.2 lbs. (19 oz.) in tank}$$

## 2. Pounds Per Acre

The label may list the recommended dosage as pounds per acre. If the job requires a full tank, you must know how many gallons your equipment applies per acre and the spray tank capacity. Use these formulas:

$$\frac{\text{Gallons in tank}}{\text{Gallons applied per acre}} = \text{Acres sprayed per tankful}$$

Acres sprayed per tank x Pounds formulation per acre = Pounds formulation needed in tank

**Example 3:**

Your sprayer applies 15 gallons per acre and your tank holds 400 gallons. The label rate is 3 pounds of formulation per acre.

$$\frac{\text{Gallons in tank (400)}}{\text{Gallons per acre (15)}} = \text{26.7 acres sprayed per tankful}$$

$400 \div 15 = 26.7$

Acres sprayed per tankful (26.7) x Pounds formulation per acre (3) = Pounds needed in tank (80.1)

$26.7 \times 3 = 80.1$

Add 80 pounds of pesticide formulation to the tank.

If the job requires less than a full tank, you must know how many acres you wish to treat and how many gallons your sprayer is pumping per acre. Figure both the number of gallons needed in the tank and the pounds of formulation to add. Use these formulas:

Gallons per acre x Acres to be treated = Gallons needed in tank

Acres to be treated x Pounds formulation per acre = Pounds formulation needed in tank

**Example 4:**

You wish to spray 3.5 acres and your equipment is applying 15 gallons per acre. The label rate is 3 pounds per acre.

Gallons per acre (15) x Acres to be treated (3.5) = Gallons needed in tank (52.5)

15 x 3.5 = 52.5

Acres to be treated (3.5) X Pounds formulation per acre (3) = Pounds formulation ( 10.5) needed in tank

3.5 x 3 = 10.5

If the recommended dosage is given as pounds of active ingredient per acre, you must first convert that figure to pounds of formulation per acre. Use the following formula:

$$\frac{\text{Pounds of active ingredient per acre x 100}}{\text{Percent of active ingredient in formulation}} = \text{Pounds formulation per acre}$$

Then follow the formulas listed above under "pounds per acre" to find the pounds of formulation to add to your tank.

**Example 5:**

You wish to apply 2 pounds of active ingredient per acre. Your formulation is 80 percent WP.

$$\frac{\text{Pounds of a.i. per acre (2) x 100}}{\text{\% a.i. in formulation (80)}} = \text{2.5 pounds formulation per acre}$$

2 x 100 ÷ 80 = 2.5

If the recommended rate is a percentage of active ingredient in the tank, another formula is necessary. First find the number of gallons of spray in the spray tank (either the tank capacity or gallons needed for job if less than tank capacity). Then:

$$\frac{\text{Gallons of spray in tank x \% a.i. wanted x Weight of carrier (lbs. per gal.)}}{\text{\% active ingredient in formulation}}$$

$= \text{Pounds formulation to add to tank}$

---

**Example 6:**

Your mist blower directions call for a spray containing 1.25 percent active ingredient. You need to mix 40 gallons of spray for the job. The pesticide is a 60 percent SP and you will use water as the carrier.

$$\frac{\text{Gallons of spray (40) x \% a. i. needed (1.25) X Weight of water/gal (8.3)}}{\text{\% active ingredient in formulation (60)}}$$

= 6.9 pounds formulation needed in tank (40 x 1.25 x 8.3 . 60 = 6.9 lbs.)

---

## B. MIXING LIQUID FORMULATIONS

Dosages for liquid formulations (EC, F, SC, etc.) are often listed as pints, quarts, or gallons per 100 gallons or per acre. Use the pounds per 100 gallons and pounds per acre (above) for making these calculations. Substitute the appropriate liquid measure for "pounds" in the formulas.

---

**Example 7:**

The label rate is 2 pints of pesticide formulation per 100 gallons of water. Your spray tank holds 300 gallons.

$$\frac{\text{Gallons per tank (300) x pints per 100 gal. (2)}}{\text{100 gallons}} = 300 \times 2 \div 100 = 6$$

6 pints formulation needed in tank

---

**Example 8:**

Your sprayer applies 22 gallons per acre and your tank holds 400 gallons. The label rate is 1.5 quarts per acre.

$$\frac{\text{Gallons in tank (400) X quarts per acre (1.5)}}{\text{Gallons per acre (22)}} = 400 \text{ X } 1.5 \div 22 =$$

27.3 qts. needed in tank

If the recommendation for the liquid formulation is listed as pounds of active ingredient per acre, you must first convert that figure to gallons of formulation to apply per acre. The label of a liquid formulation always tells how many pounds of active ingredient are in one gallon of the concentrated formulation (4 EC has 4 pounds of active ingredient per gallon; 6 EC contains 6 pounds per gallon, etc.).

$$\frac{\text{Pounds of active ingredient needed per acre}}{\text{Pounds of active ingredient per gallon of formulation}}$$

$$= \text{Gallons of formulation per acre}$$

**Example 9:**

The recommendation is for 1 pound of active ingredient per acre. You purchased an 8 EC, which contains 8 pounds of active ingredient per gallon. Your tank holds 500 gallons and is calibrated to apply 25 gallons per acre.

$$\frac{\text{Pounds a.i. to apply per acre (1)}}{\text{Pounds a.i. per gallon (8)}} = 1 \div 8 = .125 \text{ (1/8)}$$

1/8 gallons (1 pint) per acre

$$\frac{\text{Gallons in tank (500)}}{\text{Gallons per acre (25)}} = 500 \div 25 = 20 \text{ Acres per tankful}$$

Acres per tankful (20) X gallons per acre (1/8 or .125) = Gallons to add to tank (2.5)

20 x .125 = 2.5

If the recommended rate is a percentage of active ingredient in the tank, use this formula:

$$\frac{\text{Gallons of spray x \% a.i. wanted x Weight of carrier (lbs/gal)}}{\text{lbs. a.i. per gallon of formulation}}$$

= Gallons of formulation to add

**Example 10:**

You wish to make 100 gallons of a 1 percent spray using water as the carrier. You have a 2 EC formulation (2 pounds active ingredient per gallon). How many gallons of the 2 EC should you add to the 100 gallons of water in the tank?

$$\frac{\text{Gallons of spray (100) x \% a.i. wanted ( 1 ÷ 100) x Weight of water (8.3)}}{\text{lbs. a.i. per gallon of formulation (2)}}$$

= 100 X .01 x 8.3 ÷ 2 = 4.15 gals of formulation to add

## C. MIXING CONCENTRATES FOR AIR BLAST SPRAYERS OR MIST BLOWERS

If the dosage recommendations are listed as pounds or gallons per acre or pounds or gallons per 100 gallons of carrier for use in boom or hydraulic sprayers, you will need to convert the dosage to the concentration factor chosen (usually 2X, 3X, 4X, 5X, or 10X). Simply follow the steps listed above for the dry or liquid formulation you are using. The last answer should be multiplied by the concentration factor.

Pounds or gallons of formulation per tank x Concentration factor = Pounds or gallons formulation per tank in concentrate form

**Example 11:**

The label lists the dosage as 4 pounds formulation per 100 gallons of water for dilute application. Your air blast sprayer tank holds 600 gallons. You wish to apply a 5X concentration.

$$\frac{\text{Gallons per tank (600) x lbs. per 100 gallons recommended (4)}}{\text{100 gallons}}$$

= lbs. needed in tank for hydraulic sprayer (24)

600 x 4 ÷ 100 = 24

Pounds formulation per tank for hydraulic sprayer (24) x Concentration wanted (5X) = Pounds of formulation to add to air blast tank (120)

24 x 5 = 120

## REFERENCES

Baker, P. B., *Arizona Agricultural Pesticide Applicator Training Manual*, Cooperative Extension, University of Arizona, Tucson, AZ, 1992.

Bohmont, B. L., *The Standard Pesticide User's Guide*, 4th ed., Prentice Hall, NJ, 1997. *Litigation Manual*, Clark Boardman Callaghan, New York, NY, 1996.

Johnson, J. M. and G. W. Ware, *Pesticide Litigation Manual*, Clark Boardman Callaghan, New York, NY, 1996.

Meister R. T., (Ed.) *Farm Chemicals Handbook '96*, Meister Publishing Co., Willoughby, OH, 1996.

U.S. Environmental Protection Agency, *Applying Pesticides Correctly: A Guide for Private and Commercial Applicators*, revised 1991.

U.S. Environmental Protection Agency, *Applying Pesticides Correctly: A Guide for Private and Commercial Applicators*, 1983.

U.S. Environmental Protection Agency, *The Worker Protection Standard for Agricultural Pesticides—How to Comply: What Employers Need to Know*, 1993.

U.S. Environmental Protection Agency, *Protective Clothing for Pesticide Users*—Poster, 1993.

# CHAPTER 5

# TOXICOLOGY AND MODE OF ACTION

Any chemical substance may evoke a toxic effect. A chemical substance that exerts an injurious effect in the majority of cases when it comes into contact with living organisms is termed a poison. Pesticides, by necessity, are poisons. Familiar chronic responses to various irritants include silicosis, lung cancer, and necrosis of the liver or kidneys. However, the toxic effects of different pesticides and their formulations differ greatly. Two types of effects can be distinguished:

1. An *acute effect* normally occurs shortly after contact with a single dose or exposure to a poison, in this case a pesticide. The magnitude of the effect depends on the innate toxicity of the substance, duration of exposure, and the method of application or exposure to a particular organism. Thus, a smaller dose of arsenious oxide than sodium chloride (table salt) will produce toxic symptoms in most animal species, and a drop of sulfuric acid is less dangerous on the skin than it is in the eye. Acute toxicity very often results from the disruption of an identifiable biochemical or physiological system and, as a consequence, acute toxic responses are more readily quantifiable.
2. A *chronic effect*, on the other hand, sometimes occurs when an organism is exposed to repeated small and non-lethal doses of a potentially harmful substance. Familiar chronic responses to various irritants include silicosis, lung cancer, and necrosis of the liver or kidneys.

## I. METHODS TO ESTIMATE THE RELATIVE TOXICITY OF PESTICIDES TO HUMANS

The toxicity value of a pesticide is a relative measure to estimate its toxic effect on an organism.

A frequently used measure of acute toxicity is the $LD_{50}$ or $LC_{50}$ values, which mean lethal dose or lethal concentration of a poison that kills half the organisms of a test population. The $LD_{50}$ is usually expressed in terms of milligrams of poison per kilogram of body weight of the experimental animals (mg/kg). An example might be a pesticide with an $LD_{50}$ of 550, which would indicate that 550 milligrams of this pesticide given to animals that weigh 1 kilogram each would kill 50% of the test population.

To determine the $LD_{50}$ of a pesticide, batches of animals are exposed to a range of doses and the percentages of organisms dying in each batch are recorded. The results can usually be represented by a curve (sigmoidal) shown in Figure 5.1.

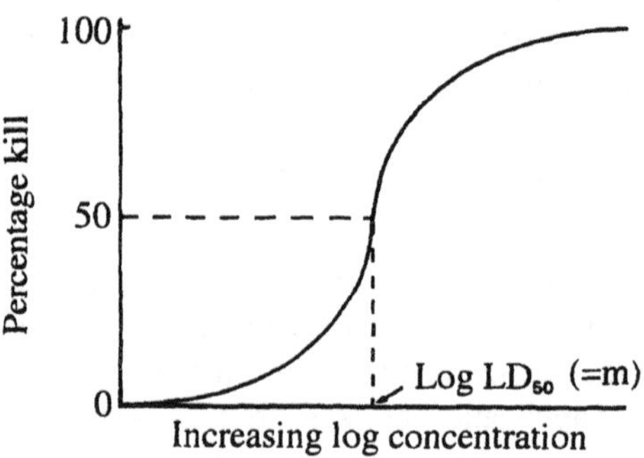


**Figure 5.1** Dose-response curve illustrating the log of the dose that causes a 50 percent mortality in the test population.

The curve is steepest in the region of 50 percent so that a small change in concentration in this region of the curve causes a large change in percentage kill. The dose that kills half of the organisms is thus a more sensitive index of toxicity than any other dose, and this is why the $LD_{50}$ is usually adopted as a standard for comparing the relative toxicity of substances.

The toxicity data based on the $LD_{50}$ values determined from tests on animals involve several problems and should not be interpreted as exact values for human toxicity. These values should be used as guides in estimating the relative toxicities of pesticides and can be also used to determine the level of protection (PPE) required to perform various tasks such as the application or handling of pesticides.

The $LC_{50}$ is a somewhat similar concept to the $LD_{50}$. It is the concentration, usually in air or water surrounding test animals, that causes 50 percent mortality. It is therefore a convenient index for studying vapors inhaled by animal or water-soluble pesticides in contact with fish.

In this book the average values of the oral toxicities ($LD_{50}$ values) as a single dose ingested by rats or similar mammals are quoted to give some indication of the mammalian toxicities of different types of pesticides. The smaller the $LD_{50}$ value, the more toxic the chemical, so that the toxicities of chemicals can be graded by the $LD_{50}$ values as follows:

| **Descriptor** | **$LD_{50}$ (mg/kg)** |
|---|---|
| Super or extremely toxic | < 5 |
| Highly toxic/hazardous | 5-50 |
| Moderately toxic/hazardous | 50-500 |
| Slightly toxic/hazardous | 500-5,000 |
| Relatively non-toxic or harmless | >5,000 |

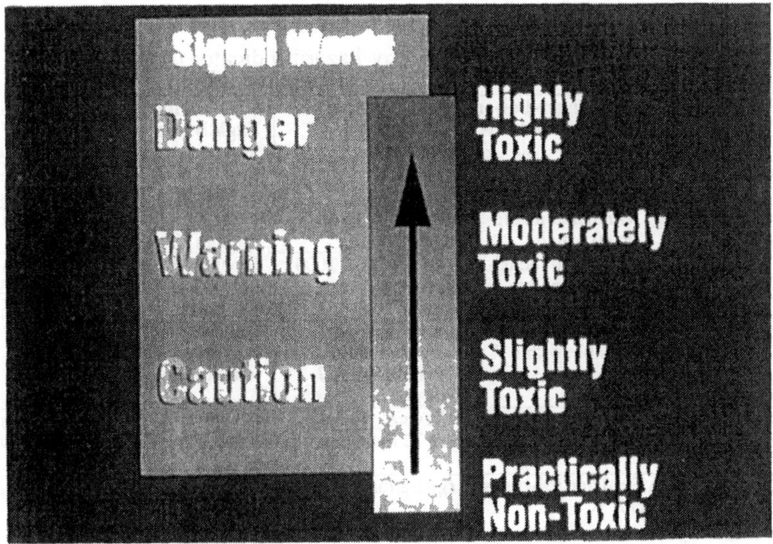


**Figure 5.2** Toxicity increases as the signal word progresses from caution to danger. Signal words are required on all labels and labeling.

## II. TOXICITY AND LABELING

All pesticide labels must contain "signal words in bold print, to attract the attention of the buyer/user of the product." The use of these signal words is mandated by EPA regulations.

All pesticide products meeting the criteria of toxicity category I shall bear on the front panel of the label the signal word DANGER. In addition, if the product was assigned to toxicity I on the basis of its oral, inhalation, or dermal toxicity (as distinct from skin and eye local effects), the word POISON shall apear in red on a background of contrasting color along with the skull and crossbones.

All pesticide products meeting the criteria of toxicity category II shall bear on the front panel the signal word WARNING.

All pesticide products meeting the criteria of toxicity category III shall bear on the front panel the signal word CAUTION.

All pesticide products meeting the criteria of toxicity category IV shall bear on the front panel the signal word CAUTION.

As an aid to those unfamiliar with signal words and toxicity values the EPA has published Figure 5.2 and the following table.

**Table 5.1** EPA toxicity categories by hazard indicator (signal words).

| | Toxicity Categories | | | |
|---|---|---|---|---|
| **Hazard Indicators** | **I—Danger-Poison Highly hazardous** | **II—Warning Moderately hazardous** | **III—Caution Slightly hazardous** | **IV—Caution Relatively nonhazardous** |
| Oral $LD_{50}$ | Up to and including 50 mg/kg | From 50 to 500 mg/kg | From 500 to 5,000 mg/kg | Greater than 5,000 mg/kg |
| Inhalation $LD_{50}$ | Up to and including 0.2 mg/l | From 0.2 through 2 mg/l | From 2.0 through 20 mg/l | Greater than 20 mg/l |
| Dermal $LD_{50}$ | Up to and including 200 mg/kg | From 200 to 2000 mg/kg | From 2,000 to 20,000 mg/kg | Greater than 20,000 mg/kg |
| Eye effects | Corrosive; corneal opacity not reversible within 7 days | Corneal opacity reversible within 7 days; irritation persisting for 7 days | No corneal opacity; irritation reversible within 7 days | No irritation |
| Skin effects | Corrosive | Severe irritation at 72 hours | Moderate irritation at 72 hours | Mild or slight irritation at 72 hours |

Source : "EPA Pesticide Programs, Registration and Classification Procedures, Part II" Federal Register 40: 28279.

Table 5.2 and Figures 5.3 and 5.4 describe how the signal words can be used to select the appropriate PPE for various pesticide formulations.

**Table 5.2** The use of signal words to select the appropriate PPE depending on the relative toxicity of the formulation.

| | Signal Word | | |
|---|---|---|---|
| **Formulation** | **Caution: Slightly toxic** | **Warning: Moderately Toxic** | **Danger—Poison: Highly toxic** |
| Dry | Long-legged trousers, no cuffs, long-sleeve shirt, shoes, socks, hat or head covering. | Long-legged trousers, long-sleeve shirt, shoes, socks, wide-brimmed hat, and gloves. | Long-legged trousers, long-sleeve shirt, shoes, socks, hat, gloves, and air purifying respirator[a] if label reads "Poisonous or fatal if inhaled." |
| Liquid | Long-legged trousers, long-sleeve shirt, shoes, socks, and hat. | Long-legged trousers, long-sleeve shirt, shoes, socks, wide-brimmed hat, and chemical-resistant gloves. | Long-legged trousers, long-sleeve shirt, or disposable coveralls, chemical-resistant boots, socks, wide-brimmed hat, chemical-resistant gloves and goggles or face shield. Air purifying respirator[a] if label reads "Do not breathe vapors or spray mists" or "Poisonous or fatal if inhaled." |
| Liquid (when mixing) | Long-legged trousers, long-sleeve shirt, shoes, socks, hat, gloves and chemical-resistant apron. | Long-legged trousers, long-sleeve shirt, shoes, socks, wide-brimmed hat, chemical-resistant gloves, goggles or face shield and chemical-resistant apron. Air purifying respirator[a] if label reads "Do not breathe vapors or spray mists" or "Poisonous or fatal if inhaled." | Long-legged trousers and long-sleeve shirt or coveralls, chemical-resistant gloves and boots, goggles or face shield. Chemical-resistant apron and air purifying respirator[a]. |
| Liquid (prolonged exposure to spray or application in enclosed area) | Long-legged trousers, long-sleeve shirt, chemical-resistant gloves and boots and water-impermeable hat. | Water-repellent, long legged trousers, long-sleeved shirt, chemical-resistant gloves, boots and apron, waterproof wide-brimmed hat, face shield and air purifying respirator[a]. | Chemical-resistant suit, boots and gloves. Waterproof hood or wide-brimmed hat, face shield, and air purifying respirator[a]. |

Adapted in part from U.S. Environmental Protection Agency and Department of Agriculture, *Applying Pesticides Correctly*.

[a] Air purifying respirator with the appropriate cartridge for dry, aqueous liquid or mist, organic liquid or vapor.

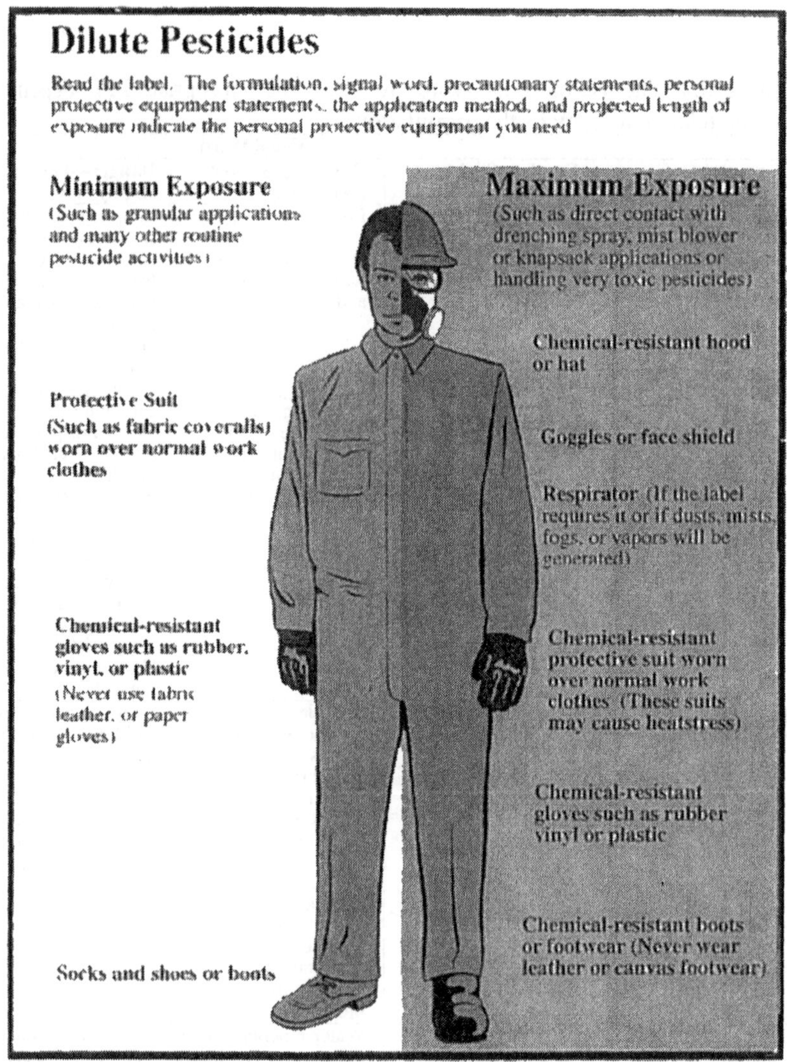


**Figure 5.3** Recommended attire for handling dilute pesticides (EPA, "Protective Clothing for Pesticide Users," poster, 1993).

## III. CLASSIFICATION OF PESTICIDES

Pesticides are any substance used directly to control pest populations or to prevent or reduce pest damage. Not all pesticides actually kill the target organism. For example, some fungicides inhibit the growth of a fungus without killing it, and attractants and repellents only lure a pest to or divert it from a particular site.

We can classify pesticides in several ways, each having its own value for a given purpose. For example, we can group pesticides according to:

- Their chemical composition.
- Use patterns (e.g., foliar vs. soil fungicides).

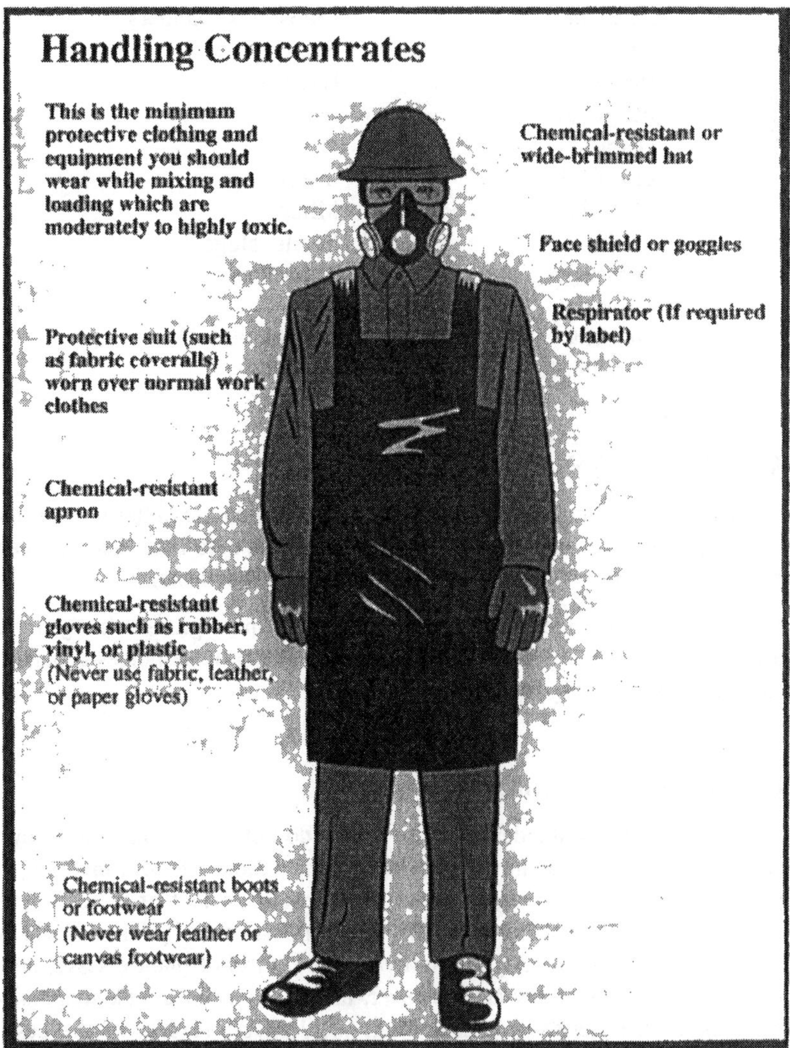


**Figure 5.4** Recommended attire for handling concentrate pesticides (taken in part from EPA, "Protective Clothing for Pesticide Users," poster, 1993).

- The target pest, which is the pest you are trying to control (e.g., gypsy moth insecticide).
- The group of pests controlled (e.g., herbicides, insecticides).

The last of these is the most common classification scheme and is based on the pest group controlled.

| **Pesticide Group** | **Pests Controlled** |
|---|---|
| Acaricide | Mites, ticks, spiders |
| Avicide | Birds |
| Bactericide | Bacteria |
| Fungicide | Fungi |
| Herbicide | Weeds |
| Insecticide | Insects |
| Miticide | Mites |
| Molluscicide | Snails, slugs |
| Nematicide | Nematodes |
| Piscicide | Fish |
| Predacide | Vertebrates |
| Rodenticide | Rodents |

## A. CHEMISTRY

This list does not include those pesticides which are either not normally used as pest control agents (plant growth regulators, defoliants, and desiccants) or are not directly toxic to the pest (attractants and repellents).

Chemical pesticides can be divided into two main groups: inorganic and organic compounds. As we will see, a third group of pesticides consists of natural disease-causing agents.

### 1. Inorganic Pesticides

The inorganic pesticides are those which do not contain carbon. They are of mineral origin and commonly contain arsenic, copper, boron, mercury, sulfur, tin, or zinc. The inorganics were the most important of the early pesticides and were the principal chemicals used for pest control prior to World War II. They are still used today, primarily for the control of plant diseases and as wood preservatives. They are, however, usually toxic to a wide range of organisms, a characteristic which is often not desirable (except in the case of wood preservatives). They also are generally less effective than many of the organic compounds. Some do have relatively low acute toxicity to humans, although compounds containing lead, mercury, and arsenic have generated widespread health and environmental concerns and their use has been either banned or severely curtailed.

## 2. Organic Pesticides

The organic pesticides contain carbon. They also contain hydrogen and often oxygen, nitrogen, phosphorus, sulfur, or other elements. Most pesticides used today are organic compounds. A few organic pesticides are either derived or extracted directly from plants. Most, however, are synthetic compounds. It is these compounds that have been responsible for the expanded use of pesticides since World War II. They are often extremely effective and easy to use, have been relatively low cost, and some are quite specific in their activity. They have, however, been the principal focus of health and environmental concerns and are the pesticides most commonly associated with problems of pesticide use and misuse.

**Figure 5.5** Assorted products containing various organic pesticides.

## 3. Biological Pesticides

A distinct group of pest control agents are the so-called microbial pesticides. These are bacteria, viruses, and fungi which cause disease in given species of pests. Although they occur naturally in certain areas, they are sometimes intentionally introduced in sufficient quantities that a relatively high level of control becomes possible. They tend to be highly specific in their activity and are often virtually harmless to non-target species. There are, however, only a few microbial pesticides registered for use at this time, and their success has been limited. The best known example is the bacterium, Bacillus thuringiensis (Bt), which has been used with some effectiveness for the control of certain caterpillars, flies, mosquitoes, worms, and borers.

## B. CLASSIFICATION OF PESTICIDES

The names of pesticides are often confusing. We often see what appears to be the same product referred to by a number of different names. Likewise, different products may have similar names.

In the formulated product, there is usually a specific component of a product that has pesticide activity. That component is the active ingredient; a particular product may contain more than one active ingredient.

It is important to read the label. Different products containing the same active ingredient(s) are not necessarily labeled for the same uses, and different formulations may have different toxicities.

Active ingredients are given chemical names which are directly related to their chemical composition. This chemical name is frequently a long name, difficult to pronounce, and difficult to use. It will normally appear on the label, often in parentheses.

Generally, active ingredients are also assigned common names which appear on the label, frequently on the same line with and just before the chemical name. A given chemical or common name always refers to a particular active ingredient, regardless of the manufacturer or formulator.

Manufacturers generally give one or more specific names to each formulation of a particular active ingredient; these names are referred to as trade names or brand names. It is the trade name which appears in large letters at the top of the label. Occasionally, the trade name is the same as the common name.

Several examples of pesticide nomenclature are given below:

1. Chemical Name: Isopropylamine salt of N-(phosphonomethyl) glycine
   Common Name: Glyphosate isopropylammonium
   Family Name: Organophosphate
   Trade Name: Accord®, Pondmaster®, Ranger®, Roundup®, etc.

2. Chemical Name: 6-chloro-$N^2$-ethyl-$N^4$-isopropyl-1,3,5-triazine-2,4-diamine
   Common Name: Atrazine
   Family Name: Triazine

Trade Names: AAtrex®, Atrazine®, Flotrazine®, etc.

3. Chemical Name: 2-chloro-N-(2,4-dimethyl-3-thienyl)-N-(2 meth-oxy-1-methylethyl) acetamide
   Common Name: Dimethenamid
   Family Name: Amide or substituted amide
   Trade Name: Frontier®

## IV. INSECTICIDES

Pesticides are often classified according to their deposition or distribution on or in treated plants; however, the following differentiation may be made:

*Insecticides with local action.* These affect the pest directly, or they have to be distributed as evenly as possible on the plant's surface, where uptake by the insect occurs. Many insecticides show a good depth of action, penetrating into the leaf and reaching even hidden insects on the underside of the leaf not directly touched by the insecticide. Depth of action is important, particularly in the control of mining developmental stages of the insects, as for example the maggots of the beet leaf miner *(Pegomyia hyoscyami).* Insecticides with a predominantly local action are to be found primarily in the chlorinated hydrocarbons, many organic phosphoric acid esters, and carbamates displaying an effective penetrating action.

*Insecticides with systemic action* are taken up relatively quickly by the plants and transported into the vascular system. According to the type of application, uptake occurs through the roots or the parts of the plant above ground. Distribution is chiefly by the xylem, but is also possible by the phloem and by diffusion from cell to cell. The persistence of activity is dependent on the type of substance, the intensity of breakdown in the plant or the soil, and environmental conditions. A much longer period of protection can be maintained if, by application of granulates at drilling or planting out, a depot of the substance is created in the soil from which the active substance is released slowly and taken up by the plants.

The following particularly important advantages of using systemic insecticides should be mentioned:

- Rapid uptake through the plants minimizes the degree of danger to beneficial insects. A few hours after application, only the insects which suck or feed on the plants are affected, whereas their natural enemies, for example, bees are not.
- Good distribution in the plants reaches even hidden pests, whereas locally acting agents affect them only slightly or not at all.
- Requirements with respect to even distribution and weather resistance are less stringent for systemic plant protection agents. Due to their

rapid uptake, they are largely impervious to the influence of external factors.

**Figure 5.6** Synthetic pyrethroid-containing products.

## A. TOXICOLOGY AND MODE OF ACTION

The way in which a pesticide affects target pest(s) is commonly referred to as its mode of action. All pesticides are alike in blocking some metabolic processes; however, how they accomplish this—their mode of action—is sometimes very difficult to determine, and in many cases is unknown or only partially understood.

In terms of their mode of action, insecticides fall into seven classes: physical toxicants, protoplasmic poisons, nerve poisons, metabolic inhibitors, cytolytic toxins, muscle poisons, and alkylating agents.

## 1. Physical Toxicants

Physical toxicants are those materials that act to block any physiological process, not biochemically or neurologically, but mechanically. Examples include oils used to control mosquito larvae by blocking or clogging their respiratory opening and heavier oils applied to fruit trees during the dormant season to control scales by clogging their respiratory openings. Other physical toxicants are the inert, abrasive dusts such as boric acid, diatomaceous earth, and silica gel. These materials kill by absorbing waxes from the insect cuticle, effecting the continuous loss of moisture from the insect body, which in turn causes desiccation and death from dehydration.

## 2. Cytoplasmic, Neurological, and Muscular Poisons, Metabolic Inhibitors, Cytolytic Toxins, and Alkylating Agents

### a. *Organophosphate Insecticides*

The following table lists current organophosphate products. They are grouped by relative toxicities.

**Table 5.3** Organophosphate products grouped by relative toxicities.

| **Highly toxic*:** | **Moderately toxic:** |
|---|---|
| tetraethyl pyrophosphate (TEPP), | quinalphos (Bayrusil), |
| imefox (Hanane, Pestox XIV) | ethion (Ethanox) |
| phorate (Thimet, Rampart, AASTAR) | chlorpyrifos (Dursban, Lorsban, Brodan) |
| disulfoton⁺ (Disyston) | edifenphos |
| fensulfothion (Dasanit) | oxydeprofos⁺ (Metasystox-S) |
| demeton⁺ (Systox) | sulprofos (Bolster, Helothion) |
| terbufos (Counter, Contraven) | isoxathion (E-48, Karphos) |
| mevinphos (Phosdrin, Duraphos) | propetamphos (Safrotin) |
| ethyl parathion (E605, Parathion, Thiophos) | phosalone (Zolone) |
| azinphos-methyl (Guthion, Gusathion) | thiometon (Ekatin) |
| fosthietan (Nem-A-Tak) | heptenophos (Hostaquick) |
| chlormephos (Dotan) | crotoxyphos (Ciodrin, Cypona) |
| sulfotep (Thiotepp, Bladafum, Dithione) | phosmet (Imidan, Prolate) |
| carbophenothion (Trithion) | trichlorfon (Dylox, Dipterex, Proxol, Negu-von) |
| chlorthiophos (Celathion) | cythioate (Protean, Cyflee) |
| fonofos (Dyfonate, N-2790) | phencapton (G 28029) |
| prothoate⁺ (Fac) | pirimiphos-ethyl (Primicid) |
| fenamiphos (Nemacur) | DEF (DeGreen, EZ-Off D) |
| phosfolan⁺ (Cyolane, Cylan) | methyl trithion |
| methyl parathion (E 601, Penncap-M) | dimethoate (Cygon, DeFend) |
| schradan (OMPA) | fenthion (mercaptophos, Entex, Baytex, Tigu-von) |
| mephosfolan⁺ (Cytrolane) | dichlofenthion (VC-13 Nemacide), |
| chlorfenvinphos (Apachlor, Birlane) | bensulide (Betasan, Prefar) |
| coumaphos (Co-Ral, Asuntol) | EPBP (S-Seven) |
| phosphamidon (Dimecron) | diazinon (Spectracide) |

**Table 5.3** continued

| **Highly toxic*:** | **Moderately toxic (continued):** |
|---|---|
| methamidophos (Monitor) | profenofos (Curacron) |
| dicrotophos (Bidrin) | formothion (Anthio) |
| monocrotophos (Azodrin) | pyrazophos (Afugan, Curamil) |
| methidathion (Supracide, Ultracide) | naled (Dibrom) |
| EPN | phenthoate (dimephenthoate, Phenthoate) |
| isofenphos (Amaze, Oftanol) | cyanophos (Cyanox) |
| endothion, bomyl (Swat) | crufomate (Ruelene) |
| famphur (Famfos, Bo-Ana, Bash) | merphos (Folex, Easy Off-D) |
| fenophosphon (trichloronate, Agritox) | pirimiphos-methyl (Actellic) |
| dialifor (Torak) | iodofenphos (Nuvanol-N) |
| cyanofenphos (Surecide) | chlorphoxim (Baythion-C) |
| dioxathion (Delnav) | propyl thiopyrophosphate (Aspon) |
| mipafox (Isopestox, Pestox XV) | bromophos (Nexion) |
| | tetrachlorvinphos (Gardona, Appex, Stirofos) |
| **Moderately toxic*:** | temephos (Abate, Abathion) |
| bromophos-ethyl (Nexagan), | fenitrothion (Accothion, Agrothion, Sumithion) |
| leptophos (Phosvel), | pyridaphenthion (Ofunack) |
| dichlorvos (DDVP, Vapona), | acephate (Orthene) |
| ethoprop (Mocap), | malathion (Cythion) |
| demeton-S-methyl⁺ (Duratox, Metasystox (i)), | ronnel (fenchlorphos, Korlan) |
| triazophos (Hostathion) | etrimfos (Ekamet), |
| oxydemeton-methyl⁺ (Metasystox-R) | phoxim (Baythion) |

* Compounds are listed in approximate order of descending toxicity. "Highly toxic" organophosphates have listed oral $LD_{50}$ values (rat) less than 50 mg/kg; "moderately toxic" agents have $LD_{50}$ values in excess of 50 mg/kg.

⁺ These organophosphates are systemic; they are taken up by the plant and translocated into foliage and sometimes into the fruit.

## Toxicology and Mode of Action

Organophosphates poison insects and mammals primarily by phosphorylation of the acetylcholinesterase enzyme (ACHE) at nerve endings. The enzyme is critical to normal control of nerve impulse transmission from nerve fibers to muscle and gland cells, and also to other nerve cells in autonomic ganglia and in the brain. Some critical portion of the tissue enzyme mass must be inactivated by phosphorylation before symptoms and signs of poisoning are manifested. At sufficient dosage, loss of enzyme function allows accumulation of acetylcholine (ACh, the impulse-transmitting substance) at cholinergic neuroeffector junctions (muscarinic effects), at skeletal nerve-muscle junctions and autonomic ganglia (nicotinic effects), and in the brain. At cholinergic nerve junctions with smooth muscle and gland cells, high ACh concentration causes muscle contraction and secretion, respectively. At skeletal muscle junctions, excess ACh may be excitatory (may cause muscle twitching), but may also weaken or paralyze the cell by depolarizing the end-plate. In the brain, high ACh concentrations cause sensory and behavioral disturbances, incoordination, and depressed motor function. Depression of respiration and pulmonary edema are the usual causes of death from organophosphate poisoning. Recovery depends ultimately on generation of new enzymes in all critical tissues.

Organophosphates are efficiently absorbed by inhalation, ingestion, and skin penetration. To a degree, the occurrence of poisoning depends on the rate at which the pesticide is absorbed. Breakdown occurs chiefly by hydrolysis in the liver; rates of hydrolysis vary widely from one compound to another.

Rarely, certain organophosphates have caused a different kind of neurotoxicity consisting of damage to the axons of peripheral and central nerves and associated with inhibition of "neurotoxic esterase" (NTE). Manifestations have been chiefly weakness or paralysis and paresthesia of the extremities, predominantly the legs, persistent for weeks to years. Most of these rare occurrences have followed an acute poisoning episode of the anticholinesterase type, but some have not been preceded by acute poisoning. Only a few of the many organophosphates used as pesticides have been implicated as causes of delayed neuropathy in humans. EPA guidelines require that organophosphate and carbamate compounds which are candidate pesticides be tested in susceptible animal species for this neurotoxic property.

Other specific properties of individual organophosphates may render them more hazardous than basic toxicity data suggest. By-products can develop in long-stored malathion which strongly inhibit the hepatic enzymes operative in malathion degradation, thus enhancing its toxicity. Certain organophosphates are exceptionally prone to storage in fat tissue, prolonging the need for antidote as stored pesticide is released back into the circulation. Animal studies have demonstrated potentiation of effect when two or more organophosphates are absorbed simultaneously; enzymes critical to the degradation of one are inhibited by the other. Whether this interaction is a significant factor in human poisonings is not known.

### *b. N-Methyl Carbamate Insecticides*

Table 5.4 lists current carbamate insecticides products. They are grouped by relative toxicities.

**Table 5.4** Carbamate insecticides products grouped by relative toxicities.

| **Highly toxic*:** | **Moderately toxic*:** |
|---|---|
| aldicarb[+] (Temik) | dioxacarb (Elocron, Famid) |
| oxamyl (Vydate L, DPX 1410) | promecarb (Carbamult) |
| methiocarb (Mesurol, Draza) | bufencarb (metalkamate, Bux) |
| carbofuran (Furadan, Curaterr, Crisfuran) | propoxur (aprocarb, Baygon) |
| isolan (Primin) | trimethacarb (Landrin, Broot) |
| methomyl (Lannate, Nudrin, Lanox), | pirimicarb (Pirimor, Abol, Aficida, Aphox, Fernos, Rapid) |
| formetanate (Carzol) | dimetan (Dimethan) |
| aminocarb (Matacil) | carbaryl (Sevin, Dicarbam) |

**Figure 5.4** continued

| **Highly toxic*:** | **Moderately toxic* (continued):** |
| --- | --- |
| cloethocarb (Lance) | isoprocarb (Etrofolan, MIPC) |
| bendiocarb (Ficam, Dycarb, Multamat, Niomil, Tattoo, Turcam). | |

* Compounds are listed approximately in order of descending toxicity. "Highly toxic" N-methyl carbamates have listed oral $LD_{50}$ values (rat) less than 50 mg/kg body weight; "moderately toxic" agents have $LD_{50}$ values in excess of 50 mg/kg.
+ This pesticide is taken up by some plants into the foliage and sometimes into the fruit.

## Toxicology and Mode of Action

The N-methyl carbamate esters cause reversible carbamylation of acetylcholinesterase enzyme, allowing accumulation of acetylcholine, the neurotransmitter substance, at parasympathetic neuroeffector junctions (muscarinic effects), at skeletal muscle myoneural junctions and autonomic ganglia (nicotinic effects), and in the brain (CNS effects). The carbamyl-acetylcholinesterase combination dissociates more readily than the phosphoryl-acetylcholinesterase complex produced by organophosphate compounds. This lability has several important consequences: 1) it tends to limit the duration of N-methyl carbamate poisonings; 2) it accounts for the greater span between symptom-producing and lethal doses than exists in the case of most organophosphate compounds; and 3) it frequently invalidates the measurement of blood cholinesterase activity as a diagnostic index of poisoning.

N-methyl carbamates are absorbed by inhalation and ingestion and some by skin penetration. Dermal absorption of particular compounds (notably carbofuran) is very slight. N-methyl carbamates are hydrolyzed enzymatically by the liver and the degradation products are excreted by the kidneys and the liver.

At cholinergic nerve junctions with smooth muscle and gland cells, high acetylcholine concentration causes muscle contraction and secretion, respectively. At skeletal muscle junctions, excess acetylcholine may be excitatory (cause muscle twitching), but may also weaken or paralyze the cell by depolarizing the end-plate. In the brain, elevated acetylcholine concentrations may cause sensory and behavioral disturbances, incoordination, and depressed motor function (rarely seizures), even though the N-methyl carbamates do not penetrate the central nervous system very efficiently. Depression of respiration combined with pulmonary edema is the usual cause of death from poisoning by N-methyl carbamate compounds. Figure 5.7 illustrates the neuralmuscular system showing its parts and chemical transmission of nerve impulse and point of action, the synapse, where the accumulation of acetylcholine occurs.

*c. Solid Organochlorine Insecticides (chlorinated hydrocarbons, chlorinated organics, chlorinated insecticides, and chlorinated synthetics)*

The following table lists current organochlorine products.

**Table 5.5** Organochlorine products.

| | |
|---|---|
| endrin (Hexadrin) | chlordecone (Kepone) |
| aldrin (Aldrite, Drinox) | terpene polychlorinates (Strobane) |
| endosulfan (Thiodan) | chlordane (Chlordan) |
| dieldrin (Dieldrite) | dicofol (Kelthane) |
| toxaphene (Toxakil, Strobane-T) | mirex (Dechlorane) |
| lindane (gamma BHC or HCH, Isotox) | methoxychlor (Marlate) |
| hexachlorocyclohexane (BHC) | dienochlor (Pentac) |
| DDT (chlorophen othane) | TDE (000, Rhothane) |
| heptachlor (Heptagran) | ethylan (Perthane) |

The United States Environmental Protection Agency has sharply curtailed the availability of many organochlorines, particularly DDT, dieldrin, heptachlor, mirex, chlordecone, and chlordane. Others, however, are still the active ingredients of various home and garden products and some agricultural, structural, and environmental pest control products. Hexachlorobenzene is used as a seed protectant fungicide. Technical hexachlorocyclohexane (misnamed benzene hexachloride, BHC) includes multiple stereoisomers; only the gamma isomer (lindane) is insecticidal. Lindane is the active ingredient of some pest control products used in the home and garden, on the farm, and in forestry and animal husbandry. It is also the active agent in the medicinal Kwell, used for human ectoparasitic disease.

## Toxicology and Mode of Action

In varying degrees, organochlorines are absorbed orally, by inhalation, and by dermal absorption. The efficiency of dermal absorption is variable. Hexachlorocyclohexane, including lindane, the cyclodienes (aldrin, dieldrin, endrin, chlordane, heptachlor), and endosulfan are efficiently absorbed across skin, while dermal absorption efficiencies of DDT, perthane, dicofol, methoxychlor, toxaphene, mirex, and kepone are substantially less. Gastrointestinal, and probably dermal, absorption of organochlorines is enhanced by fat and fat solvents. While most of the solid organochlorines are not highly volatile, pesticide-laden aerosol or dust particles trapped in respiratory mucous and subsequently swallowed may be vehicles leading to significant gastrointestinal absorption.

Following exposure to some organochlorines (notably DDT), a significant part of the absorbed dose is stored in fat tissue as the unchanged parent compound. Most organochlorines are in some degree dechlorinated, oxidized, then conjugated. The chief route of excretion is biliary, although nearly all organochlorines yield measurable urinary metabolites. Unfortunately, many of the unmetabolized pesticides are efficiently reabsorbed by the intestine (enterohepatic circulation), substantially retarding fecal excretion. Metabolic dispositions of DDT and DDE (a DDT degradation product), the beta isomer of hexachlorocyclohexane, dieldrin, heptachlor epoxide, mirex, and kepone tend to be slow, leading to storage in body fat. Storable lipophilic compounds are likely to be excreted in maternal milk.

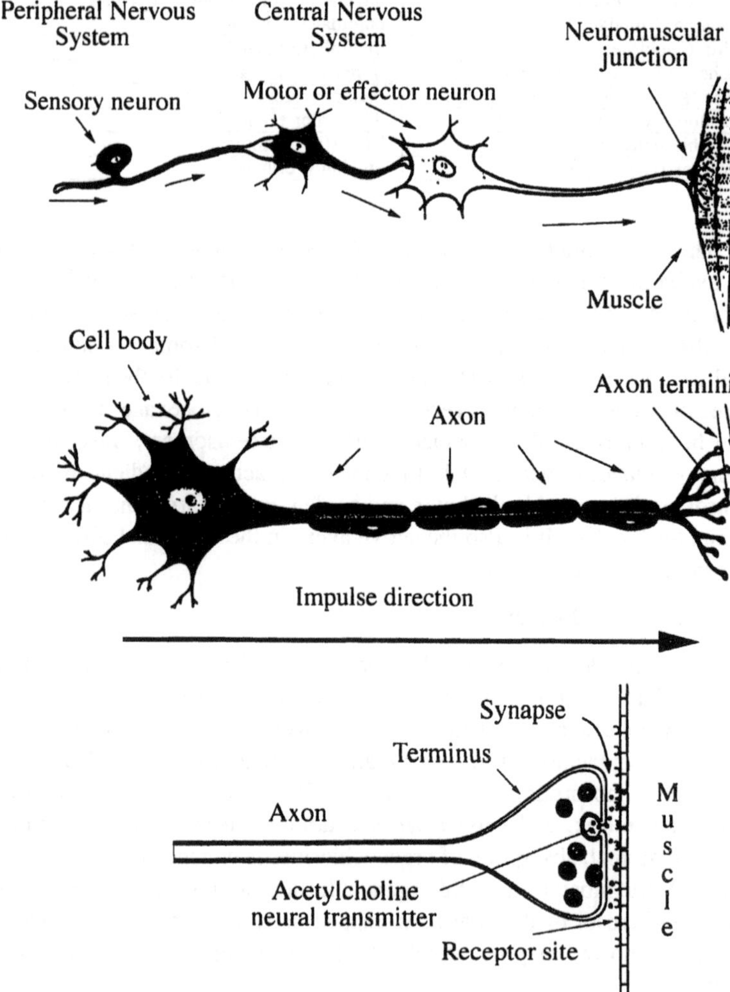


**Figure 5.7** Diagram of the neuralmuscular system showing its parts and chemical transmission of nerve impulse.

The chief toxic action of the organochlorine pesticides is on the nervous system, where these compounds interfere with fluxes of cations across nerve cell membranes, increasing neuronal irritability. This effect is manifested mainly as convulsions, sometimes limited to myoclonic jerking, but often expressed as violent seizures. Convulsions caused by the more slowly metabolized cyclodienes may recur over periods of several days. Convulsions may cause death by interfering with pulmonary gas exchange and by generating severe metabolic acidosis. Various disturbances of sensation, coordination, and mental function are also characteristic of acute organochlorine poisoning. High tissue concentrations of organochlorines increase myocardial irritability, predisposing to cardiac arrhythmias. When tissue organochlorine concentrations drop below threshold levels, recovery from the poisoning occurs. Organochlorines are not cholinesterase inhibitors.

High tissue levels of some organochlorines (notably DDT, DDE, cyclodienes, mirex, and kepone) have been shown to induce hepatic microsomal drug-metabolizing enzymes. This tends to accelerate excretion of the pesticides themselves, but may also stimulate biotransformation of critical natural substances, such as steroid hormones and therapeutic drugs, occasionally necessitating reevaluation of required dosages in persons intensively exposed to organochlorines. Human absorption of organochlorine sufficient to cause enzyme induction is likely to occur only as a result of prolonged intensive exposure.

Hexachlorobenzene (a fungicide) has caused porphyria cutanea tarda in humans. It does not cause convulsions. Lindane, chlordane, and dieldrin have been associated anecdotally with certain rare hematologic disorders, including aplastic anemia; the incidence of these effects appears to be extremely low.

Poisoning by endosulfan has caused blindness in sheep. Mirex at high dosages produces cataracts in rats and mice. The DDT analogue known as DDD is selectively concentrated in adrenal tissue, where high levels have an inhibitory effect on corticosteroid synthesis, and a damaging effect on the cells. Certain other organochlorines are also bioconcentrated in the adrenal cortex.

### d. *Insecticides of Biological Origin*

This section will cover several widely used insecticidal products of natural origin, and also a growth promoting agent, gibberellic acid. It discusses, in order, pyrethrum and pyrethrins, nicotine, rotenone, sabadilla, bacillus thuringiensis, and gibberellic acid.

#### 1. Pyrethrum and Pyrethrins

Pyrethrum is an extract of dried chrysanthemum flowers. The extract contains about 50% active insecticidal ingredients known as pyrethrins. The ketoalcoholic esters of chrysanthemic and pyrethroic acids are known as pyrethrins, cinerins, and jasmolins. These strongly lipophilic esters rapidly penetrate many insects and paralyze their nervous systems. Both crude pyrethrum extract and purified pyrethrins are contained in various commercial products, commonly dissolved in petroleum distillates. Some are packaged in pressurized containers ("bug-bombs"), usually in combination with the synergists piperonyl butoxide and n-octyl bicycloheptene dicarboximide. The synergists retard enzymatic degradation of pyrethrins. Some commercial products also contain organophosphate or carbamate insecticides. These are included because the rapid paralytic effect of pyrethrins on insects ("quick knockdown") is not always lethal.

Pyrethrum and pyrethrin products are used mainly for indoor pest control. They are not sufficiently stable in light and heat to remain as active residues on crops. The synthetic insecticides known as pyrethroids (chemically similar to pyrethrins) do have the stability needed for agricultural application.

Toxicology and Mode of Action

Crude pyrethrum is a dermal and respiratory allergen, probably due mainly to noninsecticidal ingredients. Contact dermatitis and allergic respiratory reactions (rhinitis and asthma) have occurred following exposures. A strong cross-reactivity with ragweed pollen has been recognized. Single cases exhibiting anaphylactic and pneumonitic manifestations have also been reported. The refined pyrethrins are probably less allergenic, but appear to retain some irritant and/or sensitizing properties.

Pyrethrins are absorbed orally and by inhalation, but only slightly across intact skin. They are very effectively hydrolyzed to inert products by mammalian liver enzymes. This rapid degradation combined with relatively poor bioavailability probably accounts in large part for their relatively low mammalian toxicity. Dogs fed extraordinary doses exhibit tremor, ataxia, labored breathing, and salivation. Similar neurotoxicity rarely, if ever, has been observed in humans, even in individuals who have used pyrethrins for body lice control (extensive contact) or pyrethrum as an anthelmintic (ingestion).

In cases of human exposure to commercial products, the possible role of other toxicants in the products should be kept in mind. The synergists piperonyl butoxide and n-octyl bicycloheptene dicarboximide have low toxic potential in humans, but organophosphates or carbamates included in the product may have significant toxicity. Pyrethrins themselves do not inhibit cholinesterase enzyme.

There are presently no practical tests for pyrethrin metabolites or pyrethrin effects on human enzymes or tissues that can be used to confirm absorption.

2. Nicotine

Nicotine is an alkaloid contained in the leaves of many species of plants, but is usually obtained commercially from tobacco. A 95% solution of the free alkaloid in organic solvent has been marketed in the past as a greenhouse fumigant. Another product used for the same purpose is a 40% aqueous solution of nicotine sulfate. Significant volatilization of nicotine occurs from both products. Commercial nicotine insecticides have long been known as Black Leaf 40. Formulations include sprays and dusts. Very little nicotine insecticide is used in the United States today; in fact, most nicotine poisonings are the result of ingestion of tobacco products.

Toxicology and Mode of Action

Nicotine alkaloid is efficiently absorbed orally, dermally, and by inhalation. The sulfate salt is absorbed orally and by inhalation, but is poorly absorbed across the skin. Extensive biotransformation occurs in the liver resulting in a residence half-life of two hours or less. Both the liver and kidney participate in the formation and excretion of multiple endproducts, which are excreted within a few hours.

Toxic action is complex, involving both stimulation and blockade of autonomic ganglia and skeletal muscle neuromuscular junctions, as well as direct effects on the central nervous system. Paralysis and vascular collapse are prominent features of acute poisoning, but death is usually due to respiratory paralysis, which may ensue promptly after the first symptoms of poisoning. Nicotine is not an inhibitor of cholinesterase enzyme.

3. Rotenone

Although this natural substance is present in a number of plants, the source of most rotenone used in the United States is the dried derris root imported from Central and South America. It is formulated as dusts, powders, and sprays (less than 5% active ingredient) for use in gardens and on food crops. Many products contain piperonyl butoxide as a synergist, and other pesticides are included in some commercial products. Rotenone degrades rapidly in the environment. Emulsions of rotenone are applied to lakes and ponds to kill fish.

Toxicology and Mode of Action

Although rotenone is toxic to the nervous systems of insects, fish, and birds, commercial rotenone products have presented little hazard to man over many decades. Neither fatalities nor systemic poisonings in humans have been reported in relation to ordinary use. Low concentration in commercial products, degradability, an intense nauseant effect in man, and poor absorption across gut and skin are probable factors accounting for the good safety record of rotenone.

Numbness of oral mucous membranes has been reported in workers who got dust in their mouths from the powdered derris root. Dermatitis and respiratory tract irritation have also been reported in occupationally exposed persons.

When rotenone has been injected into animals, tremors, vomiting, incoordination, convulsions, and respiratory arrest have been observed. These effects have not been reported in occupationally exposed humans.

Commercial products currently available include Foliefume, Noxfish, Noxfire, Nusyn-Noxfish, PB-Nox, Prentox, Chem-Fish, Rotacide, and Rotenone Solution FK-11.

4. Sabadilla

Sabadilla consists of the powdered ripe seeds of a South American lily. It is used as a dust, with lime or sulfur, or dissolved in kerosene, mainly to kill ectoparasites on domestic animals and humans. Insecticidal alkaloids are those of the veratrin type. The concentration of alkaloids in commercial sabadilla is usually less than 0.5%. Little or no sabadilla is used in the United States today, but there is probably some used in other countries.

Toxicology and Mode of Action

Sabadilla dust is very irritating to the upper respiratory tract, causing sneezing and irritation of the skin.

Veratrin alkaloids are apparently absorbed across the skin and gut, and probably by the lung as well. Veratrin alkaloids have a digitalis-like action on the heart muscle (impaired conduction and arrhythmias).

Although poisoning by medicinal veratrin preparations may have occurred in the remote past, systemic poisoning by sabadilla preparations used as insecticides has been very rare or nonexistent.

5. Bacillus thuringiensis (Bt)

Several strains of the Bacillus thuringiensis are pathogenic to some insects. The bacterial organisms are cultured, then harvested in spore form for use as insecticides. Production methods vary widely. Proteinaceous and nucleotide-like toxins generated by the vegetative forms (which infect insects) are responsible for the insecticidal effect. The spores are formulated as wettable powders, flowable concentrates and granules for application to field crops and for control of mosquitoes and black flies.

Toxicology and Mode of Action

The varieties of Bacillus thuringiensis used commercially survive when injected into mice, and at least one of the purified insecticidal toxins is toxic to mice. Infections of humans have been extremely rare (two recognized cases) and no occurrences of human toxicosis have been reported. From studies involving deliberate ingestion by human subjects, it appears possible, but not likely, that the organism can cause gastroenteritis. Bacillus thuringiensis products are exempt from residual tolerance limits on raw agricultural commodities in the United States. Neither irritative nor sensitizing effects have been reported in workers preparing and applying commercial products.

**Table 5.6** Commercial products containing Bacillus thuringiensis.

| Variety or Strain | Product Trade Name |
|---|---|
| kurstaki | Bactur, Bactospeine, Dipel, Futura, Sok-Bt, Thuricide, Tribactur |
| israelensis | Bactimos, Skeetal, Teknar, Vectobac |
| aizawai | Certan |

*e. Other Insecticides, Acaricides, and Repellents*

This section will discuss insecticides, acaricides, and repellents having toxicological characteristics distinct from the insecticides discussed in the previous section. It discusses pyrethroids, fluorides, borates, chlordimeform, propargite, substituted haloaromatic urea compounds, chlorobenzilate, cyhexatin, methoprene, sulfur, diethyltoluamide, alkyl phthalates, and benzyl benzoate.

1. Synthetic Pyrethroids

These modern synthetic insecticides are chemically similar to natural pyrethrins, but modified to increase stability in the natural environment. They are now widely used in agriculture, in homes and gardens, and for treatment of ectoparasitic disease. Figure 5.6 shows several different products containing various synthetic pyrethroids.

The following table list includes the names of several products that are not currently in commercial production. These are included because they may be marketed in the future, if not in the United States, then possibly in other countries. The active ingredient are the pyrethroids that are in commercial production.

**Table 5.7** Commercial products containing pyrethroids.

| Active Ingredients | Product Trade Names |
|---|---|
| allethrin | Pynamin |
| alphametrin | |
| barthrin | |
| bifenthrin | Capture, Talstar |
| bioresmethrin | |
| biopermethrin | |
| cismethrin | |
| cyclethrin | |
| cyfluthrin | Baythroid |
| cypermethrin | Ammo, Barricade, CCN52, Cymbush, Cymperator, Cyperkill, Folcord, KafilSuper, NRDC 149, Polytrin, Siperin, Ripcord, Flectron, Ustaad, Cyrux |

**Table 5.7** continued.

| Active Ingredients | Product Trade Names |
|---|---|
| deltamethrin | Decamethrin, Decis |
| dimethrin, | |
| esfenvalerate | Asana |
| fenpropathrin | Danitol, Herald, Meothrin, Ortho Danitol, Rody |
| fenvalerate | Belmark, Fenkill, Pydrin, Sumicidin, Tribute |
| flucythrinate | AASTAR, Pay-off |
| fluvalinate | Mavrik, Mavrik Aquaflow, Spur |
| furethrin, | |
| indothrin, | |
| permethrin | Ambush, BW-21-Z, Ectiban, Eksmin, Kafil, Permasect, Perthrine, Pounce, Pramex, Outflank, Talcord, Tor |
| phthalthrin | Neopynamin |
| resmethrin | Benzofuroline, Chrysron, Pynosect, Synthrin |
| tetramethrin | Neopynamin, Phthalthrin |
| tralomethrin | Scout |

Pyrethroids are formulated as emulsifiable concentrates, wettable powders, granules, and concentrates for ultralow volume application. They may be combined with additional pesticides (sometimes highly toxic) in the technical product or tank mixed with other pesticides at the time of application. AASTAR is a combination of flucythrinate and phorate. Phorate is a highly toxic organophosphate.

Nix is a 1% permethrin creme applied to control human ectoparasites.

Toxicology and Mode of Action

Although certain pyrethroids exhibit striking neurotoxicity in laboratory animals when administered by intravenous injection and some are toxic by the oral route, systemic toxicity by inhalation and dermal absorption is low. There have been very few systemic poisonings of humans by pyrethroids. Although limited absorption may account for the low toxicity of some pyrethroids, rapid biodegradation by mammalian liver enzymes (ester hydrolysis and oxidation) is probably the major factor responsible. Most pyrethroid metabolites are promptly excreted, at least in part, by the kidney.

Extraordinarily high absorbed doses may rarely cause incoordination, tremor, salivation, vomiting, diarrhea, and irritability to sound and touch. Extreme doses have caused convulsions in laboratory animals.

Apart from systemic neurotoxicity, some pyrethroids do cause distressing paresthesia (abnormal sensations) when liquid or volatilized materials contact human skin. Sensations are described as stinging, burning, itching, and tingling, progressing to numbness. The skin of the face seems to be most commonly affected, but the hands, forearms, and neck are sometimes involved. Sweating, exposure to sun or heat, and application of water enhance the disagreeable sensations. Sometimes the effect is noted within minutes of exposure, but a 1-2 hour delay in appearance of symptoms is more common. Sensations rarely persist more than 24 hours. Little or no inflammatory reaction is apparent where the paresthesia are reported; the effect is presumed to result from pyrethroid contact with sensory nerve endings in the skin. Not all pyrethroids cause a marked paresthetic reaction; it is prominent following exposure to pyrethroids whose structures include cyano-groups: fenvalerate, flucythrinate, cypermethrin, and fluvalinate. The paresthetic reaction is not allergic in nature; sensitization does not occur. Neither race, skin type, nor disposition to allergic disease affect the likelihood or severity of the reaction.

Persons treated with permethrin for lice or flea infestations sometimes experience itching and burning at the site of application, but this is chiefly an exacerbation of sensations caused by the parasites themselves and is not typical of the paresthetic reaction described above.

The manifestations of neurologic disorder seen in laboratory animals given the more toxic pyrethroids in large doses are salivation, irritability, tremors, ataxia, choreoathetosis (writhing convulsions), fall in blood pressure, and death. Severe metabolic acidosis is characteristic.

Due to the inclusion of unique solvent ingredients, certain formulations of fluvalinate are corrosive to the eyes.

Pyrethroids are not cholinesterase inhibitors.

2. Antibiotics

Antibiotics are natural products derived from fungi or bacteria with antibacterial activity. A new group of products is derived from the fermentation of *Streptomyces avermitilis*. The product (a mixture of the fermentation) is abamectin, the common name for products such as Avermectin®, Affirm®, and Avid®.

Toxicology and Mode of Action

These products containing abamectin are effective against nematodes, and insects, ticks, and mites.

The mode of action has been shown to be the blocking of the neurotransmitter gamma aminobutyric acid (GABA) at the neuromuscular junctions of insects and mites.

3. Insecticidal Soaps

These soaps are the alkali salts of fatty acids. These soaps can be made from potassium, sodium, or ammonia.

Toxicology and Mode of Action

The mode of action of these fatty acid salts is the reduction of surface tension of water which allows water to penetrate the insect spiracles and reduce the oxygen availability leading to suffocation and death. These products are effective against soft-bodied pests such as aphids, spider mites, mealybugs, and whitefly.

These products have no toxicity to mammals, and most are biodegradable.

Commercial products include M-Pede, DeMoss, and Savona.

4. Fluorides

Toxicology and Mode of Action

The probable mode of action of the fluorides is the inhibition of a large number of metal-containing enzymes with which the fluoride forms complexes. These enzymes include phosphatases and phosphorylases. Because sodium fluoride may complex with many key enzymes, it is highly toxic to all plant and animal life. An example of a commercial product is Florocid.

Sodium fluoride

Sodium fluoride is a crystalline mineral once widely used in the United States for control of larvae and crawling insects in homes, barns, warehouses, and other storage areas. It is highly toxic to all plant and animal life.

Commercial product is Florocid.

Sodium fluosilicate

Sodium fluosilicate (sodium silico fluoride) has been used to control ectoparasites on livestock as well as crawling insects in homes and work buildings. It is approximately as toxic as sodium fluoride.

Commercial products include Satsan (dust formulation) and Prodan (bait formulation).

Sodium fluoaluminate

Sodium fluoaluminate (sodium aluminofluoride, cryolite) is a stable mineral containing fluoride. It is used as an insecticide on some vegetables and fruits. Cryolite has very low water solubility, does not yield fluoride ions on decomposition, and presents very little toxic hazard to mammals, including man.

Commercial product is Kryocide.

Hydrofluoric acid

Hydrofluoric acid is an important industrial toxicant, but is not used as a pesticide.

Fluoroacetate

Sodium fluoroacetate is a highly toxic rodenticide, and even extremely small amounts can cause severe or fatal poisoning. Fluoroacetate acts after its metabolic conversion to fluorocitric acid as a competitive inhibitor of the enzyme aconitase, competing with citric acid for the enzyme and, therefore, blocking the tricarboxylic acid cycle.

Commercial products include Compound 1080, Fratol, and Yasoknock.

Toxicology

Sodium fluoride and fluosilicate used as insecticides present a serious toxic hazard to humans because of high inherent toxicity and the possibility that children crawling on floors of treated dwellings will ingest the material.

Absorption across the skin is probably slight and methods of pesticide use rarely include a hazard of inhalation, but uptake of ingested fluoride by the gut is efficient and potentially lethal. Excretion is chiefly in the urine; renal clearance of fluoride from the blood is rapid. However, large loads of absorbed fluoride poison renal tubule cells. Functional tubular disturbances and sometimes acute renal failure result.

5. Boric Acid and Borates

Commercially available as boric acid, sodium tetraborate decahydrate (borax), sodium pentaborate, boron trioxide, and sodium biborate.

Commercial products include Polybor and Pyrobor.

Products are formulated as tablets and powder to kill larvae in livestock confinement areas and cockroaches in residences. Solutions are rarely sprayed as a nonselective herbicide.

Toxicology and Mode of Action—Borate

Borax dust is moderately irritating to skin. Inhaled dust causes irritation of the respiratory tract, cough, and shortness of breath.

There have been few poisonings from the pesticidal uses of borates, although powders and pellets scattered on the floors of homes do present a hazard to children. Most poisonings have resulted from injudicious uses in human medicine aimed at suppressing bacterial growth, such as compresses for burns. Many poisonings of newborns occurred in the 1950s and 1960s.

6. Formamidines

Formamidines comprise a small group of promising insecticides. Chlordimeform, formetanate, and amitraz are examples of this group. These compounds are effective against most stages of mites and ticks. Thus they are classified as ovicides, insecticides, and acaricides. Formulations are emulsifiable concentrates and water-soluble powders.

Commercial products include Bermat, Fundal, Galecron, and Ovatoxin.

Toxicology and Mode of Action

The mode of action of the members of this group appears to be the inhibition of the enzyme monoamine oxidase. This inhibition leads to an accumulation of toxic intermediates.

In a reported episode of occupational exposure to chlordimeform, several workers developed hematuria. Hemorrhagic cystitis, probably due to chloraniline biodegradation products, was the source of blood in the urine. Symptoms reported by the affected workers were: gross blood in the urine, painful urination, urinary frequency and urgency, penile discharge, abdominal and back pain, a generalized "hot" sensation, sleepiness, skin rash and desquamation, a sweet taste, and anorexia. Symptoms persisted for 2-8 weeks after exposure was terminated.

7. Organosulfurs

Toxicology and Mode of Action

Propargite is an acaricide with residual action. Formulations are wettable powders and emulsifiable concentrates.

Commercial products include Omite, Comite, and Uniroyal D014.

Propargite exhibits very little systemic toxicity in animals. No systemic poisonings have been reported in humans. However, many workers having dermal contact with this acaricide have experienced skin irritation and possibly sensitization in some cases. Eye irritation has also occurred. For this reason, stringent measures should be taken to prevent inhalation or any skin or eye contamination by propargite.

There is no readily available method for detecting absorption of propargite.

8. Substituted Ureas (Acylureas)

Toxicology and Mode of Action

These are haloaromatic substituted ureas which control insects by impairing chitin deposition in the larval exoskeleton. They are formulated as wettable powders, oil dispersible concentrates, and granules for agriculture and forestry and in settings where fly populations tend to be large, such as feedlots.

**Table 5.8** Commercial products containing substituted ureas.

| Active ingredient | Products |
|---|---|
| Diflubenzuron | Dimilin, Micromite |
| Teflubenzuron | Nomolt, Dart, Diaract |
| Triflumuron | Alsystin |
| Chlorfluazuron | Atabron, Jupiter |
| Hexaflumuron | Trueno, Consult |

**Table 5.8** continued

| Active ingredient | Products |
|---|---|
| Flufenoxuron | Cascade |
| Flucycloxuron | Andalin |
| Novaluron | GR-572 |

There is limited absorption across the skin and intestinal lining of mammals, after which enzymatic hydrolysis and excretion rapidly eliminates the pesticide from tissues. Irritant affects are not reported, and systemic toxicity is low. Methemoglobinemia is a theoretical risk from chloroaniline formed hydrolytically, but no reports of this form of toxicity have been reported in humans or animals from diflubenzuron exposure.

9. Organotins

Toxicology and Mode of Action

The organotins are a group of pesticides that have both acaricide and fungicide activities. Of particular interest is cyhexatin (tricyclohexyl tin hydroxide), a highly specific acaricide. Later, fenbutatin-oxide was introduced and has proven to be highly effective against mites on fruits, greenhouse crops, and ornamentals.

Tricyclohexyl tin hydroxide is formulated as a 50% wettable powder for control of mites on ornamentals, hops, nut trees, and some fruit trees.

It is moderately irritating, particularly to the eyes. While information on the systemic toxicity of this specific tin compound is lacking, it should be assumed that cyhexatin can be absorbed to some extent across the skin and that substantial absorbed doses would cause nervous system injury.

Commercial products include Plictran and Vendex.

10. Insect Growth Regulators

Toxicology and Mode of Action

Methoprene is a long chain hydrocarbon ester active as an insect growth regulator. It is effective against several insect species. Formulations include slow-release briquets, sprays, foggers, and baits.

Commercial products include ZR-515, Altosid SR-10 and CP-10, Apex SE, Diacon, Dianex, Kabat, Minex, Pharorid, and Precor.

Methoprene is neither an irritant nor a sensitizer in humans or laboratory animals.

Systemic toxicity in laboratory animals is very low. No human poisonings or adverse reactions in exposed workers have been reported.

11. Inorganic Sulfur

Elemental sulfur is an acaricide and fungicide widely used on orchard, ornamental, vegetable, grain, and other crops. It is prepared as dust in various particle sizes and applied as such, or it is formulated with various minerals to improve flowability, or is applied as an aqueous emulsion or wettable powder.

Commercial products that are currently available include Brimstone, Lacco Sulfur, Clifton Sulfur, Sul-Cide, Cosan, Kumulus S, Sofril, Sulfex, Thiolux, Thiovit, Magnetic 6, Liquid Sulfur, Thion, Zolvis, and Golden Dew.

Toxicology and Mode of Action

Elemental sulfur is moderately irritating to the skin, and airborne dust is irritating to the eyes and the respiratory tract. In hot sunny environments, there may be some oxidation of foliage-deposited sulfur to irritating gaseous sulfur oxides, which are very irritating to the eyes and respiratory tract.

Ingested sulfur powder induces catharsis and has been used medicinally (usually with molasses) for that purpose. Some hydrogen sulfide is formed in the large intestine and this may present a degree of toxic hazard. However, an adult has survived ingestion of 60 grams.

Ingested colloidal sulfur is efficiently absorbed by the gut and is promptly excreted in the urine as inorganic sulfate.

12. Insect Repellants

a. Diethyltoluamide (DEET)—This chemical is one of the most widely used liquid insect repellents used today. It is suitable for application to skin or to fabrics. It is formulated with ethyl or isopropyl alcohol, usually in pressurized containers.

Commercial products that are currently available include Detamide, Metadelphene, MGK, OFF, and Repel.

Toxicology and Mode of Action

For many years, diethyltoluamide has been effective and generally well tolerated as an insect repellent applied to human skin, although tingling, mild irritation, and sometimes desquamation have followed repeated applications. In some cases, DEET has caused contact dermatitis and exacerbation of preexisting skin disease. It is very irritating to the eyes, but not corrosive.

Serious adverse effects have occurred when used under tropical conditions, when it was applied to areas of skin that were occluded during sleep. Under these conditions, the skin became red and tender, then exhibited blistering and erosion, leaving painful weeping denuded areas that were slow to heal. Permanent scarring resulted from most of these severe reactions.

DEET is efficiently absorbed across the skin and by the gut. Blood concentrations of about 3 mg per liter have been reported several hours after dermal application in the prescribed fashion. Toxic encephalopathic reactions have apparently occurred in rare instances following dermal application, mainly in children who were intensively treated. The more frequent cause of systemic toxicity has been ingestion, deliberate in adults, accidental in young children.

Discretion should be exercised in recommending DEET for persons who have acne, psoriasis, an atopic predisposition, or other chronic skin condition. It should not be applied to any skin area that is likely to be opposed to another skin surface for a significant period of time (antecubital and popliteal fossse, inguinal areas).

Great caution should be exercised in using DEET on children. Only the products containing the lower concentrations (usually 15%) should be used, and application should be limited to exposed areas of skin, using as little repellent as possible. If continuous repellent protection is necessary, DEET should be alternated with a repellent having another active ingredient. If headache or any kind of emotional or behavioral changes occur, use of DEET should be discontinued immediately.

b. Dimethyl phthalate has been widely used as an insect repellent applied directly to the skin. Dibutylphthalate is impregnated into fabric for the same purpose. It is more resistant to laundering than dimethyl phthalate.

The only commercial product currently available is DMP.

Toxicology and Mode of Action

Dimethyl phthalate is strongly irritating to the eyes and mucous membranes. It has caused little or no irritation when applied to skin, and dermal absorption is apparently minimal. It has not caused sensitization.

Tests in rodents have indicated low systemic toxicity, but large ingested doses cause gastrointestinal irritation, central nervous system depression (to coma), and hypotension. One accidental ingestion by a human resulted in coma, but recovery was prompt.

13. Benzyl Benzoate

Benzyl benzoate is incorporated into lotions and ointments. This agent has been used for many years in veterinary and human medicine against mites and lice.

Toxicology

Apart from occasional cases of skin irritation, adverse effects have been few. The efficiency of skin absorption is not known. Absorbed benzyl benzoate is rapidly biotransformed to hippuric acid which is excreted in the urine. When given in large doses to laboratory animals, benzyl benzoate causes excitement, incoordination, paralysis of the limbs, convulsions, respiratory paralysis, and death. No human poisonings have been reported.

*f. Arsenical Pesticides*

1. Inorganic Trivalent

a. Arsenic trioxide—"White arsenic," arsenous oxide. These are active ingredients in some ant pastes and veterinary preparations.

b. Sodium arsenite— Chem Pels C, Chem-Sen 56, Kill-All, Penite, Prodalumnol Double. These can be used in aqueous solution for weed control; limited use as insecticide.

c. Calcium arsenite— Mono-calcium arsenite. Formulated as a flowable powder for insecticidal use on fruit.

d. Copper arsenite (acid copper arsenite)—Formulated as a wettable powder, for use as insecticide, wood preservative.

e. Copper acetoarsenite— Insecticide. Examples include Paris Green, Schweinfurt Green, Emerald Green, French Green, and Mitts Green. Products are no longer used in the U.S. However, they are still used outside the U.S.

2. Inorganic Pentavalent

a. Sodium arsenate—Disodium arsenate. Examples are Jones Ant Killer, Terro Ant Killer.

b. Lead arsenate—Gypsine, Soprabel, Talbot. This is of limited use in the U.S. Used as wettable powder insecticide outside the U.S.

c. Zinc arsenate—This powder is used in the U.S. as an insecticide on potatoes and tomatoes.

Toxicology of Arsenic

Toxicity of the various arsenic compounds in mammals extends over a wide range, determined in part by unique biochemical actions of each compound, but also by absorbability and efficiency of biotransformation and disposition. Overall, arsines present the greatest toxic hazard, followed closely by arsenites (inorganic trivalent compounds). Inorganic pentavalent compounds are somewhat less toxic that arsenites, while the organic (methylated) pentavalent compounds incur the least hazard of the arsenicals that are used as pesticides.

The pentavalent arsenicals are relatively water soluble and absorbable across mucous membranes, while trivalent arsenicals, having greater lipid solubility, are more readily absorbed across the skin. However, poisonings by dermal absorption of either form have been extremely rare. Ingestion has been the usual basis of poisoning; gut absorption efficiency depends on the physical form of the compound, its solubility characteristics, the gastric pH, gastrointestinal motility, and gut microbial transformations. Once absorbed, many arsenicals cause toxic injury to cells of the nervous system, blood vessels, liver, kidney, and other tissues. Two biochemical mechanisms of toxicity are recognized: 1) reversible combination with thiol groups contained in tissue proteins and enzymes, and 2) substitution of arsenic anions for phosphate in many reactions, including those critical to oxidative phosphorylation. Because there are many uncertainties regarding the biotransformation of various arsenicals in the gut and in the body, and because the toxic potentials of the biotransformation products are not well established, it is generally safest to manage cases of arsenic ingestion as though all forms of arsenic are highly toxic.

Mammals, including humans, detoxify inorganic arsenic by methylation, yielding cacodylic acid (dimethylarsinic acid) as the chief urinary excretion product. Disposition by urinary excretion is usually prompt. Elimination of the arsonic acid (monomethyl) compounds has not been extensively studied, but urinary excretion of the unaltered compound and/or a further methylated form would seem likely.

## V. HERBICIDES

Weeds may be defined as plants growing out of place or where man does not wish them to grow. Organized agriculture dates back some 10,000 years and the extensive cultivation of crops is only about 5,000 years old. Weeds have caused problems for at least 5,000 years by competing with crops for moisture, nutrients, and light.

The idea of controlling weeds with chemicals is not new; for more than a century chemicals have been employed for weed control. The removal of plants from such places as railway tracks, fence lines, recreation areas, timber yards, roadsides, and power lines with crude chemicals such as rock salt, salt solutions, crushed arsenical ores, creosote, oil wastes, sulfuric acid, and copper salts with massive doses dates back to the nineteenth century and continued into the early 1900s.

Herbicides have provided a more effective and economical means of weed control than mechanical cultivation. Together with fertilizers and improved varieties of plants, herbicides have made an immense contribution to increased yields and reduced costs.

## A. CLASSIFICATION OF HERBICIDES

Every type of herbicide classification has overlap, and this is true especially when the assignment is made according to the type of use and mode of action. It is possible to differentiate herbicides by the following parameters:

The timing of application and sowing or planting time of the plants grown:

- pre-sowing
- preemergence
- postemergence herbicides.

The uptake area:

- soil herbicides
- leaf herbicides.

The mode of action:

- contact herbicides
- systemic herbicides.

The range of use:

- nonselective herbicides
- selective herbicides.

In this section, we emphasize the use of chemical classification of herbicides. The two major classifications are inorganic and organic herbicides.

### 1. Inorganic Herbicides

As mentioned earlier in Chapter 1, the first chemicals used in weed control were the inorganics. Until the 1960s, arsenic-containing compounds were commonly used and therefore we will start with them.

a. *Organic and Inorganic Arsenical Compounds*

1. Inorganic Pentavalent

a. Arsenic acid—Dessicant L-10, Hi-Yield Dessicant H-10, Zotox. Water solutions were used as defoliants and herbicides.

2. Organic Pentavalent

a. Cacodylic acid (sodium cacodylate)—Nonselective herbicide, defoliant, silvicide. Products include Boiate, Bolls-Eye, Bophy, Dilie, Kack, Phytar 560, Rad-E-Cate 25, and SaNo.

b. Methane arsonic acid— MAA. Nonselective herbicide.

c. Monosodium methane arsonate—MSMA. Nonselective herbicide, defoliant, silvicide. Several products include Ansar 170, Arsonate Liquid, Bueno 6, Daconate 6, Dal-E-Rad, Drexar 530, HerbAll, Merge 823, Mesamate, Target MSMA, Trans-Vert, Weed-E-Rad, and Weed-Hoe.

d. Disodium methane arsonate—DSMA. Selective postemergence herbicide, silvicide. Several products include Ansar 8100, Arrhenal, Arsinyl, Crab-E-Rad, Di-Tac, DMA, Methar 30, Sodar, and Weed-E-Rad 360.

e. Monoammonium methane arsonate—MAMA. Selective postemergence herbicide.

f. Calcium acid methane arsonate—CAMA. Selective postemergence herbicide. Calar, Super Crab-E-Rad-Calar, Super Dal-E-Rad.

Arsonates are absorbed and translocated to underground tubers and rhizomes, making them extremely useful against Johnson grass and nut sedges when applied as spot treatments.

Toxicology of Arsenic (see section IV, *f.*, Toxicology and Mode of Action).

### *b. Ammonium Salts*

Ammonium salts were also used for weed control. These salts, ammonium thiocyanate, ammonium nitrate, ammonium sulfate, along with ammonium sulfamate were used as foliar sprays. The modes of action of these ammonium salts along with the use of other salts such as iron sulfate and copper sulfate cause desiccation and plasmolysis.

### *c. Borates*

The borate herbicides were another group of inorganics. These included sodium tetraborate sodium metaborate, tetraborate decahydrate (borax), and sodium borate.

Commercial products that are currently available include Polybor and Pyrobor.

Toxicology and Mode of Action

Borates are absorbed by the plant roots and translocated throughout the plant. They are nonselective and persistent herbicides. Borates accumulate in the reproductive structures of the plant, but the precise mode of action is unclear.

Formulated as powders and solutions and often mixed with other herbicides; they are spread or sprayed as nonselective herbicides.

Dust formulations are moderately irritating to skin. Inhaled dust causes irritation of the respiratory tract, cough, and shortness of breath.

As mentioned eariler in Section IV, Insects, there have been few poisonings from the pesticidal uses of borates. Most poisonings have resulted from injudicious uses in human medicine aimed at suppressing bacterial growth, such as compresses for burns. Many poisonings of newborns occurred in the 1950s and 1960s.

*d. Clorates*

The chlorates are unstable molecules and strong oxidizing agents. They are highly flammable when in contact with organic materials and sometimes ignite spontaneously. Some formulations contain flame retardants, such as sodium metaborate.

Sodium chlorate has been used widely as a soil sterilant. It is readily absorbed by the roots and leaves and may be translocated throughout the plant. Its mode of action is not well understood and may be related to the high oxidizing capacity of the chlorate ion.

Magnesium chlorate is used as a defoliant and dessicant.

## 2. Organics

*a. Aliphatic acids*

Mode of Action

*TCA,* the sodium salt of trichloroacetic acid, is primarily employed to control monocotyledons. Uptake of the highly water-soluble trichloroacetate takes place mainly via the roots or rhizomes. The active substance is concentrated in the growth-active areas, and symptoms of wilting, dark green discoloration, or even of leaf deformation may occur. The plants finally begin to die after several days.

*Dalapon,* the sodium salt of dichloropropionic acid, acts similarly, mainly against monocotyledons. Uptake is possible via the roots as well as the leaves. Transport takes place via both the xylem and phloem. The most powerful action is obtained by spraying in the spring on young growing plants. Deposition in the growth-active zones is possible as breakdown in the plants occurs only very slowly. Little is known about the mechanism of action, although the protein-precipitating properties, inhibition of pantothenic acid synthesis, and other reactions contribute to it.

The very low toxicity of Dalapon to fish makes its application possible in the elimination of reeds, sedges, and rushes, etc., on and in drainage ditches, but only after permission has been granted by the appropriate water authorities and not in water protection areas. Table 5.9 lists several commerical products containing aliphatic acids.

*Glyphosate,* due to its excellent systemic properties and broader spectrum of activity, has greatly replaced the use of herbicides based on TCA and Dalapon in previous years. Uptake occurs through the leaves, and transport predominantly via the phloem, so that root or rhizome weeds are particularly severely affected. Some days after its use, its effect is visible in brightening of the leaves and inhibited growth; treated plants die after about 2-3 weeks. Glyphosate blocks the synthesis of aromatic amino acids, which are particularly important as precursors of proteins and secondary plant compounds. Figure 5.8 shows a package of Roundup.

*Glufosinate* is a postemergent, nonselective, partially systemic contact herbicide which acts on leaves as well as in the plant after uptake and transport. Its action is delayed at lower temperatures. It causes inhibition of photosynthesis and interferes with amino acid metabolism of the plant, thereby causing accumulation of $NH_4$. The most important uses are in the control of seed and root weeds in vineyards and in fruit growing. A relatively quick degradation with a half-life between 30 and 40 days occurs in the soil.

**Table 5.9** Commercial products containing aliphatic acids.

| Active Ingredient | Product—Trade Name |
|---|---|
| TCA (trichloroacetic acid) | Konesta, NaTA |
| Dalapon | Dalacide, Dalapon-Na, Gramevin, Unipon |
| Glyphosate | Accord, Ranger, Rodeo, Roundup |
| Glufosinate | Basta, Dash, Finale, Liberty |

Toxicity

Several members of this family of herbicides are severely irritating and can cause permanent damage to the eyes.

*b. Acetanilide*

Toxicology and Mode of Action

The acetanilides (propachlor, butachlor, alachlor and metolachlor) are used as preemergence herbicides against annual broad-leaved weeds and grasses. The acetanilides owe their herbicidal activity to the inhibition of photosynthetic electron transport and energy production.

Several formulations of Alachlor, namely Crop Star, Lasso, Micro-Tech C, and Partner are RUP products due to their oncogenicity.

Dual is a product with the active ingredient metolachlor. It has a signal word *caution* and is therefore a slightly toxic to relatively nontoxic product.

**Figure 5.8** Roundup is a contact herbicide containing glyphosate. This container contains liquid concentrate which is diluted prior to use.

**Table 5.10** Commercial products containing acetanilides.

| Active Ingredient | Trade Name |
|---|---|
| Alachlor | Crop Star, Lasso, Micro-Tech C, Partner |
| Propachlor | Ramrod |
| Metolachlor | Dual |
| Metazachlor | Butisan S |

### *c. Amides*

Mode of Action

There are no consistent modes of action for this group. They are similar in activity to the acetanilides. These soil-applied preemergent herbicides have been reported to inhibit root and shoot elongation but are not trasnslocated in the xylem.

Modes of action of this group are severalfold. Naptalam interferes with cell growth and development by affecting auxin transport, while the mode of action of butam diphenamid is unknown. Propylzamide interferes with cell division and diflufenican causes chlorosis resulting from inhibition of carotenoid biosynthesis. Carbetamide, naproamide, and propyzamide are thought to act by inhibiting protein synthesis.

Toxicology

Amides are moderately to mildly irritating. Several formulations may contain solvents which can cause injury by both inhalation and oral ingestion.

Among the amides are the products listed in Table 5.11.

**Table 5.11** Commercial products containing amides.

| **Active Ingredient** | **Trade Name** |
|---|---|
| Dimethenamide | Frontier |
| Napropamide | Devrinol, Naproguard |
| Propanil | Stam, Stampede, Strel |
| Bensulide | Betasan, Prefar |
| Naptalam | Alanap-L, Naptro |
| Propyzamide (pronamide) | Kerb |
| Carbetamide | Carbetamex |

### *d. Benzoic or Arylaliphatic Acid Herbicides*

Mode of Action

The herbicides resemble and mimic auxins in the activity. They compete with natural auxins, cause abnormal elongation at the growing terminals, tissue proliferation, induce adventitious roots, and modify the arrangement of leaves and other organs. These herbicides act in the same way as the chlorophenoxy acids (2,4-D and relatives); however, the specific mechanism of action of the benzoic acid herbicides is unknown.

**Table 5.12** Commercial products containing benzoic or arylaliphatic acid.

| **Active Ingredient (common name)** | **Trade Name** |
|---|---|
| Chloramben | Amiben, Vegiben |
| Dicamba | Banvel, Metambane, Scotts Proturf, Veteran |
| 2,3,6-Trichlorobenozic acid (2,3,6-TBA) | Tribac |

*e. Benzonitrile Herbicides*

Dichlobenil, bromoxynil, and ioxynil are contact herbicides used to control hard-to-kill weeds.

Mode of Action

The modes of action of this group are not similar. Dichlobenil is applied to the soil as granules and prevents the germination of annual weeds and shoot growth in perennials. Bromoxynil and ioxynil induce foliar chlorosis and necrosis. The degradation products of these herbicides also inhibit other complex cellular processes, and their precise modes of action are unknown.

**Table 5.13** Commercial products containing benzonitriles.

| Active Ingredients | Trade Names |
|---|---|
| Dichlobenil | Barrier, Casoron, Dyclomec, Norosac |
| Bromoxynil | Bromotril, Buctril, Certrol, Combine Emblem, Koril, Korilene, Merit, Pardner, Sabre, Sanoxynil |
| Ioxynil | Not available in the U.S. |

*f. Chlorophenoxy Herbicides*

These herbicides are highly selective for broadleaved weeds. 2,4-D was the first chlorophenoxy acid herbicide developed. There are several compounds that belong to this group in addition to 2,4-D and they include 2,4-DB, Dichlorprop, MPCA, MPCB, MCPP, 2,4,5-T, and 2,4,5-TP (Silvex).

Mode of Action

These herbicides cause auxin-like responses (hormone-induced growth) in broadleaf plants. They cause rapid cell division, resulting in unusual growth, activated phosphate metabolism, and increased RNA and protein synthesis. The mode of action thus appears to consist of a large number of structural and biochemical reactions causing prolonged abnormal growth and the failure of the plant to undergo changes characteristic of maturity and senescence.

Typical symptoms of the action of growth regulators manifest themselves as crookedness, deformation, leaf or stem malformations, and adventitious root formation. Increased distension as well as signs of wilting can appear. The weed's growth is arrested, and it dies as a result of generalized damage. The extent and length of time within which these effects occur are dependent on the type of plant, the dose adsorbed, and environmental conditions.

Commercial Products

Several hundred commercial products contain chlorophenoxy herbicides in various forms, concentrations, and combinations. Following are names of widely advertised formulations. In some cases, the same name is used for products with different ingredients. Exact composition must therefore be determined from the product label.

**Table 5.14** Commercial products containing chlorophenoxy herbicides.

| Chlorophenoxy Acid Compound | Commercial Products |
|---|---|
| 2,4-D (2,4-dichlorophenoxyacetic acid) | Agrotect, Amoxone, AquaKleen, BH 2,4-D, Chipco Turf Herbicide D, Chloroxone, Crisalamina, Crisamina, Crop Rider, D50, Dacamine, Debroussaillant 600, Ded-Weed SULV, Desormone, Dinoxol, Emulsamine BK, Emulsamine E-3, Envert DT, Envert 171, Super D Weedone, Weedone, Estone, Farmco, Fernesta, Fernimine, Fernoxone, Ferxone, Formula 40, Gordon's Amine 400, Gordon's LV 400 2,4-D, Hedonal, Herbidal, Lawn-Keep, Macondray, Miracle, Netagrone 600, Pennamine D, Planotox, Plantgard, Salvo, Spritz-Hormit/2,4-D, Superormone Concentre, Transamine, Tributon, Tuban, U 46, U 46 D-Ester, U 46 DFluid, Weed-B-Gon, Weedar, Weedatul, Weed-Rhap, Weed Tox, Weedtrol, Gordon's Dymec Turf Herbicide Amine 2,4-D, Gordon's Phenaban 801, Acme Amine 4, Acme Butyl Ester 4, Acme LV 6, Acme LV 4, Gordon's Butyl Ester 600, DMA 4, Dormone |
| **Chlorophenoxy Acid Compound** | **Commercial Products** |
| 2,4-DP (2,4-dichlorophenoxypropionic acid) | BH 2,4-DP, Desormone, Hedonal, Hedonal DP, Kildip, Polymone, Seritox 50, U 46, U 46 DPFluid, Weedone DP, Weedone 170 |
| 2,4-DB (2,4-dichlorophenoxybutyric acid) | Butoxon, Butoxone, Butyrac, Embutox |
| 2,4,5-T (2,4,5-trichlorophenoxyacetic acid) | Amine 2,4,5-T for rice, Dacamine, Ded-Weed, Farmco Fence Rider, Forron, Inverton 245, Line Rider, Super D Weedone, T-Nox, Trinoxol, U 46, Weedar, Weedone |
| MCPA (2-methyl, 4-chlorophenoxy aliphatic acids and esters) | metaxon, Agroxone, Weedone |
| MCPB (2-methyl, 4-chlorophenoxy aliphatic acids and esters) | Can-Trol, PDQ, Thistrol |
| MCPP (2-methyl, 4-chlorophenoxy aliphatic acids and esters) | mecoprop, Methoxone M, Mecopex, Gordon's Mecomec |

Sodium, potassium, and alkylamine salts are commonly formulated as aqueous solutions, while the less water-soluble esters are applied as emulsions. Low molecular weight esters are more volatile than the acids, salts, or long-chain esters.

Chlorophenoxy compounds are sometimes mixed into commercial fertilizers to control growth of broadleaf weeds.

## Toxicology

Some of the chlorophenoxy acids, salts, and esters are moderately irritating to skin, eyes, and respiratory and gastrointestinal linings. In a few individuals, local depigmentation has apparently resulted from protracted dermal contact with chlorophenoxy compounds.

The chlorophenoxy compounds are absorbed across the gut wall, lung, and skin. They are not significantly fat-storable. Excretion occurs almost entirely by way of the urine. Apart from some conjugation of the acids, there is limited biotransformation in the body. Figure 5.9 pictures a retail package containing 2,4-D (Weed-B-Gon®).

**Figure 5.9** Trigger pump sprayer containing 2,4-D (Weed-B-Gon®).

*g. Dinitrophenols*

In 1933, 2-methyl-4,6-dinitrophenol (DNOC) was introduced. DNOC, along with the related compound dinseb is used on a limited scale because of its toxicity to plants, animals and the environment.

Toxicity and Mode of Action

These are contact herbicides which, in high concentrations, cause foliar desiccation and collapse. Their mode of action appears to be their ability to uncouple oxidative phosphorylation and hence inhibit ATP synthesis. Since this process is common in plants and mammals, it is not surprising that dinitrophenols have high mammalian toxicities.

Nitroaromatic compounds are highly toxic to humans and animals. Most nitrophenols and nitrocresols are absorbed efficiently from the gastrointestinal tract, across the skin, and by the lung when fine droplets are inhaled. Fatal poisonings have occurred as a result of dermal contamination. Except in a few sensitive individuals, the compounds are only moderately irritating to the skin and mucous membranes.

*h. Bipyridyliums or Dipyridyls*

Mode of Action and Toxicology

These compounds act as contact herbicides, rapidly killing all green plant growth on which they fall. They are rapidly translocated from foliage, but not from the plant roots, because they are deactivated on contact with soil.

In solution, bipyridiums almost completely dissociate into ions, and in chloroplasts, during photosynthesis, the positive ion is reduced to a stable free radical. In the presence of oxygen, the free radicals are reorganized to the original ion and hydrogen peroxide, which destroys the plant tissue.

Particularly in concentrated form, paraquat causes injury to tissues with which it comes into contact. It leaves the skin of the hands dry and fissured, sometimes resulting in loss of fingernails. Prolonged contact with skin may cause blistering and ulceration, with subsequent absorption of paraquat in sufficient dosage to cause systemic poisoning. Prolonged inhalation of spray droplets may cause nosebleed. Eye contamination results in severe conjunctivitis and sometimes protracted and even permanent corneal opacification. When ingested in adequate dosage, paraquat has life-threatening effects on the gastrointestinal tract, kidney, liver, heart, and other organs.

Diquat is somewhat less damaging to skin than paraquat, but irritation effects may appear following dermal contamination. Diquat has severe toxic effects on the central nervous system that are not typical of paraquat.

Paraquat is a synthetic nonselective contact herbicide, usually marketed as the dichloride salt. Dimethyl sulfate salts are also produced. Liquid technical products range from 20% to 50% concentration. Trade names of liquid concentrates are Ortho Paraquat CL, Ortho Paraquat Plus, Cekuquat, Crisquat, Herbaxon, Herboxone, Dextrone, Esgram, Gramocil, Gramoxone, Goldquat

276, Sweep, Osequat Super, Gramonol, Toxer Total, Pillarxone, and Pillarquat.

Paraquat is commonly formulated in combination with other herbicides:

With diquat: Actor, Preeglone, Preglone, Priglone, Weedol (a 2.5% soluble granule formulation).

With monolinuron: Gramonol

With diuron: Gramuron, Para-col, Tota-col, Dexuron

With simazine: Terraklene, Pathclear

Diquat is usually prepared as the dibromide monohydrate salt, 20% to 25% in liquid concentrates. Diquat and reglon are alternative common names.

Commercial products are Ortho Diquat, Aquacide, Dextrone, Reglone, Reglox, and Weedtrine-D. Combinations with paraquat are listed above.

Diquat is still used as a water herbicide but is now applied as a dessicant and terrestrial herbicide as well.

*i. Carbamate, Aromatic or Phenyl Carbamate, and Thiocarbamate Herbicides*

The carbamates are esters of carbamic acid. They exhibit fungicideal, insecticidal as well as herbicidal activity.

Mode of Action and Toxicology

This group of herbicides has many diverse modes of action. The mode of action of most of the members of this group is by interfering with photosynthesis, or by interfering with cell division in meristematic tissues (growing termini). This latter action is effected by impairing nucleic acid metabolism and protein synthesis.

Most of these herbicides are absorbed by the plant roots and are translated in the xylem.

There is evidence suggesting that many of the thiocarbamates reduce the production of cuticular wax and interfere with lipid biosynthesis which probably represents the primary mode of action of these herbicides.

As a group, these herbicides exhibit moderate to low acute toxicity. The carbamates and their relatives present no particular hazard in normal use.

**Table 5.15** Commercial products containing carbamates or related compounds.

| Active Ingredient | Trade Name |
|---|---|
| Asulam | Asulox |
| EPTC | Eptam, Eradicane |
| Butylate | Sutan |

**Table 5.15** continued.

| Active Ingredient | Trade Name |
|---|---|
| Cycloate | Ro-Neet, Sabet |
| Molinate | Ordram |
| Pebulate | Tillam |
| Vernolate | Vernam |
| Diallate | Avedex, Di-allate |
| Triallate | Far-Go |
| Benthiocarb (thiobencarb) | Bolero, Saturn |
| Phenmedipham | Betanal |
| Desmedipham | Betanex |

*j. Dinitroanilines or Nitroanilines*

Mode of Action and Toxicology

The dinitroanilines are some of the most widely used herbicides. They are used almost exclusively as soil-incorporated, preemergent, selective herbicides. Uptake occurs from the soil via germinating seeds or the roots.

This group of pesticides has several modes of action. Most members inhibit cell division in the meristematic tissues. In others, the development of several enzymes is inhibited, while others may also effect uncoupling of oxidative phosphorylation.

These herbicides are moderately to mildly irritating and have only minimal toxic effects.

**Table 5.16** Commercial products containing dinitroanilines.

| Active Ingredient | Trade Name |
|---|---|
| Trifluralin | Treflan, TRI-4 |
| Benfluralin | Benefin, Balan, Quilan |
| Profluralin | Tolban |
| Ethalfluralin | Sonalan, Curbit |
| Fluchloralin | Basalin, Flusul |
| Oryzalin | Dirimal, Surflan |
| Nitralin | Planavin |
| Butralin | Amex, Tabamex, Tamex, Tobago |
| Pendimethalin | Accotab, Prowl, Stomp |
| Dipropalin | Dipropalin |
| Isopropalin | Paarlan |
| Flumetralin | Prime+ |

*k. Diphenyl Ethers*

The diphenyl ethers are generally used as preemergent or early postemergent herbicides. They are formulated as granules or wettable powders and are rapidly absorbed from the soil by plant roots. Some of the more successful members of this group do not translocate, and their activity is enhanced by high-volume spraying, along with the use of surfactants to increase their coverage and penetration.

Mode of Action

The diphenyl ether herbicides are active only in the presence of light and cause chlorosis of leaf tissue. They inhibit the Hill reaction in photosynthesis and photophosphorylation. However, the primary mode of action probably involves the photosynthetic reduction to form radicals, which initiate destructive reactions in lipid membranes leading to cell leakage.

Toxicology

The toxicity of the product is dependent on the formulation. Formulations such as emulsifiable concentrates contain inert additives or solvents which can severely irritate or damage the skin and eyes. Before handling, read the label carefully.

**Table 5.17** Commercial products containing diphenyl ether herbicides.

| Active Ingredient | Trade Name |
|---|---|
| Fluoroglycofen-ethyl | Compete |
| Fomesafen | Flexstar, Reflex |
| Oxyfluorfen | Goal |
| Acifluorfen | Blazer |
| Lactofen | Cobra Herbicide |
| Nitrofen | Nitrofen, Trizilin |
| Fluorodifen | Fluorodifen |
| Bifenox | Modown |

*l. Benzothiadiazoles*

Mode of Action and Toxicity

The only representative of this group is bentazon. It is used as a selective postemergence control of many broadleaf weeds. It is usually formulated as a soluble concentrate of the sodium salt of bentazon.

Its mode of action is similar to other photosynthesis inhibitors in that its activity is dependent on light. It inhibits photosynthetic carbon dioxide fixation and photosynthetic electron transport.

Basagran is mildly irritating to eyes and respiratory tract and is therefore designated to Toxicity Class III.

One of the only commercial products with bentazon as the sole active is Basagran. Storm is a product that contains aciflurofen along with bentazon.

*m. Cyclohexenones*

This group includes sethoxydim, clethodim, cycloxydim, and tralkoxydim. These compounds are used as systemic postemergence herbicides to control annual and perennial grass weeds.

Mode of Action and Toxicity

These compounds are inhibitors of lipid biosynthesis and auxin activity.

These compounds are moderately to mildly irritating. Several formulations include solvents which may be irritating.

**Table 5.18** Commercial products containing cyclohexenones.

| Active Ingredients | Trade Names |
|---|---|
| Sethoxydim | Poast |
| Clethodim | Select |
| Cycloxydim | Focus, Laser, Stratos |
| Tralkoxydim | Grasp |

*n. Imidazoles*

Several of these compounds are popular, nonselective, broad spectrum, systemic herbicides. However, imazamethabenz-methyl is used as a selective postemergent herbicide.

Mode of Action and Toxicity

They are absorbed through the roots and foliage and translocated to the growing or meristematic regions of the plant. Several members of this group have been shown to inhibit branched-chain amino acid biosynthesis.

Several members of this group are mildly irritating to skin and eyes.

**Table 5.19** Commercial products containing imidazole compounds.

| Active Ingredient | Trade Names |
|---|---|
| Imazapyr | Arsenal, Chopper |
| Imazaquin | Image, Scepter |
| Imazethapyr | Passport, Pursuit |
| Imazamethabenz-Methyl | Assert |

*o. Oxyphenoxy Acid Esters*

Mode of Action

This is a relatively new class of preemergent herbicides active against most grass at economical rates, with very little activity towards broadleaf crops. The compounds are translocated from the point of uptake to the growing points where they inhibit normal cell growth. They are known to inhibit lipid synthesis and auxin activity.

Toxicity

Several members of this group are formulated as emulsifiable concentrates and oil in water emulsions. The inert ingredients can cause slight irritation to eyes, skin, and respiratory tract.

**Table 5.20** Commercial products containing oxyphenoxy acid esters.

| Active Ingredients | Trade Names |
|---|---|
| Diclofop-methyl | Hoelon 3EC, Illoxan |
| Fenoxaprop-ethyl | Acclaim, Horizon, Whip1EC |
| Fenoxaprop-p-ethyl | Option, Whip 360 |
| Fluazifop-butyl | Fusilade 2000, |
| Fluazifop-p-butyl | Fusilade DX |

### p. *Substituted or Phenyl Urea Herbicides*

These urea derivatives are a large group of important herbicides. The herbicidal efficacy of the first described member of this group, diuron, was shown in 1951.

Most of the ureas are relatively nonselective and are usually applied to the soil as preemergent herbicides; some have postemergent uses, while others are active when applied to the foliage. Tebuthiuron is used on noncropland areas, rangeland, rights-of-way, and industrial sites. Ureas are predominately absorbed by the roots and translocated to the rest of the plant.

Mode of Action and Toxicity

The mode of action of the substituted ureas is relatively well known. It results in an inhibition of photosynthesis by blocking photosynthetic electron transport and photophosphorylation.

Several members of this group are formulated as dry flowables, flowables, water dispersible granules, and wettable powders. Improper application can cause moderate irritation of the eye, skin, nose, and respiratory tract.

**Table 5.21** Commercial products containing substituted ureas.

| Active Ingredients | Trade Names |
|---|---|
| Diuron | Karmex, Seduron |
| Fenuron | Fenuron, Dozer |
| Fluometuron | Cotoran |
| Linuron | Lorox, Afalon |
| Siduron | Tupersan |
| Tebuthiuron | Spike |
| Thidiazuron | Dropp |

*q. Pyridazinones*

Most members of this group are used as preemergent and early postemergent, broad spectrum herbicides. Since the mode of action of this group is varied and complex, effecting many biosynthetic processes that are involved in the photosynthetic pathway, there is no single easily identifiable mode of action. The results of their activities is an inhibition in chlorophyll synthesis and a bleaching of the foliage.

Several members of this group are formulated as emulsifiable concentraters with solvents which can cause severe to moderate injury to the eyes, skin, and respiratory tract.

**Table 5.22** Commercial products containing pyridazinone compounds.

| Active Ingredient | Trade Names |
|---|---|
| Clomazone | Command, Gamit, Merit |
| Pyrazon (Chloridazon) | Pyramin |
| Norflurazon | Evital, Solicam, Zorial Rapid 80 |
| Oxadiazon | Chipco, Ronstar |

*r. Phthalic Acid Herbicides*

There are only two significant compounds in this group, and these are endothall and chlorothal (DCPA). Chlorothal is a selective, preemergent herbicide used to control many annual grasses and broadleaf weeds. Endothall has several different uses such as a preemergent and postemergent herbicide, a defoliant, desiccant, aquatic algicide, and growth regulator.

Toxicity

Chlorothal is slightly irritating to the eyes, while endothall has a "Warning" classification because of its moderate oral toxicity.

Mode of Action

Endothall has been shown to interfere with photosynthesis, while chlorothal inhibits cell division in the meristemic tissue.

**Table 5.23** Commercial products containing phthalic acid herbicides.

| Active Ingredient | Trade Name |
|---|---|
| Chlorothal | Dacthal |
| Endothall | Accelerate, Aquathol K, Des-I-Cate, Endothal Turf herbicide, Herbicide 273, Hydrothol |

*s. Pyridinoxy and Picolinic Acids*

This group of herbicides consists of clopyralid, fluroxypr, picloram, and triclopyr. These compounds have both soil absorption and foliar activity.

They are used as selective preemergent and postemergent herbicides to control broadleaf weeds and woody plants.

Mode of Action

These compounds act by inhibiting nucleic acid synthesis, which in turn interferes with protein and enzyme synthesis and metabolism.

Toxicity

Some formulations may cause severe to moderate skin and eye irritation and permenent damage.

**Table 5.24** Commercial products containing pyridinoxy and picolinic acid herbicides.

| Active Ingredient | Trade Names |
|---|---|
| Clopyralid | Lontrel, Reclaim, Stinger |
| Fluroxypyr | Starane |
| Picloram | Grazon, Tordon |
| Triclopyr | Garlon, Grandstand, Rely |

*t. Triazines*

The triazines were developed by the Swiss firm, J. R. Geigy Limited, in 1952. Two well-known examples of this group of heterocyclic compounds are simazine and atrazine. These are persistent soil-acting herbicides which can be applied in large concentrations as total weed killers on rights-of-way or industrial sites. At lower concentrations, they can be used for selective control of germinating weeds in a variety of crops. They are taken up by the roots of emerging weed seedling, causing it to turn yellow and die, but owing to their low solubility in water, they do not penetrate to the deeper levels of the soil and consequently have little effect on the deep rooted crops.

Triazines are used selectively on certain crops and nonselectively on others. They are used in greatest quantity in crop production and are used nonselectively on industrial sites.

Prometryne is used for selective pre- and postemergence control of annual weeds in vegetables, cotton, and sunflowers. Desmetryn provides postemergence control of annual weeds. Methoprotryne, ametryn, dipropetryn, ethiozin, terbutryn, cyanazine, and eglinazine-ethyl provide effective control of grass weed in cereals. Aziprotryne is used for weed control in a wide range of vegetables and acts through both the soil and foliage. Metribuzin and metamitron are used for pre- and postemergence weed control.

Altogether, some 25 triazines are currently used as commercial herbicides, and the most important is atrazine.

Mode of Action

Triazines act by interfering with photosynthesis, and it seems clear that, like the substituted urea herbicides, the primary site of action is inhibition of the Hill reaction of photosynthetic electron transport.

Toxicity

The triazines listed below are nonirritating to mildly irritating to eyes or skin. Cyanazine has a designation of Toxicity Category II, "Warning," because of its moderate oral $LD_{50}$.

**Table 5.25** Commercial products containing triazines.

| Active Ingredient | Trade Names |
|---|---|
| Ametryn | Evik, Gesapax |
| Atrazine | AAtrex, Atratol 90, Atrazine |
| Aziprotryne | Brasoran, Mesoranil |
| Cyanazine | Bladex, Fortrol |
| Dipropetryn | Sancap |
| Ethiozin | Tycor |
| Metamitron | Goltix |
| Metribuzin | Lexone, Sencor, Sencoral |
| Prometryne | Caparol |
| Simazine | Aquazine, Caliber 90, Princep |
| Terbuthylazine | Gardoprim |
| Terbutryn | Ternit |

*u. Uracils and Substituted Uracils (Pyrimidines)*

Unsubstituted uracils have no herbicidal activity but certain derivatives substituted in the 3,5,6-positions are active. The most important examples are bromouracil, terbacil, and lenacil. These compounds were introduced by DuPont in 1963. These herbicides are applied to the soil and are used for selective weed control. Bromacil and terbacil are pre- and postemergence herbicides; they control a wide spectrum of grass and broadleaf weeds when applied early in the growing season. These compounds must be carried to the roots by soil moisture, where they are absorbed and translocated to the rest of the plant. This family of herbicides is used for asparagus, sugar cane, pineapple, apple, and citrus crops.

Mode of Action

Uracils and substituted uracils owe their herbicidal activity to inhibition of photosynthesis by blocking the Hill reaction. This activity is also present in the ureas and triazines.

Toxicity

Liquid formulations of bromacil are designated to Toxicity Category II because they can irritate the nose, throat, and skin if inhaled as mist or dust.

Formulations of terbacil currently available are listed as relatively nontoxic, Toxicity Category IV.

**Table 5.26** Commercial products containing uracil compounds.

| Active Ingredient | Trade Name |
|---|---|
| Bromacil | Hyvar, Uragan |
| Terbacil | Sinbar |
| Lenacil | Adol, Venzar (lenacil is used primarily in Europe) |

*v. Sulfonylureas*

Introduced in 1982 by DuPont, this group has made a major impact on weed control technology. They are remarkable active compounds, selectively controlling many dicotyledonous weeds in cereals at very low dose rates, and they have very low mammalian toxicity.

Mode of Action

The sulfonylureas are potent inhibitors of plant growth; seed germination is generally affected, but subsequent root and shoot growth is severely inhibited in sensitive seedlings. Death follows chlorosis and necrosis.

Sulfonylureas are generally formulated as wettable powders or water dispersible granules, and the compounds are readily absorbed by both the roots and foliage of plants. They are translocated by both the phloem and xylem.

Toxicity

All the products below, with the exception of Permit, are relatively nontoxic (Toxicity Category IV). Permit has an oral $LD_{50}$ high enough to be in Toxicity Category III.

**Table 5.27** Commercial products containing sulfonylurea herbicides.

| Active Ingredient | Trade Name |
|---|---|
| Bensulfuron-methyl | Londax |
| Chlorsulfuron | Glean, Telar |
| Chlorimuron-ethyl | Classic |
| Halosulfuron-methyl | Battalion, Manage, Permit |
| Metsulfuron-Methyl | Ally, Escort |
| Nicosulfuron | Accent |
| Primisulfuron | Beacon |
| Sulfometuron-Methyl | Oust |
| Thifensulfuron | Pinnacle |
| Triasulfuron | Amber |

*w. Plant Growth Regulators*

Plant growth regulators (PGRs) are phytochemicals, which in minute amounts alter the behavior and/or growth of plants, blossoms, or fruit. These chemicals are regulated as pesticides, and besides being called PGRs, they are referred to as plant hormones and plant-regulating substances.

1. Auxins

Auxins are compounds that stimulate various growth processes. Auxins occur naturally and can be manufactured.

Auxins are used to thin apples and pears, increase yields of vegetables, and promote rooting in cuttings and fruit setting.

The mechanism of action is not completely clear, but auxins may act by controlling the types of enzymes produced by the cell. In any event, auxins induce cell enlargement, water uptake, and cell wall expansion.

Toxicity

Amid-Thin (naphthalene acetamide) is labeled "Danger" because of the potential of severe irreversible eye damage, whereas Fruitone N, NAA 800, and Tree-Hold Sprout Inhibitor (NAA) are relatively nontoxic (Category III). See Table 5.28 for a listing of commerical products.

2. Gibberellic Acid (Gibberellin, $GA_3$)

Toxicity and Mode of Action

Gibberellic acid is not a pesticide, but it is commonly used in agricultural production as a growth promoting agent. Gibberellins are compounds that promote cell division and cell elongation. They are metabolic products of a cultured fungus, formulated into tablets, granules, and liquid concentrates for application to soil beneath growing plants and trees.

More than 50 gibberellins have been isolated and identified. They are termed $GA_1$, $GA_2$, $GA_3$, etc. The $GA_3$ is the most commonly used gibberellin.

Experimental animals tolerate large oral doses without apparent adverse effect. No human poisonings have been reported. Sensitization has not been reported, and irritant effects are not remarkable. See Table 5.28 for a listing of commerical products.

3. Cytokinins

Mode of Action

The third type of plant hormone is known to control the induction of cell division in plants, and these substances are known collectively as cytokinins.

Cytokinins occur naturally and are manufactured. Most of the cytokinins are derivatives of adenine.

Cytokinins also delay senescence and hence can be used to prolong the shelf life of fresh vegetables and cut flowers and to promote seed germination. Cytokinins appear to be involved with transfer RNA (t-RNA) and, therefore, they possibly play a role in regulating the incorporation of specific amino acids into proteins.

Toxicity

Some formulations may cause severe reversible eye irritation and have a signal word, "Warning," on the label (ProShear). See Table 5.28 for a listing of commerical products.

4. Ethylene and Ethylene Generators

In contrast to complex auxins, gibberellins, and cytokinins, ethylene is a simple gaseous compound. In 1932, it was discovered that ethylene would promote flowering in pineapples. The subsequent discovery of the role of ethylene in fruit ripening has been of considerable commercial importance in the banana and citrus industries.

Mode of Action

Ethylene appears to influence the enzymes of the physiological state of the absorbing tissue. Ethylene may serve as a trigger, or be induced by auxins, or act as a synergist resulting in a chain of biochemical and/or physiological events such as the ripening of bananas or the prevention of the opening of flower buds of carnations and fruit trees.

Ethephon, introduced in 1967, is an ethylene generator. It generates ethylene when added to solutions with a pH >4, which are then sprayed onto fruit trees to promote ripening. Another active ingredient, 2-chloroethylmethyl-bis(benzoxy) silane, was introduced by Ciba-Geigy in 1972. Like ethepon, this silane decomposes in the presence of water to release ethylene. This compound is used as a chemical abscission or as a thinning agent in apples, peaches, oranges, and other fruits.

The exact mode of action of ethylene is unclear but probably involves the binding of the molecule to a specific receptor.

Toxicity

The products Cerone and Prep are classified in the Toxic Category I because of the potential of severe irreversible eye damage. They are labeled with the signal word "Danger." The product Ethrel is labeled with the signal word "Warning" because it can cause severe reversible eye irritation, whereas the product Florel has a formulation which is labeled with the word "Caution." See Table 5.28 for a listing of commerical products.

**Table 5.28** Commercial products containing plant growth regulators.

| Active Ingredients | Trade Name |
|---|---|
| Indolacetic Acid (IAA) | |
| Indoleacetaldehyde (IAAld) | |
| 2,4-dichlorophenoxyacetic Acid (2,4-D)* | See chlorophenoxyacetic acids |
| 4[(4-Chloro-*o*-tolyl)oxy]butyric Acid (MCPB*) | See chlorophenoxyacetic acids |
| Naphthalene Acetamide (NAD) | Amid-Thin |
| β-Naphthaleneacetic Acid (NAA) | Fruitone N, NAA 800, Tree-Hold Sprout Inhibitor |
| β-Naphthoxyacetic Acid (BNOA) | |
| Gibberellic acid | Activol, Berelex, Cekugib, Gibberellin, Giberl, Grocel, Pro-Gibb Plus, Regulex |
| Cytokinins | ProShear, Paturyl, Burst, Cytex |
| Ethephon | Cerone, Ethrel, Florel, Prep |

*Classified as a chlorophenoxyacetic acid herbicide with auxin-like properties

## VI. FUNGICIDES

Fungicides are extensively used in industry, agriculture, the home, and garden for:

1. protection of seeds during storage, shipment, and germination
2. protection of mature crops, fruits, seedlings, flowers, and grasses in the field, in storage, and during shipment
3. suppression of mildews that attack painted surfaces
4. control of slime in paper pulps
5. protection of carpet and fabrics in the home.

Fungicides vary enormously in their potential for causing adverse effects in humans. Historically, some of the most tragic epidemics of pesticide poisoning occurred because of mistaken consumption of seed grain treated with organic mercury or hexachlorobenzene. However, most fungicides currently in use are unlikely to cause frequent or severe systemic poisonings for several reasons.

1. Many have low inherent toxicity in mammals and are inefficiently absorbed.
2. Many are formulated as suspensions of wettable powders or granules, from which rapid, efficient absorption is unlikely.
3. Methods of application are such that relatively few individuals are intensively exposed.

Apart from systemic poisonings, fungicides as a class have probably caused disproportionate numbers of irritant injuries to skin and mucous membranes, as well as some dermal sensitizations.

Fungicides may be differentiated according to their mode of action into protective or surface acting and curative or systemic (internal) acting. The classification principle used in this text is based on the chemical of the fungicides in order to clearly summarize specific groups or families of compounds. It will not be possible to discuss all the agents or formulations commercially available today and, therefore, this overview must be restricted to the most important groups and their members.

The following discussion considers the mode of action and the toxicity of the more widely used fungicides and some formulated products.

## A. PROTECTANT OR SURFACE ACTIVE FUNGICIDES

Many commercial fungicides currently in use belong to the class known as protectant or surface fungicides. These are usually applied to plant foliage as dusts or sprays. Such materials do not appreciably penetrate the plant surface of cuticle and are not translocated within the plant, whereas the more recent systemic fungicides are absorbed by the plant via the roots, leaves, or seeds and are translocated within the plant.

Most pathogenic fungi penetrate the cuticle and ramify through the plant tissue. If one chooses to use a protectant fungicide, it must be applied before the fungal spores reach the plant if it is to be effective. Only a few fungi are restricted to the surface of the leaf (e.g., powdery mildews), and in these cases, protectant fungicides may also possess eradicant action.

For protectant fungicides to be effective, they must satisfy the following conditions:

1. They should possess a low phytotoxicity and, therefore, they should not cause damage the host plant during application.
2. They should be toxic to the fungal spore or undergo rapid conversion to a toxic form within the spore before the fungus penetrates the cuticle.
3. They must be capable of penetrating the spore and reach their site of action.
4. Since most fungicides are applied as foliar sprays, they must be formulated so that when deposited on the plant, they are resistant to the effects of weathering.

None of the protectant fungicides in use today are completely nonphytotoxic. Since fungi and plants both possess the same or similar metabolic or physiological processes, and the fungus is an infection within the plant, it is almost impossible to discover a completely nonphytotoxic fungicide that would specifically block a process or reaction in one without affecting the other.

Fungicides may be applied to seeds, foliage, or fruits as sprays or dusts. For a dust to be effective, it must provide uniform coverage, and this requires a small particle size. Spraying is the more widely used method of application. The fungicide may be applied in solution or as a fine suspension and, therefore, the reduction in particle size increases the effectiveness of the fungicide.

### 1. Inorganic and Organic Metal Fungicides

The earliest fungicides were inorganic chemicals like sulfur, copper, and mercury.

#### a. *Sulfur*

Sulfur alone and in combination as lime-sulfur was one of the most important fungicides. In the 1800s and early 1900s, sulfur dust and spray were used against mildew on fruit trees. In 1958, the weight of sulfur used against fungi was four times that of all other fungicides. However, the recent development of organic fungicides has reduced the use of sulfur along with other inorganic chemicals.

There are three formulations of sulfur used predominantly today. The first is finely ground sulfur mixed with 1 to 5 percent talc or clay to assist in the dusting effectiveness. The sulfur in this form may be used as a carrier for additional pesticides. The second form is colloidal sulfur that is so fine that it must be formulated as a paste to enable it to be mixed with water. The third form is wettable sulfur. In this formulation, the sulfur is finely ground with a wetting agent so that it will mix readily with water for spraying.

Mode of Action

There has been much speculation regarding the mode of action of sulfur. The current view is that sulfur itself ($S_8$) may be toxic to fungi. Recent studies indicate that sulfur is not biologically inert and probably becomes involved in biological redox reactions. These studies also suggest that sulfur can more easily penetrate spores in the presence of urea, hydrocarbons, surfactants, and lime.

Toxicity

Sulfur in many of its formulations is relatively nontoxic. However, some formulations may slightly irritate the skin.

Commercial products include Microthiol Special, Sulfur Alfa, and Thiolux.

*b. Copper Compounds*

Some of the oldest and most extensively used fungicides belong to this group of compounds. In 1982, the French botanist Millardet discovered the effectiveness of a mixture of copper sulfate and lime on the downey mildew blight of grape vines. This copper-lime liquid, termed the "Bordeau Mixture," was of great importance at that time and marked the beginning of the commercialization of fungicides.

Since copper is toxic to plants, it must be used at low levels or in the insoluble form. For this reason, the relatively insoluble or "fixed" copper salts are used. These compounds release copper ions at very low rates that are adequate for fungicidal activity but not at concentrations that would harm or kill the host plant.

Since copper compounds are relatively insoluble in water, they resist the effects of weathering and provide longer protection against fungal infection.

Mode of Action

The current theory on the mode of action of copper compounds is that the copper ion, $Cu^{2+}$, is the active component which is released from the different salts on the leaf surfaces. The copper ions in the presence of $CO_2$ from the air and the organic acids excreted from the plant and/or fungal spores interact together to produce the resultant activity. The copper ions and complex-bound copper are capable of penetrating the spores and lead to the inhibition of enzyme reactions. This can occur by the removal of other important metals from their compounds by chelation and also by blocking or interacting with the sulfhydryl groups of the spore enzymes.

The use of copper compounds has declined sharply in recent years due to the development of new organic fungicides. However, they still are important as cost effective agents for the control of downy mildews.

Toxicology

The dust and powder preparations of copper compounds are irritating to the skin, respiratory tract, and particularly to the eyes. The soluble copper salts (such as the sulfate and acetate) are corrosive to mucous membranes and the cornea. Limited solubility and absorption probably account for generally low systemic toxicities of most compounds. The more absorbable organic copper compounds exhibit the greatest systemic toxicity in laboratory animals.

The principal features of poisoning by ingested copper compounds have been 1) gastrointestinal irritation (vomiting and burning pain in the mouth, esophagus and stomach, abdominal pain and diarrhea, sometimes with blood), 2) headache, sweating, weakness, and sometimes shock, 3) liver enlargement and jaundice, 4) hemolysis and methemoglobinemia, and 5) albuminuria, hemoglobinuria, and sometimes acute renal failure.

Inorganic copper compounds include cuprous oxide; cupric oxide; copper hydroxide; copper carbonate; basic; copper ammonium carbonate; copper acetate; copper sulfate; copper sulfate, tribasic (Bordeau Mixture); copper oxychloride; copper silicate; copper lime dust; and copper potassium sulfide. Figure 5.10 shows a package of Kocide 101, a copper-containing product.

**Figure 5.10** Fungicides. Benlate contains the carbamate fungicide benomyl. Kocide 101 contains copper hydroxide, which is an inorganic copper-containing fungicide.

Organic copper compounds include copper linoleate, naphthenate, oleate, phenyl salicylate, and resinate.

Insoluble compounds are formulated as wettable powders and dusts. Soluble salts are prepared as aqueous solutions. Some organometallic compounds are soluble in mineral oils.

A great many commercial copper-containing fungicides are available. Some are mixtures of copper compounds. Others include lime, other metals, and other fungicides.

There are several copper-arsenic compounds, such as Paris Green, still used in some agricultures. Toxicity of these is chiefly due to arsenic content.

c. *Mercury Compounds*

All organic and inorganic mercurial fungicides have lost their registrations and are no longer available for any purpose.

Mode of Action

The fungicidal activity organic of mercury compounds cannot be attributed to the Hg ion or metallic Hg in the vapor phase. Since the organic residue of the molecule affects the disinfectant activity very strongly, it may be a contributing factor. As with copper, the mercurial ion reacts with enzymes containing iron and /or having a reactive sulfhydryl group.

Toxicity

The mercurial fungicides are among the most toxic pesticides ever developed in terms of chronic as well as acute hazard. Epidemics of severe, often fatal, neurologic disease have occurred when indigent residents of less-developed countries consumed methyl mercury-treated grain intended for planting of crops. Poisoning has also occurred when meat from animals fed mercury-treated seed was eaten. Most of what is known of poisoning by organic mercurial fungicides has come from these occurrences.

d. *Organic Tin (Organotin) Compounds*

Mode of Action

Most of the organotin fungicides tested were too phytotoxic to be used as foliar fungicides. These included the trialkyktin compounds. However, the triphenyltin compounds are much less phytotoxic and still possess strong fungicidal activity. These compounds interfere with fungal growth by their inhibition of oxidative phosphorylation (energy production).

Toxicity

Most of the organotin salts are moderately toxic, and some are labeled "Poison" (triphenyltin chloride). They have a low oral $LD_{50}$ and require the signal word "Warning" on the label.

These agents are irritating to the eyes, respiratory tract, and skin. They are probably absorbed to a limited extent by the skin and gastrointestinal tract. Manifestations of toxicity are due principally to effects on the brain: headache, nausea, vomiting, dizziness, and sometimes convulsions and loss of consciousness.

**Table 5.29** Commercial products containing phenyltin compounds.

| Active Ingredients | Trade Names |
|---|---|
| Fentin hydroxide | Du-Ter, Duter, Haitin, Phenostat-A H, Suzu-H, TPTH, TPTOH, Triple Tin, Tubotin |
| Fentin chloride | Aquatin, Phenostat-C, Tinmate |
| Fentin acetate | Batasan, Brestan, Phenostat-A, Phentinoacetate, Suzu, Tinestan, TPTA |

All are formulated as wettable and flowable powders for use mainly as fungicides to control blights on field crops and orchard trees. Fentin chloride is also prepared as an emulsifiable concentrate for use as a molluscicide (Aquatin 20 EC).

Tributyltin salts are used as fungicides and antifouling agents on ships. They are somewhat more toxic by the oral route than triphenyltin, but toxic actions are otherwise probably similar.

### 2. Organic Fungicides

Many organic fungicides have been developed during the past 50 years to replace the more toxic, nonselective inorganic predecessors. The first group of organics developed were the dithiocarbamates, developed in the 1930s and 1940s.

*a. Dithiocarbamates (ethylenebis-dithiocarbamates—EBDCs)*

The development of purely organic fungicides began with the discovery of the fungicidal activity of the dithiocarbamates. These compounds and their derivatives comprise one of the most important groups of fungicides for controlling plant diseases.

Thiram, mancozeb, maneb, ferbam, ziram, and zineb are among the most popular used fungicides. They are employed for every use known for fungicides.

Mode of Action

The active agent of the dithiocarbamates is the dithiocarbamate ion, which can interfere with the fungal metabolism in several different ways.

1. They can react with metal-containing enzymes (e.g., polyphenoloxidase and ascorbic acid oxidase) to form a copper-dithiocarbamate complex with high fungicidal activity (thiram and ziram probably owe their fungicidal activity to this type of reaction).

2. They can react directly with sulfhydryl groups of enzymes, e.g., glucose-6-phosphate dehydrogenase or redox compounds such as cysteine or glutathione.
3. By reacting with specific substances of plants, they can promote the formation of or create different forms of transport for the active substance or even for stronger fungicidal compounds. This type of reaction is observed with sodium dimethyldithiocarbamate. As such, this compound has only a slight fungicidal activity, but is more effective in combination with the amino acid alanine or a glucoside (nabam, maneb, and zineb probably act through this type of mechanism).

Toxicity— Thiram

Thiram dust is moderately irritating to human skin, eyes, and respiratory mucous membranes. Contact dermatitis has occurred in occupationally exposed workers. A few individuals have experienced sensitization to thiram.

Systemic human poisonings by thiram itself have been very few, probably due to limited absorption in most circumstances involving human exposure. See Table 5.30 for a listing of the commercially available products containing thiram.

Thiocarbamates are commonly formulated as dusts, wettable powders, or water suspensions. They are used to protect seeds, seedlings, ornamentals, turf, vegetables, fruit, and apples.

Toxicity—Ziram and Ferbam

Ziram and Ferbam dusts are irritating to the skin, respiratory tract, and eyes. Prolonged inhalation of ziram is said to have caused neural and visual disturbances, and, in a single case of poisoning, a fatal hemolytic reaction.

See Table 5.30 for a listing of the commercially available products containing ziram and ferbam.

These are formulated as flowable and wettable powders used widely on fruit and nut trees, apples, vegetables, and tobacco.

Toxicity—EBDC Compounds

These fungicides may cause irritation of the skin, respiratory tract, and eyes. Both maneb and zineb have apparently been responsible for some cases of chronic skin disease in occupationally exposed workers, possibly by sensitization.

See Table 5.30 for a listing of the commercially available products containing maneb, zineb, nabam, and mancozeb.

Maneb and Zineb are formulated as wettable and flowable powders. Nabam is provided as a soluble powder and in water solution.

Mancozeb is a coordination product of zinc ion and maneb. It is formulated as a dust and as wettable and liquid flowable powders.

**Table 5.30** Commercial products containing dithiocarbamates (ethylenebis-dithiocarbamates—EBDCs).

| Active Ingredients | Trade Names |
|---|---|
| Thiram | AAtack, Aules, Chipco Thiram 75, Fermide 850, Fernasan, Hexathir, Mercuram, Nomersam, Polyram-Ultra, Pomarsol forte, Spotrete-F, Spotrete WP 75, Tetrapom, Thimer, Thioknock, Thiotex, Thiramad, Thirasan, Thiuramin, Tirampa, TMTD, TMTDS, Trametan, Tripomol, Tuads, Vancide TM |
| Ziram | Carbazinc, Corozate, Cuman, Drupina 90, Fungostop, Hexazir, Mezene, Prodaram, Tricarbamix Z, Triscabol, Zerlate, Vancide MZ-96, Zincmate, Ziram Technical, Ziram F4, Ziram W76, Ziramvis, Zirasan 90, Zirberk, Zirex 90, Ziride, Zitox |
| Ferbam | Carbamate, Ferbam, Ferberk, Hexaferb, Knockmate, Trifungol |
| Maneb | Akzo Chemie Maneb, BASF-Maneb Spritzpulver, Dithane M22, Kypman 80, Manex 80, Maneba, Manesan, Manex, Manzate, Manzate D, M-Diphar, Polyram M, Remasan Chloroble M, Rhodianebe, Sopranebe, Trimangol, Tubothane, Unicrop Maneb |
| Zineb | Aspor, Chem Zineb, Devizeb, Dipher, Dithane Z-78, Hexathane, Kypzin, Lonacol, Parzate, Parzate C, Polyram Z, Tiezene, Tritoftorol, Zebtox, Zineb 75 WP, and Zinosan |
| Nabam | Chem Bam, DSE, Nabasan, Parzate, Spring Bak |
| Mancozeb | Manzeb, Manzate 200, Akzo Chemie Mancozeb, Mancozin, Manzin, Dithane M-45, Nemispor, Penncozeb, Policar MZ, Policar S, Vondozeb Plus, Ziman-Dithane |

*b. Thiazoles*

This group of fungicides contains a five-membered ring that is cleaved to form the active fungicidal agent, -N=C=S, or a dithiocarbamate, depending upon the structure of the parent compound. Etridiazole is a soil fungicide which has been shown to undergo the ring cleavage mentioned above.

Mode of Action

The probable mode of action is the inhibition of respiration and mitochondrial destruction.

Toxicity

The formulation of the active ingredient is important to its toxicity. Emulsifiable concentrates (Truban EC) are designated to Toxicity Category I and have the signal word "Danger" on the label. Several wettable powder formulations are in Toxicity Category II and have the signal word "Warning" on the label and, in general, granulated formulations are in Toxicity Category III and have the signal word "Caution" on the label.

The importance of the inert ingredients in the above formulations is evident. The end-user of the product should be aware of the toxicity of these ingredients.

Commercial products include Etridiazole, Asterra, Koban, Pansoil, Terrazole, and Truban.

*c. Triazines*

The only member of this group of pesticides with fungicidal activity is anilazine. The manufacturer (Miles, Inc.) has canceled all uses in the U.S. Labelling requires the signal word "Warning."

Anilazine is supplied as wettable and flowable powders and is used on vegetables, cereals, coffee, ornamentals, and turf.

Toxicity

This product has caused skin and eye irritation in exposed workers. Acute oral and dermal toxicities in laboratory animals are low. Human systemic poisonings have not been reported.

Commercial products include Anilazine, Dyrene, Kemate, and Triasyn.

*d. Substituted Aromatics (substituted benzenes and phenols)*

This group of fungicides contains several different subgroups, the chlorobenzenes, phenols, and the dicarboximide fungicides.

Pentachloronitrobenzene (PCNB) was introduced in the 1930s and is widely used as a soil fungicide for seed treatment and foliar applications. Hexachlorobenzene was introduced in 1945 to treat seeds and foliage (HCB, Anticarie, Ceku C.B., No Bunt). Chlorothalonil was introduced in 1964; it is a broad spectrum foliar or soil-applied fungicide used on many crops. Dicloran is a popular and widely used fungicide for the protection of a variety of fruits and vegetables.

The majority of phenols, especially those containing chlorine, are too phytotoxic to permit their use as agricultural fungicides. They are widely used as industrial fungicides. Cresols contribute to the fungicidal action of creosote which is used as a timber preservative. Pentachlorophenol, another chlorinated phenol, and its esters are widely used as industrial biocides for the protection of such materials as wood and textiles.

Mode of Action

The mode of action of this group of compounds is diverse and depends on the parental compound. The fungicidal activity of most phenols depends on their ability to uncouple oxidative phosphorylation and therefore prevent the production of ATP which is required for growth. Other substituted aromatics reduce growth rates by reacting with the amino or sulfhydryl groups of essential metabolic compounds.

Toxicity

PCNB, hexachlorobenzene (HCB), and dichloran have only slight irritant effects and relatively low single-dose toxicity. Several formulated products of chlorothalonil are classified in Toxicity Category I and are corrosive to eyes and skin (Bravo, ClortoCaffaro, Daconil, Echo). Other formulations are classified in Toxic Category II as severe irritants to eyes and skin.

**Table 5.31** Commercial products continuing substituted aromatics compounds.

| Active Ingredients | Trade Names |
|---|---|
| Hexachlorobenzene | HCB, Anticarie, Ceku C.B., No Bunt |
| Pentachloronitrobenzene | PCNB, Quintozene, Terraclor, Turfcide, Avicol, Botrilex, Earthcide, Folosan, Kobu, Kobutol, Pentagen, Tilcarex, Tri-PCNB |
| Chlorothalonil | Bravo, ClortoCaffaro, Clortosip, Daconil 2787, Exotherm Termil, Tuffcide |
| Dichloran | DCNA, Ditranil, Botran, Allisan, Kiwi Lustr 277, Resisan |

Formulations

PCNB—emulsifiable concentrates, wettable powders, and granules.

Hexachlorobenzene—dusts and powders.

Dichloran—wettable powder, dusts, and flowable powders.

Chorothalonil—wettable powder, water dispersible granules, and flowable powders.

### *e. Dinitrophenols*

The majority of phenols, especially those containing chlorine, are toxic to microorganisms. However, the majority of phenols are too phytotoxic to permit their use as agricultural fungicides. They are widely used as industrial fungicides. Dinocap, a mixture of several compounds, was first introduced in 1946 as a non-systemic aphicide and contact fungicide. Dinocap is used to control powdery mildew on major horticultural crops. Unlike the herbicide DNOC and other chlorinated phenols, dinocap has relatively low mammalian toxicity.

Mode of Action

The fungicidal activity of the various phenols depends on their ability to uncouple oxidative phosphorylation and thus prevent the incorporation of inorganic phosphate into ATP without effecting electron transport. The result of this inhibition is cell death due to the lack of energy for cellular metabolism.

Toxicity

As stated above, diocap has a relatively low mammalian toxicity with an oral $LD_{50}$ (rat) of 980 mg/kg and is assigned to Toxicity Category III with the signal word "Caution."

### *f. Thiophthalimides (sulfenimides)*

Introduced in 1949, captan has become one of the most popular and versatile fungicides for foliar treatment of fruits, vegetables, and ornamentals, for soil and seed treatments, and for postharvest applications. Captafol and folpet were introduced later and were also used as dusts and sprays on fruits, vegetables, and ornamentals. Captafol and folpet were later discontinued by the manufacturers because of their long-term health hazards.

Mode of Action

The fungicidal action of captan results from the toxic moiety or toxophore N—S—$CCl_3$. Studies suggest that captan interacts with the sulfhydryl groups within fungal cells to produce thiophosgene. This highly reactive chemical then reacts with free sulfhydryl, amino and hydroxyl groups of enzymes and causes the inhibition of metabolic processes.

Toxicity

Captan has been classified in Toxic Category I because of the potential for severe eye damage. Dermal sensitization may occur; captafol appears to have been responsible for several episodes of occupational contact dermatitis. No systemic poisonings by thiophthalimides have been reported in man (captan has a very high oral $LD_{50}$ in rats of 9,000 mg/kg).

**Table 5.32** Commercial products containing thiophthalimide compounds.

| **Active Ingredient** | **Trade Names** |
|---|---|
| Captan | Captanex, Captaf, Drexel, Merpan, Orthocide, Vondcaptan |
| Captafol | Not marketed in the U.S. |
| Folpet | Not marketed in the U.S. |

### *g. Quinones*

The first member of this group, chloranil, was first used as a fungicide in 1938. It was found to be useful as a seed protectant and foliage fungicide for downey mildew. The more popular, related derivative known as dichlone was used on a number of fruit and vegetable crops and for the control of blue-green alge in ponds.

Mode of Action

Quinones react with the sulfhydryl groups of enzymes such as amylase and carboxylase. This reaction interferes with their activity and, indirectly, with other cellular processes such as oxidative phosphorylation. Chloranil and dichlone have been discontinued and are no longer available in the U.S.

Toxicity

The quinone fungicides are of low to moderate acute toxicity with oral $LD_{50}$ values (rat) of: chloranil 4,000, dichlone 1,300, and dithianon 638 mg/kg. Under some conditions, these compounds may cause severe skin irritation in sensitive individuals.

**Table 5.33** Commercial products containing quinone fungicides.

| Active Ingredient | Trade Names |
|---|---|
| Chloranil | Spergon |
| Dichlone | Phygon |
| Dithianon | Delan |

## B. SYSTEMIC FUNGICIDES

### 1. Organic Fungicides

Systemic fungicides are absorbed by the plant through the roots, leaves, or seeds and translocated thoughout the plant. In this way the whole plant, including new growth, is protected against fungal attack or an established fungal infestation.

Phytotoxicity is a much more difficult problem to overcome with systemics, because they are brought into intimate contact with the host plant.

Most of the early systemic fungicides (benomyl) showed translocation through the xylem. However, pyrimidine systemic fungicides, e.g., ethirimol and dimethirimol, appear to move through the phloem.

#### a. *Oxathiins and Related Carboxanilide Compounds*

Oxathiins are a group of heterocyclic compounds with interesting systemic properties. Carboxin and oxycarboxin are examples of this group.

Carboxin was the first seed disinfection agent with systemic activity used in cereals to control loose smut in barley. Today, newer fungicides have displaced carboxin.

Oxycarboxin has good activity against rust fungi and is particularly useful for their control in ornamental plants. It is also used against some soil-borne seedling diseases in cotton growing.

Fenfuram and methfuroxam are important components in the control of smut in wheat, barley, and oats as well as stripe disease in barley and snow mold.

Pyracarbolid acts against rust fungi and is important in the control of coffee rust.

Mode of Action

The fungicidal properties of carboxin and related carboxanilides are due to their inhibition of glucose and acetate oxidative metabolism and RNA and DNA synthesis, although the latter may be caused by lack of cellular energy due to the inhibition of respiration.

Toxicity

The toxicity of these products is dependent on the formulation and the inert ingredients. Several formulations have severe to moderate inhalation and eye hazards. However, most products have only slight oral and dermal hazards. Fenfuram is classified as an irritant.

Formulations

Carboxin is available as flowable concentrate, true solutions and wettable powder, and oxycarboxin is available as emulsifiable concentrate and wettable powder.

**Table 5.34** Commercial products oxathiins and related carboxanilide compounds.

| **Active Ingredient** | **Trade Names** |
|---|---|
| Carboxin | Enhance, Pro-Gro, Vitavax 30C |
| Oxycarboxin | Plantvax, Carbexsin |
| Fenfuram | Pano-ram |
| Pyracarbolid | Sicarol (discontinued) |

*b. Benzimidazole Compounds*

Since 1968, this group of compounds has been widely accepted and used as a systemic, curative, and protective action. Benomyl, with its broad spectrum of activity against many injurious fungi, also has exhibited ovicidal and miticidal actions against spider mites. Figure 5.10 shows the product with the trade name Benlate, which contains benomyl.

## Mode of Action

Uptake occurs via the roots and leaves, and transport occurs by way of the xylem. The active compound which is transported is the methylbenzimidazole carbamate (MBC or BCM). In plants the active compound is readily formed by hydrolysis. It is not clear whether a further cleavage occurs or if this cleavage occurs in the plant or fungal cells.

Benzimidazoles primarily act on cell and nuclear division. Nuclear division is inhibited by the binding of the MBC to the microtubular proteins involved in the synthesis of the mitotic spindle apparatus.

Benomyl has an extremely wide spectrum of activity and can be used as a leaf fungicide, seed disinfectant, and soil treatment.

Carbendazim, which is the common name used for the active agent MBC or BCM, is also marketed as a fungicide. It exhibits a wide spectrum of activity, is used to control snow molds, and is used in stem injections to inhibit Dutch elm disease.

Fuberidazol is used to combat snow mold, and mainly in combination with other substances for cereal seed stock disinfection. It was one of the first organic fungicides able to replace organic mercury as a seed protectant.

Thiabendazol is also used to disinfect cereal seed stock in combination with other fungicides. It is used to control snow mold, bunt, and loose smut in wheat and barley, as well as stripe disease in barley. It is also used as a spraying agent against eye-spot in cereals and against ear mildew and *Septoria.* In potatoes, thiabendazol is used as a spraying agent to control Fusarium dry rot. Like benomy, this compound has a very wide spectrum of uses.

## Toxicity

Benomyl is a synthetic, organic fungistat having little or no acute toxic effect in mammals. No systemic poisonings have been reported in humans. Although the molecule contains a carbamate grouping, benomyl is not a cholinesterase inhibitor. It is poorly absorbed across skin; that which is absorbed is promptly metabolized and excreted.

Although injuries to exposed individuals have been few, dermal sensitization has occurred in agricultural workers exposed to foliage residues.

Thiabendazole is widely used as an agricultural fungicide, but most experience with its toxicology in humans has come from medicinal use against intestinal parasites. Oral doses administered for this purpose are far greater than those likely absorbed in the course of occupational exposure. Thiabendazole is rapidly metabolized and excreted in the urine, mostly as a conjugated hydroxy-metabolite. Symptoms and signs that sometimes follow ingestion are dizziness, nausea, vomiting, diarrhea, epigastric distress, lethargy, fever, flushing, chills, rash and local edema, headache, tinnitus, paresthesia, and hypotension.

Most formulations of benomyl are classified as Toxicity Category IV and are considered relatively nontoxic. Arbotect, a trade name of a product containing thiabendazol, is classified as Toxicity Categrory III and is considered a slight eye and skin irritant. Fuberidazol is classified as a Toxicity Category II fungicide.

**Table 5.35** Commercial products containing benzimidazole fungicides.

| Active Ingredients | Trade Names |
|---|---|
| Benomyl | Benlate, Tersan, Benex |
| Thiabendazole | Apl-Luster, Arbotect, Mertect, TBZ, Tecto, Thibenzole |
| Carbendazim | Bavistin, Delsene |
| Fuberidazol | Fuberidazole |

c. *Pyrimidine Compounds*

In the late 1960s, the pyrimidines were introduced as fungicides. This group includes dimethirimol, ethirimol, bupirimate, and fenarimol. They are used to control powdery mildew in several plants.

Mode of Action

As with almost all systemic fungicides, after foliar or soil application, absorption and translocation via the xylem follows. The mechanisms of action of pyrimidines are not very clear; however, information suggests that spore germination and mycelial growth are inhibited. In addittion, studies with ethirimol suggest that it inhibits RNA synthesis as well as adenosine metabolism by blocking adenosine deaminase.

Toxicity

Most of the pyrimidines are classified in Toxicity Category III and are only slightly irritating. However, several pyrimidine formulations (e.g., Rubigan) are in Category II. In general, it is the properties of the inert ingredients present in the formulations that increase the toxicity of the product.

**Table 5.36** Commercial products containing pyrimidines.

| Active Ingredients | Trade Names |
|---|---|
| Dimethirimol | Milcurb |
| Ethirimol | Ethirimol, Milcurb Super |

**Table 5.36** continued

| Active Ingredients | Trade Names |
|---|---|
| Bupirimate | Nimrod |
| Fenarimol | Rubigan, Fenal |

*d. Phenylamines*

The phenylamine group of fungicides consists of the acylalanines, butyrolactones, and oxazolidinones. This group of fungicides includes metalaxyl, furalaxyl, benalaxyl, and oxadixyl. These compounds show protective and systemic activity against downy mildew, late blight, and other agents causing disease on fruit trees, cotton, hops, soybeans, peanuts, ornamentals, and grasses.

Mode of Action

The primary effect of these compounds on metabolism of sensitive fungi lies in their inhibition of ribosomal-RNA synthesis. The secondary effects of this inhibition is possible decrease in protein synthesis which would ultimately kill the fungus.

Toxicity

These compounds exhibit low acute oral and dermal toxicity and are considered non-irritating to skin and eyes.

**Table 5.37** Commercial products containing phenylamine fungicides.

| Active Ingredients | Trade Names |
|---|---|
| Metalaxyl | Apron, Ridomil, Subdue |
| Furalaxyl | Fongarid, Fonganil |
| Benalaxyl | Galben, Tairel, Trecatol |
| Oxadixyl | Recoil, Ripost, Sandofan, Wakil |

*e. Triazole Compounds*

This group of effective systemic fungicides exhibit both protective and curative properties. The group consists of triadimefon, triadimenol, bitertanol, hexaconazole, and propiconazol. They are broad spectrum fungicides effective against mildews and rusts on cereals, vegetables, deciduous fruits, grapevines as well as ornamentals.

Mode of Action

The mechanism of action of this group is well studied. These compounds effect fungal metabolism by inhibiting ergosterol synthesis. This compound is an important component of the cell membrane, and its absence leads to alterations of the membrane stucture and function and, among other things, to disturbances in the division and development of fungal cells.

Toxicity

Formulations vary in their toxicity depending on the inert ingredients present in the product. Several formulations, e.g., emulsifiable concentrates, are considered as moderately hazardous and are classified in Toxicity Category II. However, most product formulations are non-irritating to skin and eyes or slightly irritating to eyes. These products are classified in Categories III and IV.

**Table 5.38** Commercial products containing triazole fungicides.

| Active Ingredients | Trade Names |
|---|---|
| Bitertanol | Baycor, Baymat |
| Cyproconazole | Alto, Atemi, Sentinel |
| Diclobutrazol | Vigil |
| Hexaconazole | Anvil, Planete Aster |
| Myclobutanil | Eagle, Nova, Rally, Systhane |
| Penconazole | Award, Omnex, Topas, Topaz |
| Propiconazol | Alamo, Banner, Bumper, Orbit, Tilt |
| Tebuconazole | Elite, Folicur, Horizon, Lynx, Raxil |
| Tetraconazole | Domark, Eminent, Lospel |
| Triadimefon | Acizol, Bayleton, Strike |
| Triadimenol | Bayfidan, Baytan |

*f. Morpholine and Piperazine Compounds*

The morpholines and piperazines exhibit both protective and curative action against powdery mildew, rust, scab, and black spot on fruits, berries, cereals, and ornamentals.

Mode of Action

These compounds affect fungal metabolism by inhibiting ergosterol synthesis at some other biosynthetic steps than those affected by the pyrimidines and triazoles. Ergosterol is an important component of the cell membrane, and its absence leads to alterations of the membrane structure and, therefore, to disturbances in the division and development of the fungus.

The only member of the piperazines is triforine. The other fungicides are morpholines.

Toxicity

Toxicity varies with the formulation (e.g., emulsifiable concentrates, wettable powders, and others), with hazardous inert ingredients which may be classified in Toxicity Category I (Danger) or II (Warning) because of their potential to cause eye and skin damage. However, the active ingredient may exhibit low acute oral and dermal toxicity.

**Table 5.39** Commercial products containing morpholines and piperazines.

| Active Ingredients | Trade Names |
|---|---|
| Triforine | Denarin, Funginex, Saprol |
| Tridemorph | Calixin |
| Fenpropimorph | Carbel, Funbas, Mildofix, Mistral |
| Dimethomorph | Acrobat, Forum |

### g. *Dicarboximide Compounds*

This group of fungicides consist of vinclozolin, iprodione, and procymidon. Only procymidone acts systemically; however, vinclozolin and iprodione both exhibit protective and curative activities.

Mode of Action

The mode of action is not clearly defined, but there are indications that it might be connected with cell division, as well as chitin biosynthesis.

Toxicity

This group of compounds is relatively nontoxic and is classified in Toxicity Categories III and IV.

**Table 5.40** Commercial products containing dicarboximide compounds.

| Active Ingredients | Trade Names |
|---|---|
| Vinclozolin | Curalan, Ronilan, Ornalin, Vorlan |
| Iprodione | Chipco, Rovral |
| Procymidon | Sumilex, Sumisclex |

### h. *Organophosphates*

More than 100 organophosphate compounds show fungicidal action; however, relatively few are of practical use as fungicides. Among these few are iprobenfos and aluminium-fosetyl, which are systemic in their activity, and tolclofos-methyl, which acts as a protectant or contact fungicide.

Mode of Action

Iprobenfos have been shown to block the synthesis of phospholipids. This reduction in phospholipids alters the membrane structure in the fungus, increasing permeability with consequent loss of vital cellular components and eventually killing the fungus. Fosetyl, or its metabolite, phosphonic acid, probably acts directly on the target fungus, slowing its growth and hence allowing the plant's natural defense mechanisms time to kill the fungus. The mode of action of tolclofos-methyl is presently unknown.

Toxicity

Toxicity varies and is dependent on the active ingredient and the formulation (e.g., emulsifiable concentrates, wettable powders). Iprobenfos have a low oral $LD_{50}$ and are classified in Toxicity Category II (Warning), whereas aluminum fosetyl (fosetyl-Al) is only slightly toxic and is classified in Toxicity Category III.

**Table 5.41** Commercial products containing organophosphate compounds.

| Active Ingredients | Trade Names |
|---|---|
| Iprobenfos | Kitazin |
| Aluminum fosetyl | Aliette, Chipco |
| Tolclofos-methyl | Risolex |

## REFERENCES

Ashton, F. M. and A. S. Crafts, *Mode of Action of Herbicides*, 2nd ed., John Wiley & Sons, New York, 1981.

Briggs, S. A., *Basic Guide to Pesticides: Their Characteristics and Hazards*, Hemisphere Publishing Corp., New York, 1992.

Brown, V. K., *Acute Toxicity in Theory and Practice*, John Wiley & Sons, New York, 1980.

Chambers, J. E. and P. E. Levi (Eds.), *Organophosphates, Chemistry, Fate, and Effects*, Academic Press, San Diego, CA, 1992.

Cheremisinoff, N. P. and J. A. King, *Toxic Properties of Pesticides*, Marcel Dekker, Inc., New York, 1994.

Corbett, J. R., *The Biochemical Mode of Action of Pesticides*, Academic Press, New York, 1974.

Corbett, J. R., Wright, K., and A. C. Baillie, *The Biochemical Mode of Action of Pesticides*. Academic Press, New York, 1984.

Hayes, W. J. Jr. and E. R. Laws Jr., *Handbook of Pesticide Toxicology*, Academic Press, 1990.

Johnson, J. M. and G. W. Ware, *Pesticide Litigation Manual*, Clark Boardman Callaghan, New York, NY, 1996.

Kearny, P. C. and D. D. Kaufman, *Herbicides—Chemistry, Degradation, and Mode of Action*, Vol. 3. Marcel Dekker, Inc., New York, 1990.

Klaassen, C. D., Amdur, M. O., and J. Doull (Eds.), Casarett and Doull's Toxicology, *The Basic Science of Poisons*, 3rd ed., Macmillan Publishing Co., New York, 1986.

Lever, B. G., *Crop Protection Chemicals*, Ellis Horwood, London, U.K., 1990.

Lu, F. C., *Basic Toxicology—Fundamentals, Target Organs, and Risk Assessment*, Hemisphere Publishing Corp., New York, 1985.

Matsumura, F., *Toxicology of Insecticides*, 2nd ed., Plenum Press, New York, 1985.

Meister R. T. (Ed.), *Farm Chemicals Handbook '96,* Meister Publishing Co., Willoughby, OH, 1996.

Milne, G. W. A. (Ed.), *CRC Handbook of Pesticides*, CRC Press, Boca Raton, FL, 1995.

U.S. Environmental Protection Agency, *Applying Pesticides Correctly: A Guide for Private and Commercial Applicators,* revised 1991.

U.S. Environmental Protection Agency, *Applying Pesticides Correctly: A Guide for Private and Commercial Applicators,* 1983.

U.S. Environmental Protection Agency, *Protective Clothing for Pesticide Users*, Poster, undated.

U.S. Environmental Protection Agency, *Recognition and Management of Pesticide Poisonings*, 4th ed., 1989.

U.S. Environmental Protection Agency, *The Worker Protection Standard for Agricultural Pesticides—How to Comply: What Employers Need to Know*, 1993.

Waxman, M.F., *Hazardous Waste Site Operations: A Training Manual for Site Professionals*, John Wiley & Sons, New York, 1996.

# CHAPTER 6

# HANDLING OF PESTICIDES AND THE USE OF PERSONAL PROTECTIVE EQUIPMENT

## I. HANDLING PESTICIDES

### A. INTRODUCTION

The factors that influence selection decisions regarding personal protective equipment (PPE) and associated workplace safety practices for pesticide handlers are particularly complex in the agricultural setting. The range of handler activities, formulation options, exposure scenarios, environmental factors, and facility amenities is vast. The term "agriculture" often conjures the image of tractors with booms traversing fields of crops neatly planted in parallel rows. This is an unfortunate misconception. In addition to traditional row-crop farming, the generic term "agriculture" also typically includes activities relating to livestock, poultry, pasture, rangeland, forests, nurseries, greenhouses, highway, lawn and garden care. To guide the selection and use of PPE for the various circumstances of these diverse industries, a clear understanding of pesticide-handling tasks is needed.

### B. HANDLING ACTIVITIES

Users of agricultural pesticides perform many activities that involve direct exposure to these chemicals. Adequate PPE selection guidelines must address the range of pesticide-handling jobs and the associated exposure risks. To simplify and focus this discussion, the various pesticide activities will be described under three main operations: mixing, loading, and application.

#### 1. Mixing pesticides

Activities associated with mixing pesticides include:

- transferring pesticide from original container to mixing or application tank or hopper and adding diluent
- conveying open pesticide containers to and from use site
- containing or cleaning up a pesticide spill

- washing contaminated equipment after use
- disposing of pesticide containers
- repairing contaminated mixing equipment such as leaking or broken hose.

**Figure 6.1** Transferring pesticide to application equipment (courtesy of American Cyanamid Company).

## 2. Loading pesticides

Loading activities include:

- transferring pesticide from mixing apparatus to application equipment
- conveying open pesticide containers to and from use site
- containing or cleaning up a pesticide spill
- washing contaminated equipment after use
- repairing contaminated loading equipment.

## 3. Applying pesticides

Applying pesticides may involve:

- On foot - using hand-held sprayers or shake cans
  - using backpack or knapsack sprayers or dusters
  - using portable, aerosol generators or foggers

**Figure 6.2** Mixing pesticides (courtesy of American Cyanamid Company).

- pushing or pulling wheel-mounted sprayers
- using high-pressure, handgun nozzles
- using hand-held, fumigation equipment
- using dipping vats or spray-dip machine
- putting plants or seeds in pesticides
- flagging for aircraft applications
- repairing contaminated application equipment such as clogged nozzles
- leaking or broken hoses, or other malfunctions that occur when the equipment is in use
- adjusting contaminated application equipment, such as boom height, nozzle pattern, or hopper openings
- containing or cleaning up a pesticide spill
- disposing of excess pesticide
- washing contaminated equipment after use.

• By vehicle, pulling or mounted with

- a low-pressure sprayer
- a boom
- a granular applicator
- a high-pressure sprayer
- an injector or incorporator
- a fumigation applicator

- a power duster
- an aerosol generator or fogger
- an airblast sprayer.

**Figure 6.3** Unloading pesticides (courtesy of American Cyanamid Company).

- By using aircraft with
  - open cockpits
  - enclosed cockpits.

## C. FORMULATION CHARACTERISTICS

The range of pesticide formulations routinely used in agricultural situations complicates the selection of appropriate PPE. Pesticide form, toxicity, concentration, and additives and inert ingredients are primary considerations in planning clothing strategies.

### 1. Form

Agricultural pesticides exist in a variety of forms: gases, liquids, emulsions, granules, powders, pellets, aerosols, encapsulates, and even impregnated materials such as wax bars and wiper wicks. Each type of formulation may require distinct strategies for protecting pesticide handlers with PPE.

Liquids may splash or slosh during mixing and loading, causing a potential exposure through ocular, dermal, and oral routes. Furthermore, until sprays have dried, such formulations have the potential to soak through non-chemical-resistant materials, creating dermal exposure. Due to perspiration, they may penetrate even after drying.

Dry formulations often cause clouds or puffs of powdery pesticide to waft upward during mixing and loading, causing potential exposure through ocular, dermal, and inhalation routes (see Figure 6.4). Persons exposed to dry formulations may be unaware that pesticides have penetrated their clothing or equipment, because they do not sense "wetting through."

Gas formulations have obvious exposure potential through the inhalation route, but some also present an exposure risk through the dermal route. Exposure may occur if the gases become trapped in clothing made of fabrics or materials having low air permeability.

## 2. Toxicity

In general, the greater the pesticide toxicity, the greater the amount of protection needed by the worker. Toxicity categories (Table 6.1) are based on the $LD_{50}$, a statistical estimate of a chemical dose which, when administered, will kill 50 percent of the test animals within a stated period of observation.

**Table 6.1** Toxicity Categories

| Category | Toxicity | Signal word |
|---|---|---|
| I | Highly toxic | Danger |
| II | Moderately toxic | Warning |
| III | Slightly toxic | Caution |
| IV | Relatively nontoxic | Caution |

Correlating PPE recommendations with the pesticide toxicity categories is helpful for the pesticide user accustomed to reading signal words on the product labels. One complication, however, is that a single pesticide may be ranked as Category I for oral toxicity and as Category II for dermal toxicity. A second complication is the duration of the exposure. Pesticide handlers may mistakenly believe they do not need protective equipment because they are using pesticides of low toxicity. In fact, long-term exposure to slightly toxic pesticides can sometimes be as dangerous as a brief exposure to a highly toxic pesticide.

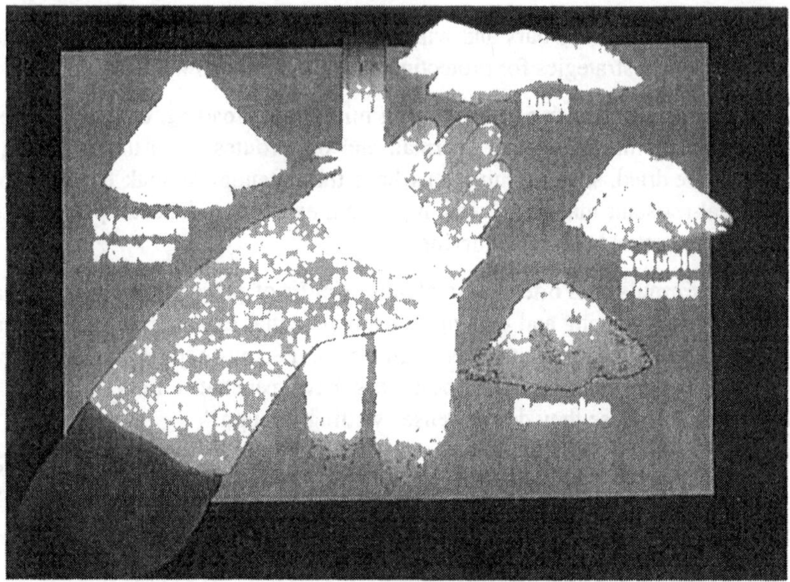


**Figure 6.4** Various types of dry formulations of pesticides.

### 3. Concentration

Agricultural pesticide handlers usually are exposed to highly concentrated pesticides during the mixing and loading process. Mixers/loaders of pesticides may be handling formulations with very high levels of active ingredients; 80 percent wettable powder formulations and 8-pounds-per-gallon emulsifiable concentrates are not unusual. Ultra-low volume concentrates and fumigant formulations may be close to 100 percent active ingredient and often have a high level of acute toxicity. Protecting handlers from exposure to these highly concentrated pesticides is clearly a priority when recommending PPE.

Many agricultural pesticides, however, are ready-to-use (RTU) formulations with very low levels of active ingredients; 5 percent granular or dust formulations are common, and low-concentrate solutions containing 10 percent or less of active ingredient are typical for some livestock formulations. Basing PPE guidelines on mixing and loading activities may not be appropriate for those low concentrations.

### 4. Additives and Inert Ingredients

Another variation among formulations of agricultural pesticides is the plethora of additives and "inert" ingredients added to the active ingredients to create each formulation. These ingredients may contribute to the exposure risk and have an impact on the selection of appropriate PPE for agricultural pesticides.

Most emulsifiable concentrate formulations and some other agricultural formulations have petroleum-based solvents (often xylene), which are a potential hazard to humans orally, dermally, ocularly, and through inhalation. The PPE requirements must reflect the toxicity not only of the active ingredient, but also of the solvents. Other additives do not present direct hazards to humans, but do increase the hazard potential of the active ingredient. The properties of some additives may increase the potential for dermal penetration; others may enhance the adherence of the pesticide to surfaces, including PPE and skin. Still others may increase the potential of the active ingredient to "spread out," thereby potentially contaminating larger areas of skin or PPE.

## D. EXPOSURE SCENARIOS

One of the challenges in establishing PPE guidelines for agricultural pesticide operations is the extensive number of distinct exposure scenarios to be found in agriculture. These scenarios are important not only in ascertaining the degree of pesticide exposure, but also in assessing the level of effort required by the user—an important factor in determining heat-stress potential.

### 1. On Foot

Persons handling pesticides while on foot are, in general, the most likely to receive direct exposure to the pesticide. They are necessarily in close proximity to the pesticide regardless of whether mixing, loading, applying, flagging, cleaning, adjusting, or repairing contaminated equipment or entering treated areas to ventilate or to remove fumigation tarps. Even minor spray drift will contact the handler. The close proximity of applicator to equipment magnifies any equipment inadequacies or imperfections. A dripping or partially clogged nozzle, an unfastened cap, a leaky hose, or a loose connection are extremely likely to result in handler exposure. Figure 6.5 shows a worker spot treating a field on foot.

In addition to direct contact with the pesticide, indirect exposure is also likely through such actions as brushing against plants, animals, or other items just treated with pesticides or from overhead fallout of pesticide, such as mists, drips, or dusts.

The potential for heat-induced illness is also likely to be highest for handlers on foot. Not only are they often exposed to direct, unshaded sunlight, they are also often carrying, pushing, pulling, or otherwise manipulating equipment or pesticide containers.

**Figure 6.5** Using a backpack sprayer for spot applications of herbicides (courtesy of University of Wisconsin-Extension).

### 2. On a Vehicle

Persons applying pesticides from a vehicle are, in general, separate from the immediate proximity of the pesticide being dispensed. Often they are located above and in front of the point of pesticide release. This factor, combined with the forward motion of the vehicle, reduces the probability that pesticide drift or runoff will contact them. Figure 6.6 shows a person in a vehicle applying a contact herbicide with a directed applicator.

In standard practice, however, the vehicle-mounted applicator sometimes must exit the vehicle and walk to the rear of the application rig to repair, adjust, or monitor the equipment or the pesticide dispersal. When this occurs, the applicator often is climbing down from and back onto a contaminated rig and/or walking through newly treated areas. Depending on the size and type of vegetation the applicator must traverse, the applicator may receive significant exposure to the pesticides, ranging from exposure to the feet and lower legs to whole-body exposure from overhead fallout. In addition, the applicator may receive further exposure due to the activity to be performed, such as unclogging nozzles or hoppers and adjusting boom height.

The heat-stress factors for pesticide users on vehicles are generally lower than those for users on foot. The obvious factor is that the equipment is replacing the energy-demanding activity of walking while simultaneously pushing, pulling, or carrying the application equipment and the pesticide. Two related factors also apply. The motion of the vehicle creates the effect of wind on the applicator with the concomitant cooling effect. Secondly, the vehicle is often equipped with a cab, awning, or other over-head shield that shades the applicator from the full force of the sun's intensity.

**Figure 6.6** Application of a contact herbicide by vehicle (courtesy of Monsanto).

Airblast/mistblower applications, unless performed while in an enclosed cab, are likely to expose the pesticide applicator to large quantities of drifting "mist," often to the point of being completely drenched in the pesticide fall-out. Such exposure cannot be avoided, even if the application is performed in conditions of little or no wind and care is taken to remain out of the direct path of the air blast.

### 3. By Air

Agricultural pesticide users who apply pesticides from aircraft are also usually distanced from the pesticide being applied. The dispersal of the pesticide is from below the aircraft, and gravity and the speed of the aircraft carry the pesticide rapidly away from the cockpit.

**Figure 6.7** Aerial application of pesticides.

Some possible exposure scenarios exist, however. When the pilot is also the mixer/loader of the pesticides, the pesticide exposure from that activity may be continued in the cockpit through contaminated PPE. The pilot also may be exposed to pesticides when climbing into or out of the cockpit of a contaminated aircraft. Finally, in some situations, a pilot making a sharp turn or flying in strong wind may fly through the pesticide swath just released from the aircraft. If the cockpit is not enclosed or if its air is not filtered, the resultant pesticide exposure could be significant.

## 4. Exposure Scenarios Deserving Special Consideration

### a. *Engineering Controls*

Some equipment, such as enclosed cabs and enclosed cockpits, greatly reduce the risk of pesticide exposure. The need for PPE is also greatly reduced, but workers should have appropriate PPE available for situations that require them to leave the enclosed areas. Other equipment, such as closed mixing and loading systems, are also designed to reduce exposure. These systems may not warrant as great a reduction in PPE use because of the high risk of exposure in case of equipment failure.

*b. Enclosed Areas.*

Agricultural pesticide handlers use some pesticides in enclosed areas such as greenhouses, silos, and barns. The use of pesticides in such areas increases the risk of inhalation exposure and may require the use of a respiratory protection device even if the identical application outdoors would not warrant such protection. The likelihood of increased potential for heat stress due to the enclosure is dependent on the effectiveness of the ventilation system.

All agricultural pesticides registered for use on crops that may be produced in greenhouses are automatically registered for use in greenhouses unless the registration and/or pesticide labeling specifically prohibits use in a greenhouse. Therefore, the increased inhalation hazard in an enclosed area should be considered when respiratory protection requirements are established for agricultural pesticides.

## E. ENVIRONMENTAL FACTORS

### 1. Weather

Weather conditions can strongly influence the PPE selection process and the work regimen of pesticide handlers. In general, agricultural pesticide use is performed under conditions where climate control or climate alteration is impossible or impractical. The advent of temperature-regulated cabs on pesticide application equipment is relatively new and not, as yet, widespread. Climate control in greenhouses consists mainly of manipulation of ventilating systems. The limited alteration in temperature that results is usually partially counteracted by the high humidity levels, which are inescapable in the enclosed, constantly moist environment. Recommendations for PPE selection should be feasible under a wide range of temperature/humidity conditions without undue risk of heat-induced illness.

Except in these and a few other instances, such as aerial application, the use of agricultural pesticides occurs at ambient air temperatures and humidity— often at the height of the summer heat. Widespread areas of the United States routinely experience temperature/humidity conditions from June through September that make any outdoor activity extremely uncomfortable even without the added burden of personal protective equipment. These climatic conditions cause strong resistance to the use of plastic or rubberized clothing, respirators, goggles or face masks, nonwoven protective suits, or coveralls worn over another layer of clothing. Under some environmental conditions, workers resist the use of long-sleeved shirts.

The undesirable effects of high heat and humidity can be moderated by the cooling effect of the wind or of the air movement created by the vehicle's motion. Too much wind, however, can increase the possibility of exposure due to drift.

At the other extreme, agricultural pesticide use in the northern United States in the early spring may take place in temperatures at or barely above freezing. Pesticide handlers working in cold weather are amenable to the use of protective suits or coveralls over another set of clothing. In these conditions, however, some rubber or plastic garments may become stiff and brittle.

Pesticide handlers also may need protection against glare, sunburn, and the harmful ultraviolet rays of the sun.

### 2. Terrain

Most agricultural pesticide use in greenhouses, on nurseries, and on row crop farms is located in areas easily traversed either on foot or by vehicle. Chemical-resistant footwear is practical and often indicated in such situations. In some forestry, pasture, and rangeland areas, however, pesticide application must be undertaken on rough, ungraded terrain not accessible by vehicles. Under these conditions, the use of chemical-resistant footwear and leg coverings may be a lower priority than reducing exposure to physical risks such as abrasions and animal bites by wearing sturdy (such as leather) footwear and leg coverings.

## F. WORKPLACE SAFETY AMENITIES

### 1. Decontamination Areas

Agricultural establishments range in complexity from the traditional one-family farm to large, complex corporations. Regardless of the size of the establishment, onsite running water for decontamination, private dressing rooms, hot showers, and professional laundry service are the exception rather than the rule. The majority of agricultural establishments do not have a controlled, "factory-like" setting. Many agricultural establishments do not have potable running water anywhere on the premises except at the "office" which may be a personal residence. There is usually no specific indoor facility provided for handlers to change into or out of PPE or to wash themselves after handling pesticides. The "decontamination facility" at the facility headquarters will often consist of cold running water, a bar of soap, and a stack of paper towels. This facility is often located in an unprotected outdoor site.

Water sources for decontamination may be even more primitive for persons handling pesticides at sites removed from the headquarters facility. Handlers may be assigned duties that are many miles from the headquarters and far removed from any source of running water. Running water is often plentiful in greenhouses and on nurseries, but it may be pumped in from a nearby supply of surface water, which is contaminated. Irrigation water on farms and nurseries may not be suitable for even emergency decontamination because pesticides have been or are being applied in the irrigation water. In these remote outdoor locations, water for routine washing of the hands and face and for emergency decontamination may be supplied in a carboy with a push-button spigot. In some remote forestry situations, the only source of wash water is a nearby stream.

### 2. Health Monitoring

The large corporate agricultural establishments may have an on-staff occupational safety officer. The officer conducts biological monitoring of individual handlers both for pesticide exposure and for heat stress symptoms. For many other agricultural establishments, however, monitoring individual handlers either for pesticide exposure or for heat-related illnesses is difficult. Many of the smaller establishments are unequipped and resistant to conducting intrusive monitoring such as blood or urine sampling of their employees. Even body weight, temperature, and blood pressure monitoring is infeasible at many of these facilities. Unlike a factory situation where employees work in close proximity to one another and to sanitation and medical facilities, pesticide handlers on agricultural establishments often work independently, totally isolated from supervisory personnel, and often miles from any modern sanitation or medical facilities. These handlers cannot rely on visual monitoring by supervisors or on biological monitoring to aid in diagnosis and intervention in the event of either a pesticide-related or heat-induced emergency. Even frequent work breaks to recover from the heat stress symptoms may be inconvenient in some agricultural situations. There may be no nearby area that is suitably shady and pesticide-free to enable a handler to remove the PPE and wash before drinking the appropriate fluids and allowing the body temperature to return to near-normal.

# II. PERSONAL PROTECTIVE EQUIPMENT PERFORMANCE AS A BARRIER TO PESTICIDES

## A. INTRODUCTION

Establishing selection guidelines for personal protective equipment and correlated work safety practices for the protection of handlers of agricultural pesticides is a challenge. A balance must be achieved between the goal of reducing the risks to the handlers and the reality of the limitations and complexities imposed by the agricultural workplace environment. Owners, operators, or supervisors must weigh the various factors and attempt to establish PPE requirements and work safety practices that are practical, sensible, and functional in the agricultural workplace and still provide the handler a significant degree of reduction in exposure to pesticides.

As discussed in previous sections, pesticides may be encountered in the form of solids, liquids, or gases. When engineering controls or work practices do not prevent the contact of the pesticide with the skin, PPE may be required. For PPE to be a barrier to a potentially harmful chemical, it must be properly designed, selected, used, and maintained. The materials of construction of the PPE are key determinants of its barrier effectiveness and, therefore, in its selection. PPE materials available to pesticides handlers include:

- woven fabrics, principally cotton and cotton/polyester blends
- nonwoven fabrics, principally of polyethylene or polypropylene
- fabrics incorporating a microporous membrane
- fabrics coated or laminated with continuous films of plastic or rubber
- free-standing films, sheets, or shaped forms of plastic or rubber.

The effectiveness of each of these types of materials as a barrier to a pesticide is dependent on a wide range of factors, including:

- chemical composition of the material
- material thickness
- material weight per unit area
- fabric treatments (finishes)
- fabric air permeability or porosity
- chemical composition of the pesticide formulation
- form of the pesticide, e.g., mist, spray, bulk liquid, dust, granules, vapor
- amount of the pesticide
- duration of the contact period
- force of contact
- temperature
- moisture presence, e.g., humidity, rain, perspiration.

The movement of pesticides through clothing fabrics may occur by one or both of two processes: penetration and permeation.

- Penetration encompasses movement through openings between fibers and yarns, stitching, seams, closures, or flaws by mechanisms such as wicking, wetting, pressure gradients, bulk flow, and so forth. For penetration to occur there must be a hole.
- Permeation involves absorption, diffusion, and desorption of the chemical within a film, coating, or an individual fiber. Permeation occurs due to a concentration gradient across a continuous (i.e., non-porous) layer of material. Standard and customized tests exist for

measuring penetration and permeation. Insight into the likely resistance of a PPE material to penetration or permeation can often be obtained by degradation testing.

With regard to chemical transit through PPE and PPE materials, it is generally true that:

- No one PPE material is resistant to all chemicals.
- One cannot predict with certainty the barrier effectiveness of a PPE material to one liquid on the basis of its performance against another liquid. Predicting chemical resistance to multi-component liquids, which many pesticides are, is particularly problematic.
- Barrier effectiveness increases with the thickness of the material.
- Barrier effectiveness increases with decreases in air permeability.
- Comfort decreases with increased barrier effectiveness.
- Worker acceptance of PPE decreases as comfort decreases.

## 1. Measures of Chemical Resistance

### a. *Penetration*

As defined above, penetration is the movement of a chemical through holes in the material or fabricated component of PPE. Standard penetration tests exist and are described below. In general, these tests have been found either not applicable or are not practiced by the research community involved in studies pertinent to protection from pesticides.

### b. *Permeation*

Permeation rates and breakthrough times are a function of temperature, the thickness and composition of the test material, and the nature of the chemical challenge. Each are discussed below.

1. Temperature

The solubility of a chemical in a polymer and the ease with which the chemical moves through the polymer network both increase with temperature. Consequently, permeation rates increase and breakthrough times decrease as temperature is increased. Most published permeation data were obtained at 20-25° C. Since worker exposures to pesticides may occur at higher temperatures, temperature effects must be considered when assessing the effectiveness of clothing. Breakthrough times for benzene in neoprene were 40 minutes at 7° C, 24 minutes at 22° C, and 16 minutes at 37° C. Over this same range the permeation rates increased by 74%. British researchers have demonstrated decreases in breakthrough times ranging from 3% to 29%, dependent on the chemical/material pair, as the temperature was increased from 23° C to 30° C.

Without permeation data for at least two temperatures, it is not possible to predict the magnitude of the temperature effect for any given chemical/material pair. Tests must be performed at the maximum (worst case) temperature and/or over the temperature range of the anticipated application of the clothing.

2. Thickness

Permeation rate is inversely proportional to the thickness of the test material. The rate decreases as the thickness increases, theoretically in a linear manner. Examples are shown in Table 6.2. Breakthrough times increase as the thickness increases. The breakthrough time relationship is not linear. Although there is no theoretical basis for it, some researchers have correlated breakthrough time with the square of the clothing material thickness. Accordingly, doubling the thickness increases the breakthrough time by a factor of four. The factor, however, may exceed four by as much as 50%. Others have found lesser increase in breakthrough time as thickness is increased. In attempting to determine such correlations, it is critical that all experimental conditions are maintained consistent for each test over the thickness range. This includes the formulation of the clothing material. As shown in the next paragraphs, specimens of the same thickness from gloves of the same generic material but from different manufacturers exhibited differences as large as an order of magnitude in breakthrough times.

**Table 6.2** Effect of thickness on the permeation of benzene through neoprene rubber at 22°C

| Thickness cm | Breakthrough time, min | Permeation rate, ug/cm²/min |
|---|---|---|
| 0.041 | 6.0 | 500 |
| 0.076 | 24.5 | 230 |
| 0.160 | 60.0 | 80 |

3. Composition of Clothing Materials

As discussed previously, PPE is fabricated from a wide variety of polymeric materials by many processes. Glove materials, for example, are multicomponent formulations containing one or more polymers and copolymers, processing aids, stabilizers, fillers, and so forth. Gloves are typically sold under the generic name of the principal polymer in the formulation (e.g., nitrile, butyl, etc.). However, the formulations for the products sold under one generic name will vary from manufacturer to manufacturer and, in some cases, within one manufacturer.

Since the barrier effectiveness of a polymeric material is determined by the specific interactions between the chemical and the material, a variety of performances may be expected for clothing sold under the same generic name. Indeed, this effect has been measured. Table 6.3 shows that for nitrile glove materials of approximately the same thickness from eight manufacturers, the breakthrough times exhibited differences of a factor of ten for perchloroethylene, a factor of six for p-xylene, and a factor of four for butyl acetate. For neoprene glove materials from five manufacturers, the breakthrough times differed by factors of eight for ethanol, five for n-hexane, and two for butyl acetate. Note: there is a greater thickness variation among the neoprene gloves than with the nitrile gloves. As will be discussed later, each of the above chemicals has been identified as one that may appear in a pesticide formulation. In particular, xylenes are widely used as an ingredient in emulsifiable concentrates of Toxicity Category I and II pesticides.

**Table 6.3** Breakthrough times (minutes) for perchloroethylene, p-Xylene, n-Butyl acetate, Ethanol, and n-Hexane through nitrile and neoprene gloves from several manufacturers (U.S. EPA, *Guidance Manual for Selecting Protective Clothing for Agricultural Pesticide Operations*).

| Manufacturer | Thickness (mm) | Perchloro-ethylene | p-Xylene | n-Butyl acetate | Ethanol | n-Hexane |
|---|---|---|---|---|---|---|
| **Nitrile** | | | | | | |
| Ansell | 0.36 | 30 | 11 | 15 | | |
| Best | 0.28 | 135 | 27 | 28 | | |
| Edmont | 0.32 | 215 | 33 | 44 | | |
| Granet | 0.32 | 90 | 24 | 32 | | |
| Intermarket | 0.36 | 125 | 28 | 39 | | |
| LRC | 0.28 | 295 | 54 | 38 | | |
| North | 0.32 | 145 | 33 | 38 | | |
| Pioneer | 0.32 | 295 | 60 | 56 | | |
| | | | | | | |
| **Neoprene** | | | | | | |
| Ansell | 0.38 | | | 14 | 500 | 106 |
| Best | 0.64 | | | 13 | 110 | 21 |
| Edmont | 0.46 | | | 20 | 480 | 82 |
| Granet | 0.55 | | | 13 | 80 | 23 |
| Pioneer | 0.55 | | | 33 | 690 | 72 |

4. Chemical Composition

Published permeation data are predominantly from tests in which the challenge chemical is neat (pure). Considerably less testing has been performed with multicomponent chemical solutions. That which has been reported represents a broad spectrum of results. In some cases, the permeation of any one component occurs at a slower rate than when the chemical was tested in neat form. In other cases, the permeation is at a greater rate, and in others there seems to be little effect. Examples of each are shown in Table 6.4. At present, there are no means for predicting the behavior of multicomponent solutions.

This situation represents a particular problem with regard to pesticides which are virtually all multicomponent formulations. Research, much of it sponsored by the EPA, is ongoing in an attempt to better understand the role of the carrier solvents, surfactants, and other components in the permeation of the active ingredients. Data from these studies are presented in detail later in this section.

**Table 6.4** Permeation of three two-component mixtures through two glove materials (U.S. EPA, *Guidance Manual for Selecting Protective Clothing for Agricultural Pesticide Operations*).

| | | Breakthrough time (minutes) | | Permeation rate ($mg/m^2/sec$) | |
|---|---|---|---|---|---|
| **Composition (A/B)/Vol%** | **Glove material (thickness, cm)** | **A** | **B** | **A** | **B** |
| Methanol/ n-butyl acetate | Nitrile (0.38) | | | | |
| 0/100 | | | 60 | | 91 |
| 10/90 | | 38 | 39 | 26 | 91 |
| 25/75 | | 24 | 28 | 50 | 71 |
| 50/50 | | 34 | 38 | 63 | 32 |
| 75/25 | | 53 | > 150 | 26 | <1 |
| 90/10 | | 103 | > 150 | 10 | <1 |
| 100/0 | | 108 | | 4 | |
| Methyl ethyl ketone/hexane | Viton (0.23) | | | | |
| 0/100 | | | >200 | | |
| 5/95 | | | 150 | | |
| 10/90 | | 39 | 33 | <25 | <25 |
| 25/75 | | 6 | 6 | 55 | 55 |
| 50/50 | | 3 | 3 | 330 | 86 |
| 75/25 | | 3 | 3 | 670 | 135 |
| 90/10 | | 3 | 3 | 1140 | 63 |
| 100/0 | | 3 | | 1100 | |
| p-Xylene /toluene | Nitrile (0.38) | | | | |
| 0/100 | | | 32 | | 26 |
| 25/75 | | 28 | 25 | 6 | 18 |
| 50/50 | | 34 | 32 | 20 | 18 |
| 75/25 | | 54 | 54 | 19 | 5 |
| 100/0 | | 53 | | 27 | |

*c. Degradation*

Degradation is any undesirable change in a physical property or the appearance of the PPE material. Examples include swelling, shrinkage, loss of puncture and tear resistance, and an increase in stiffness due to exposure to pesticides. The degradation effects of a pesticide on PPE materials can be assessed in the laboratory or in the field.

In a typical degradation test, a specimen from the PPE or, in some cases (e.g., a glove), the entire item is immersed in the pesticide for some period of time. At predetermined intervals, the specimen is removed, tested, and observed. In the laboratory, strength, puncture, and tear properties are commonly measured using a machine with which the specimen is stretched and the force required to stretch the specimen recorded. Significant increases or decreases (±20%) in any of the properties would be indicative of poor resistance to the pesticide formulation. Losses in these properties would generally result from the sorption of one or more components of the pesticide formulation by the glove material.

Increases in properties are often associated with the extraction of components from the glove material by the pesticide formulation. The stiffening of polyvinyl chloride (PVC) gloves due to extraction of plasticizer is an example.

Care must be taken when testing clothing materials that derive their physical properties from a supporting fabric but derive their barrier properties from the coating. The supporting fabric may not be affected but the barrier coating may be degraded by the chemical.

Sorption or extraction may also be evidenced by changes in the weight and/or dimensions of the test specimen. Measurement of such effects are relatively easy to perform in either the laboratory or the field. For this reason, until recently most chemical resistance charts published by suppliers of polymeric materials and polymer-based PPE were based on weight change or swelling measurements. Again, weight or dimensional changes of ±20% for rubber materials or ±3 % for plastic materials should be considered significant and indicative of poor resistance to the pesticide formulation.

## B. ARM AND HAND PROTECTION

### 1. Introduction

As discussed in the previous section, there are scores of polymeric materials from which gloves are fabricated. All of these materials are subject to some degree of chemical attack by pesticide formulations. The degree to which the pesticide formulation will attack any given polymeric material is dependent on the duration of the exposure, the temperature, the condition of the material, and the specific interactions between the polymer and the particular components of the pesticide formulation. A possible consequence of chemical attack is that the glove will no longer protect the worker's hands and lower arms from the pesticide.

Parameters for determining the chemical resistance and barrier effectiveness of glove materials to chemicals were discussed above. Many decades of testing have shown that:

- No one glove material is a barrier to all chemicals.
- Any given chemical may severely attack one glove material and not affect another material.
- Certain chemicals or combinations thereof have severe effects on all commonly available glove materials.

The effectiveness of gloves as chemical barriers is the subject of the remainder of this subsection. However, before beginning the discussion, it is important to note that chemical resistance is also fundamental to the issue of decontamination. Clothing materials that are not resistant to a chemical will absorb the chemical, making decontamination more difficult. Decontamination will then require removal of an absorbed as well as surface chemical.

### 2. Pesticide Formulation Types

Pesticide formulations are designed to optimize the cost, handling, application, and effectiveness of the active ingredient. Some fumigants are used in undiluted form (commonly referred to as "neat"), whereas some herbicide formulations may have 15 to 20 ingredients in addition to the active ingredient. The other ingredients include diluents, surfactants, stabilizers, thickeners, solid substrates, solvents, and tackifiers. The resulting formulations are typically characterized as one of the following types:

- aqueous solutions and concentrates
- emulsifiable concentrates
- wettable powders
- dry flowables (water dispersible granules)
- granules
- others—including controlled release, dusts, and aerosols.

From the perspective of protective clothing selection and chemical resistance, the pesticide formulation types may be more generally categorized as:

- solids
- solutions and dispersions that contain an organic solvent

- aqueous solutions and dispersions, which contain no organic solvents.

PPE materials having a continuous polymeric film are likely to resist permeation and penetration by all solid formulations and thus act as a barrier to solids contact with the skin. For solids, glove features such as fit, gauntlet length, durability, etc. as discussed previously, are the key issues for glove selection decisions.

Chemical resistance factors are more important for aqueous solution formulations and most important for the formulations that contain an organic solvent. Most common glove materials, with the exceptions of polyvinyl alcohol and very thin (<0.2 mm) natural rubber gloves, are highly resistant to water and would be expected to provide good barriers to permeation by formulations that contain no organic solvents.

The solutions that contain an organic solvent usually in the form of emulsifiable concentrates and field application dilutions of such concentrates, represent the greatest potential challenge to the barrier effectiveness of polymeric gloves. Many of the solvents can severely degrade or permeate the glove materials when in undiluted form.

Because most pesticide formulations are proprietary, and because until recently their components other than the active ingredient were not considered important from a health perspective, there is little published information available on the types and concentrations of organic solvents in pesticide formulations. However, information from the literature and discussions with persons knowledgeable on pesticide formulations provides some insight into the identity of the solvents. Categorized according to functional group type, solvents that are representative of those used in pesticide formulations are listed in Table 6.5.

The remainder of this section describes the results of permeation tests performed using such carrier solvents in undiluted form and using commercially available pesticide formulations containing these types of solvents. The results indicate that the permeation of various pesticide formulation components through common glove materials is often controlled by the type and amount of the solvent. Thus, the relative permeation of the pesticide active ingredient and other components can often be predicted based on information regarding the permeation of the carrier solvent in neat form. This finding can greatly simplify protective glove selection decisions, because permeation tests using the solvents alone or focused on the solvent are less complex to perform than those using actual pesticide formulations. Also a large database of information is now available for the types of organic liquids used as solvents in neat form.

## 3. Solvent Permeation

Since the early 1980s, the EPA has compiled all the available information from the government laboratories, equipment manufacturers, and suppliers into a database. Permeation data for several of the solvent liquids listed in Table 6.5 are available for common glove materials from this database.

**Table 6.5** Organic liquids used as carrier solvents in non-aqueous solution pesticide formulations (U.S. EPA, *Guidance Manual for Selecting Protective Clothing for Agricultural Pesticide Operations*).

| Functional group classification | Organic carrier solvent |
|---|---|
| Acetate | Amyl acetate<br>Butyl acetate |
| Alcohols | Diacetone alcohol<br>Diethanol amine<br>Ethanol<br>Hexylene glycol<br>Isopropanol<br>Methanol<br>Propylene glycol<br>Tetrahydrofurfuryl alcohol |
| Halogenated hydrocarbons | Methylene chloride |
| Ketones | Acetone<br>Cyclohexanone<br>Isophorone<br>Methyl isobutyl ketone |
| Petroleum distillates, aliphatic | Diesel fuel (paraffins up to 600•f b.p.)<br>Gasoline (c4-c12 paraffins)<br>Kerosene (paraffins up to 500•f b.p.)<br>Mineral oil<br>Mineral spirits<br>Naphtha<br>Petroleum oil |
| Petroleum distillates, aromatic | Xylenes<br>Xylene range solvents |

These data were reviewed and a qualitative rating was assigned to each chemical/material pair for which data were available in Table 6.6.

Table 6.6 is provided as a final aid for protective glove selection decisions. Table 6.6 identifies the preferred glove types as well as those not recommended for use with pesticide formulations containing various carrier solvent types. The glove types identified as preferred are those that consistently provide the best permeation resistance to pesticide formulations and undiluted carrier solvents within each chemical classification identified. The glove types identified as not recommended are those that consistently provide very poor permeation resistance. Also note that the recommendations in Table 6.6 are based solely on chemical resistance considerations. Other factors such as glove sizing, comfort, durability, and cost must also be considered as discussed elsewhere in this text. And finally, the proper use of chemical protective gloves is as important as their proper selection for limiting worker exposures to agricultural pesticide formulations. Figure 6.8 shows a worker selecting nitrile gloves for a handling job involving contact with a pesticide formulation containing a petroleum distillate.

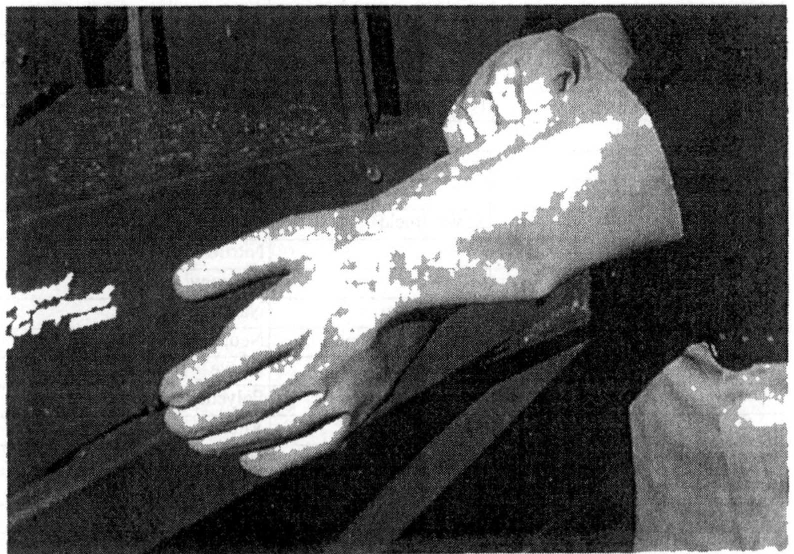

**Figure 6.8** Pesticide handler putting on nitrile gloves for mixing an emulsifiable concentrate containing a petroleum distillate (courtesy of University of Wisconsin-Extension).

## C. BODY PROTECTION

### 1. Introduction

Body protection includes shirts, pants, coveralls, aprons, hats, and the like. As mentioned earlier, these items are fabricated from three general groups of materials:

- Woven cotton and cotton/polyester fabrics
- Nonwoven fabrics, predominantly polyolefins (see Figure 6.9), but including microporous polytetrafluoroethylene (PTFE)
- Woven or nonwoven fabrics to which a continuous film of plastic or rubber has been laminated or coated.

**Table 6.6** Glove selection guidance for protection from pesticide formulations that contain organic solvents (U.S. EPA, *Guidance Manual for Selecting Protective Clothing for Agricultural Pesticide Operations*).

| | Glove Recommendations | |
|---|---|---|
| **Carrier solvent type** | **Preferred** | **Not recommended** |
| Acetates | Butyl rubber | Natural rubber |
| | Polyvinyl alcohol | Polyvinyl chloride |
| | Silver Shield | |
| Alcohols | Butyl rubber | Neoprene + natural blend |
| | Neoprene | |
| | Nitrile rubber | |
| | Nitrile + PVC blend | |
| Alcohol mixtures with: | | |
| halogenated | Neoprene | Nitrile rubber |
| hydrocarbons, or | | Polyvinyl chloride |
| aromatic petroleum | Neoprene | Nitrile rubber |
| distillates, or | | Polyvinyl chloride |
| halogenated | Silver Shield | Natural rubber |
| hydrocarbons | | Nitrile rubber |
| | | Polyvinyl chloride |
| Ketones | Butyl rubber | Natural rubber |
| | Polyvinyl alcohol | Neoprene + natural blend |
| | Silver Shield | Nitrile + PVC blend |
| | | Polyethylene |
| | | Viton |
| Ketone mixtures with: aromatic | Butyl rubber | Natural rubber |
| petroleum distillates | Neoprene | Polyvinyl chloride |
| | Nitrile rubber | |
| | Silver Shield | |
| Petroleum distillates, aliphatic mixtures with: aromatic petroleum distillates | Butyl rubber<br>Nitrile rubber | Natural rubber<br>Polyvinyl chloride |
| Petroleum distillates, aromatic | Butyl rubber | Natural rubber |
| | Nitrile rubber | Polyethylene |
| Xylenes (<40%) | Polyvinyl alcohol | Natural rubber |
| | Silver Shield | |
| | Viton | |
| | Nitrile rubber | |
| Xylenes (>40%) | Polyvinyl alcohol | Neoprene |
| | Silver Shield | Polyethylene |
| | Viton | Polyvinyl chloride |

These fabrics exhibit levels of barrier effectiveness that span many orders of magnitude. As discussed earlier, a definitive comparison of the fabrics is difficult, because the agricultural clothing research community has no standard test for measuring barrier effectiveness. Furthermore, the difficulty of this situation is compounded by varied practices among the researchers for describing the fabrics that they tested. Examples include:

- Not reporting properties such as thickness, weight, and air permeability.
- Not identifying trademarked fabrics, referring to them rather by generic descriptions. Since no two researchers use the same generic description, persons attempting to use the literature cannot ascertain whether, in fact, the same material was or was not tested by two researchers. For example, the same fabric might be referred to as "synthetic nonwoven," "nonwoven polyolefin," "nonwoven polyethylene," and "Tyvek®."
- Not identifying trademarked fabric treatments/finishes, referring to them rather by generic descriptions. For example, the descriptions of the same treatment may range from "soil repellent finish" to "fluorocarbon" to "a 3M fluorocarbon" to " Scotchgard®."

Overall recommendations derived from the information present in the EPA's *Guidance Manual for Selecting Protective Clothing for Agriculture Pesticides Operations* are presented in Table 6.7.

**Table 6.7** Body protection considerations.

| Considerations | Description |
|---|---|
| 1. | Long-sleeved shirts and long-legged pants are preferred to short sleeves and short pants. |
| 2. | For woven fabrics, heavier fabrics are better barriers than lighter fabrics. 250 g/m$^2$ has been suggested as a minimum. |
| 3. | For fabrics of the same type and weight, higher resistance to penetration was generally correlated with higher cotton content. The performance of 100% polyester is usually significantly below even 35/65 cotton/polyester blends. Although supporting data are not extensive, it seems that woven fabrics should be preferred to knits and, within the wovens, twills to plain weave. |
| 4. | In general, low air permeability, synthetic nonwovens and microporous PTFE are better barriers than woven fabrics. |
| 5. | Fluorocarbon treatment generally improves the penetration resistance of porous fabrics, both woven and nonwoven, to liquids and particles. The fluorocarbon reduces the differences between the resistances of 100% cotton and various cotton/polyester blends. Durable press finishes reduce penetration resistance to liquids. Studies of soil releasing treatments are inconclusive. Laundering reduces the effectiveness of finishes; they must be renewed periodically. |
| 6. | The highest levels of protection can only be achieved through the use of a fabric that contains a coating or lamination of a continuous film of plastic or rubber. |
| 7. | Any item of clothing, to be effective, must be properly designed, constructed, fitted, worn, and maintained. |

## D. EYE AND FACE PROTECTION

### 1. Introduction

Splashes, errant sprays, mists, clouds, etc., of pesticide vapors, aerosols, liquids, dusts, and granules are potential sources of facial contact with the pesticides. Although facial contacts would be expected to occur less frequently than, for example, contact with the hands, the health consequences of the contacts are potentially more serious, especially for contacts with the eyes.

In this section, the characteristics of eye and face protection products useful for protection from chemicals and pesticides are discussed. Occupational Safety and Health Administration's (OSHA) 29 CFR 1910.133 and American National Standard Institute's (ANSI) Z87.1 are used as references for this discussion. The regulation and the standard are useful consolidations of information and guidance on the subject area, although they may not be directly applicable to the agricultural setting.

### 2. Selection Recommendations

The process of selecting eye and face protection for pesticide handlers is difficult to generalize because of the wide range of exposure conditions that are present in agriculture. In some situations, workers are in close proximity to highly toxic pesticides where the danger of splash, irritating mists, or nuisance dusts might occur (at other times, the worker may only peripherally encounter the above dangers, yet requires appropriate protection). Common sense and fundamental technical principles should be used in determining the correct protective eyewear.

Historically, many individuals have used either no eye protection or safety glasses as the primary method of protection in this environment. There are currently several different types of safety glasses for eye protection: no sideshield, non-removable lens, half sideshield, lift front, full sideshield, detachable sideshield, and headband temple. Safety glasses are designed to protect the wearer primarily from direct impact and heat. The sideshields provide additional protection from side impacts. Of the seven types of spectacles available, the models with full sideshields provide the greatest protection from flying objects.

Shielded safety glasses are comfortable, do not cause fogging or sweating, and give good eye protection in many situations. Safety glasses must have both brow and side shields.

Other types of protective eyewear appropriate for use in environments containing pesticides include:

- Goggles
- Face shield
- Shielded safety glasses

Goggles are available with both rigid and flexible frames, the majority of users favoring the more comfortable flexible frame. Typically, the flexible frames are translucent and thus increase the field of vision.

**Figure 6.9** Worker dressed in a hooded coverall made from DuPont Tyvek®. Tyvek® is a DuPont registered trademark for its brand of spunbonded olefin. Only DuPont makes Tyvek®.

Goggles may have three types of ventilation: direct, indirect, or none. Directly ventilated goggles create the possibility of a liquid or aerosol entering through the ventilation port. To reduce and minimize the goggle lenses from fogging, indirect vented models are preferred. The indirect models exclude the direct passage of dust, liquids, light, or particles. Figure 6.10 shows a worker doning goggles and other PPE.

**Figure 6.10** Pesticide handler donning goggles, gloves, and chemical-resistant apron (courtesy of University of Wisconsin-Extension).

Face shields are protective devices intended to shield the wearer's face, or portions thereof, in addition to the eyes. The face shield would be used when handling pesticides where it is highly likely that splashing will occur. Face shields are available with or without removable windows. Figures 6.1, 6.2, and 6.3 show handlers wearing face shields.

Employees should be cautioned about wearing contact lens in certain, if not all, pesticide environments. Dusty or chemical environments may represent particular hazards to contact lens wearers. Soft contact lenses can be especially problematic. Their design makes them hydrophilic (absorb fluids); as a result they will also absorb chemical mists. If an employee is working in these conditions, it is best that he or she not wear contact lenses.

Guidelines for selecting eye and face protection follow:

- Define the potential hazard, considering the need for protection from chemicals as well as from impact from flying objects.
- Compare the potential hazards with the performance capabilities of various eye and face protective equipment.
- Make a judgment in selection of the appropriate protective equipment so that the protection is greater than the estimated hazard.
- Fit the user with the protective device, taking extra care to ensure that an adequate seal exists between the goggle and the wearer's face to prevent any pesticide from entering.
- Give instructions on care and use of the equipment.

Table 6.8 lists the types of eye protection that satisfy eyewear requirements on pesticide labels.

**Table 6.8** Selection guidance for eye and face protection.

| Label Requirement | Acceptable Eye Protection |
|---|---|
| Protective eyewear | Safety glasses with brow, front, and temple protection, face shield, goggles, or a full-face respirator |
| Goggles | Goggles, or full-face respirator |
| Full-face respirator | Full-face respirator |

## E. PPE SELECTION AND USE

### 1. Introduction

The process of selecting and using chemical-resistant PPE encompasses multiple steps and decision points, starting with a definition of the hazard and ending with the disposal of used PPE. Figures 6.11 and 6.12 picture assorted PPE.

Briefly, those persons responsible for the health of a pesticide handler must:

- Define the potential threat in terms of the toxicity of the pesticide to be handled and the conditions under which it will be handled (i.e., the exposure scenario).
- On the basis of the threat, establish the level of protection that is required.
- Select items of PPE that combine to meet the requirements for the level of protection. The selection process must take into account the effectiveness and costs of the available clothing.

- On the basis of the selected PPE, adjust the work regimen to take into account reduced productivity associated with the PPE, the potential for heat stress, and the possibility that the available PPE may not fully meet all the protection requirements.

**Figure 6.11** Assorted PPE including chemical goggles and face shield (courtesy of University of Wisconsin-Extension).

- Instruct workers in the use of the PPE and its importance to promote worker acceptance of the PPE.
- Distribute the PPE and oversee its use.
- After its use, either dispose of the PPE by appropriate means or decontaminate, repair, store, and reissue it for another use.
- Replace PPE periodically, even though it does not show wear.

## 2. Toxicity

Pesticides vary significantly in their potentials for harming humans and their modes of action. The pesticides having the most potential for acute adverse effects are classified as Category I compounds. Less than a teaspoon of some of these compounds, if taken orally, could cause death. Other Category I compounds may be absorbed through the skin with lethal or severely debilitating consequences. Still other compounds in Category I could cause severe damage to the eyes or skin. Compounds in Category II are less toxic, i.e., higher doses are required to cause the same type of effects as relatively smaller doses of Category I compounds. Category II compounds are generally described as moderately toxic. Category III are considered slightly toxic and Category IV are considered relatively nontoxic.

The acute toxicity category of the pesticide is an important factor for the level of PPE that is required. Consideration must also be given to providing PPE for pesticides that may have delayed adverse effects. For a given pesticide handling activity, a higher level of protection would be required for a more highly toxic compound. The mode of action of the pesticide also influences the level of protection. For example, compounds that are severely corrosive to the eyes would require particular attention to the selection and use of eye protection. Compounds that are readily absorbed through the skin or are highly corrosive to the skin would require specific attention to body protection, particularly for the hands. In addition to providing dermal protection, glove use is important for compounds that have high oral toxicities since there is the possibility that pesticide handlers might put food or tobacco (chewing or smoking) into their mouths before thoroughly washing their hands.

## 3. Exposure Scenarios

As outlined below, many factors combine to define the exposure scenario. In turn, the scenario combined with the toxicity of the pesticide determines the level of protection that should be used.

### a. *Pesticide Form*

From the perspective of defining the type of PPE that might be required, pesticides may be conveniently divided into four categories:

1. solids with little or no solubility in water
2. solids with high solubility in water
3. aqueous liquids and suspensions that contain no organic solvents
4. liquids and suspensions that contain organic solvents.

Common woven and nonwoven fabrics with sufficiently small pore sizes would be expected to provide a reasonable level of protection from solids. There is concern, however, that perspiration could dissolve water soluble solids present on the outside of the fabric and that the perspiration-wet fabric would provide a route for the pesticide to contact the skin. For aqueous liquids and suspensions that contain no organic solvents or active ingredients that behave as solvents, any PPE item that is a barrier to water would be expected to provide protection. Of concern would be the integrity of the seams, interfaces, edges, and closures. Those formulations that contain active ingredients that behave as solvents must be treated similarly to those that contain an organic solvent. It is for those pesticide formulations that contain an organic solvent that the selection process becomes complex, requiring consideration of the potential for chemical permeation and degradation.

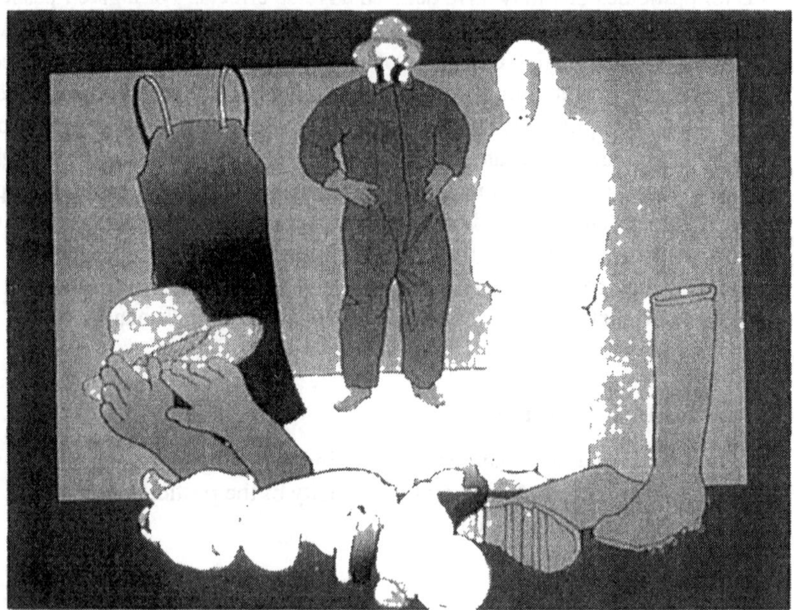

**Figure 6.12** Assorted PPE that can be used for various tasks involving hazardous chemicals and pesticides.

*b. Control of Exposure*

Pesticide formulations and handling situations vary in the degree of exposure that a pesticide handler can expect. For example, some solids are in the form of wettable powders. In one scenario, the pesticide handler might be required to open and pour a bag of such a powder into a mixing tank containing water. Depending on the wind, the rate of pouring, and size and shape of the tank into which the powder is poured, the worker may be exposed to a cloud of the powder. Alternatively, that same powder might be available in a water-soluble bag. In this case, the bag does not have to be opened and can be added directly into the mixing tank; the potential for exposure is greatly reduced. Similarly, some liquid handling systems are completely closed and there is very little potential for exposure due to splashes and spills.

The level of PPE required for persons using engineering controls to reduce exposure as illustrated above, may be significantly less than for situations in which there are multiple opportunities for worker exposure.

### c. *Duration and Frequency of Exposure*

Both duration and frequency of pesticide handling and exposure affect the level of PPE that might be required. The type of PPE that might be required for a person who mixes pesticide for one 1/2-hour period each morning and afternoon would differ from that for a person who is exposed to mists from airblast sprays for the entire work day. PPE should be removed when the pesticide handling activity is completed.

### d. *Work Task*

The work task is another important determinant of the level of PPE that is required for worker protection. The task largely defines the direction from which the potential exposure may occur (in some tasks, the potential exposure is mostly from overhead, in others from the front, in others the back and shoulders, and in still others from below the knees). Thus, the required types of PPE for the same pesticide may vary, depending on the mode of application and the specific activity of the pesticide handler. For example, weed sprayers riding a boom may require thorough PPE coverage of the lower parts of the body, but little or no protection above the waist. Mixers may only require frontal coverage—a need that could be satisfied by a combination of gloves, sleeves, apron (or sleeved-apron), and boots, leaving the back open for the dissipation of metabolic heat.

## 4. Level of Protection

Level of protection is that combination of gloves, boots, garment, goggles, etc., that is necessary to provide protection from the potential hazard. The goal is always to use the minimum level of PPE that is necessary. By so doing, costs, effect on productivity, and potential for heat stress are minimized, and the likelihood for worker acceptance (compliance) is increased.

Five levels of protection are described in Table 6.9.

**Table 6.9** Protective clothing ensembles.

| Protection Level | Description |
|---|---|
| 1 | Long-sleeved shirt and long-legged pants (cotton or cotton/polyester overall), and chemical-resistant gloves and shoes. |
| 2 | Cotton or cotton/polyester coveralls worn over a long-sleeved shirt and long-legged pants, chemical-resistant gloves, chemical-resistant hat or hood, and chemical-resistant boots. |
| 3 | Cotton or cotton/polyester coveralls worn over long-sleeved shirt and long-legged pants, chemical-resistant gloves, chemical-resistant boots, chemical-resistant apron, and goggles or face shield. |
| 4 | Nonwoven, polyolefin coverall (worn over long-sleeved shirt and long-legged pants), chemical-resistant gloves, chemical-resistant hat or hood, chemical-resistant boots, and goggles or face shield. |
| 5 | Rubber or plastic-coated coverall (worn over underwear) with hood, chemical-resistant gloves, chemical-resistant boots, and goggles and half-face respirator (or full-face respirator). |

The five levels are suggestive of the levels of PPE that might be worn for typical pesticide handler activities, as summarized in Table 6.10. Note, however, that other combinations of PPE items may be used and, in many situations, may be more appropriate than any one of the five levels. There are too many extenuating circumstances in agriculture for rigid specification of the composition of the protective ensembles.

## 5. PPE Performance

After the types of PPE components necessary for protection have been established, one must pick specific PPE items that will satisfy the requirements. The task begins by reviewing the information available for chemical resistance and for physical properties. From this information, one selects the materials of construction for the PPE items. For example, if the need is for hand protection and foot protection for >4 hours from a pesticide concentrate containing aliphatic petroleum distillates as the carrier solvent, then, with reference to Table 6.6, nitrile might be selected as the material of construction.

## 6. Availability

After the types of PPE items and their materials of construction have been determined, sources for the PPE must be identified. This step is accomplished by reference to the appendices found in EPA's *Guidance Manual for Selecting Protective Clothing for Agriculture Pesticides Operations.* Each appendix contains an extensive listing of commercially available PPE and is organized under the subheadings of the materials of construction. The description of each product includes its source, style number, sizes, and other key features of importance to the purchasing decision.

**Table 6.10** Association of PPE levels with pesticide handler activities.

| Activity | Level of Clothing* | |
|---|---|---|
| | Toxicity Categories I & II | Toxicity Category III |
| Mixing /Loading | 4 | 1 |
| Applying | | |
| -Airblast spraying | 5 | 1 |
| -Boom spraying | 2 or 4 | 1 |
| -Greenhouse | 2 or 4 | 1 |
| -Airplane | 2 | 1 |

*As defined in Table 5.9.

## 7. Cost

An individual item of PPE may pass through several vendors between its point of manufacture and the point of sale to the person responsible for the health and safety of a pesticide handler. Along the chain, there is an opportunity for mark-ups and discounts; the magnitudes of which are highly dependent on the volume of the purchase and the market conditions. Vendors and distributors are constantly changing prices in order to reflect changing raw material costs, labor costs, and competition. For all these reasons, item-by-item price information is not included in this discussion.

## 8. Impact on Productivity

### a. *Job function*

The effect of PPE on task performance and, therefore, productivity was discussed briefly in previous sections. In virtually all cases (with the possible exception of eye protection), the use of PPE will increase the time needed to complete a task. Certain PPE, furthermore, may preclude the performance of certain tasks. For example, the glove that provides chemical protection for >4 hours may reduce dexterity to a point where the worker cannot repair a clogged nozzle. For situations such as this, the preferred glove may be one that provides protection for considerably shorter times but has less negative effect on dexterity. Protection would be obtained by replacing the glove more frequently.

The negative effect of PPE on worker productivity combined with its recurring costs are leading many persons to pursue engineering controls and other alternatives to PPE as the route to worker protection.

### *b. Heat Stress*

Similar to functional impacts, all PPE (with the possible exceptions of gloves and eye and face protection) to a greater or lesser extent increases the potential for heat stress when working in warm environments. As higher levels of PPE are prescribed, persons in positions of responsibility must consider the potential for heat stress. If the potential is significant, the expectations for worker productivity must be reduced. Humans simply cannot work as long or as hard under conditions in which the body is hindered (by PPE) in its attempt to maintain thermal balance. Under some environmental conditions, certain activities cannot be performed without jeopardizing the health of the worker unless auxiliary cooling is provided to the worker. Chapter 7 will discuss this topic in more detail.

## 9. Worker Training and Compliance

To be effective, PPE must be used and maintained properly. Training is required. A myriad of federal and state programs are in place and being planned to promote the correct use of PPE by pesticide handlers. Compliance comes when workers understand the importance of the clothing to their personal well-being and that of persons with whom they may come into contact. For example, workers need to know why they should wear gloves, that there are major performance differences between glove types, that one glove does not work in all applications, that gloves must be decontaminated on at least a daily basis, and that a glove with a hole in it may be worse than no glove. Pesticide that reaches the inside of a glove does not readily evaporate and, thus, is available for skin exposure during the entire time that the glove is worn. Compliance also results when supervisors monitor and encourage the *proper* use of PPE as an essential element of the work routine.

## 10. Personal Protective Equipment Use

PPE use begins the moment the worker inspects the item for defects prior to donning it. It ends when the item has been safely doffed and made ready for thorough decontamination or disposal without transferring contamination to other persons and things. The proper use between these two points in the process requires training, monitoring, and encouragement.

Garment closures must be kept closed. Sleeves must be kept rolled down, and the integrity of the glove-sleeve and other interfaces maintained. PPE that becomes damaged must be replaced. PPE that becomes severely contaminated must be doffed immediately. Pant legs must be kept on the outside of the boots. Aprons must be kept tied and in place. Goggles and face shields must be kept in position, not pushed up on the head or hung around the neck.

Once clothing has been used for handling pesticides, it should never be used for any other activity. To do so would risk transferring contamination to other persons, animals, and things outside the pesticide handling environment.

### 11. Decontamination

The decontamination of PPE is discussed extensively later in Section IV under the heading Disposables and Reusables, and in Section V., "Spill Management." The subject is complex; there are many unanswered questions and there are many situation-specific considerations. Again, education and training are the keys here.

## F. SPECIAL PRECAUTIONS FOR HANDLING FUMIGANTS

Fumigants are pesticides that are applied as a gas or that readily form a gas when they are applied. Their pesticidal action is in the gaseous form. Fumigants are highly toxic to plants and animals, including humans. Use extreme caution and wear appropriate personal protective equipment whenever handling fumigants. Personal protective equipment requirements for protection from fumigants are often very different from the requirements for other types of pesticides. Follow labeling directions exactly for each fumigant.

Inhaling even small amounts of some fumigant gases can be fatal or cause severe injury. You must wear the respirator listed on the fumigant labeling during any handling activity, including removing tarps or other coverings, when exposure to the gas is likely.

Never work alone with fumigants, especially in enclosed areas. Arrange to be monitored at all times by another handler who has immediate access to an appropriate respirator, in case rescue is needed.

While handling a fumigant indoors or in any enclosed area, use an air-supplying respirator. In enclosed areas such as greenhouses, ship holds, railcars, bins, vaults, and chambers, there may not be enough oxygen for you to breathe. Cartridge and canister respirators will not protect you in these situations.

Some fumigants readily penetrate plastic, rubber, and leather. These fumigants may be trapped inside gloves, boots, or tightfitting coveralls and cause severe skin irritation or lead to poisoning through skin absorption. The labeling on these fumigants state the appropriate personal protective equipment to wear while handling them. Such labeling often will tell you to wear loose-fitting clothes and "breathable" footwear such as canvas or other fabric. The labeling may tell you not to wear any gloves or to wear cotton or other absorbent gloves.

## III. RESPIRATORY PROTECTION

### A. PROTECTING YOUR RESPIRATORY TRACT

The respiratory tract—the lungs and other parts of the breathing system—is much more absorbent than the skin. You must wear a respirator when the pesticide labeling directs you to do so. Even if the labeling does not require it, you should consider wearing a respiratory protective device:

- If you are in an enclosed area and the pesticide you are handling has a labeling precautionary statement such as "do not breathe vapors or spray mist," or "harmful or fatal if inhaled."
- If you will be exposed for a long time to pesticides that are in or near your breathing zone.

Some fumigants and a few other pesticide formulations contain an additive that will warn you if you begin to inhale the pesticide. Such warning agents often are used when the active ingredients in the pesticide are highly toxic ones that you would otherwise not be able to detect. The additive may have a characteristic odor or be a mild irritant to alert you to put on a respirator or that your respirator is no longer protecting you. The warning agent can help determine when you should use a respirator for products whose labeling does not require respiratory protection in all situations.

Some pesticide labeling lists the type of respirator you should wear when handling the product. Other labeling requires the use of a respirator, but does not specify the type or model to be used. NIOSH (National Institute of Safety and Health) and MSHA (Mine Safety and Health Administration) approve respirators as adequate for certain types of uses. When the pesticide labeling requires you to use a respirator, wear one that is approved by NIOSH and MSHA. If the respirator has more than one part, all the parts must be approved.

Studies have shown that many pesticide handlers do not use respirators correctly and so are not being well protected. Before using a respirator, you should be trained in the correct procedures for selecting, fitting, cleaning and sanitizing, inspecting, and maintaining respiratory protective equipment.

There are two basic types of respirators:

- **Air-supplying respirators,** which supply clean, uncontaminated air from an independent source.
- **Air-purifying respirators,** which remove contaminants from the air around you.

## 1. Air-Supplying Respirators

Air-supplying respirators are used in a few specialized situations where other types of respirators are not protective enough. Use an air-supplying respirator when the pesticide labeling tells you to do so. In addition, you should use one when handling pesticides:

- when the oxygen supply is low
- during fumigation in enclosed areas, such as greenhouses or other buildings, railcars, ship holds, or grain bins.

### a. *Supplied-Air Respirators*

These respirators pump clean air through a hose to the face mask. You are limited to working within the distance between the hose and the supply of clean air.

### b. *Self-Contained Breathing Apparatus*

This type of respirator supplies clean air from cylinders that you carry, usually on your back. This lets you move more freely and over a wider area than you can with a supplied-air respirator. Get training from competent instructors before using any type of respiratory protection, especially self-contained breathing equipment. These devices contain a limited air supply (usually about 30 minutes), which may be used up even more quickly in high temperatures or with excessive exertion. Figure 6.13 shows the different air-supplying respirators, the airline and self-contained breathing apparatus.

## 2. Air-Purifying Respirators

In most situations where pesticide handlers need to use a respirator, some type of air-purifying respirator provides enough protection. Air-purifying respirators will not protect you from fumigants, from extremely high concentrations of vapor, or when the oxygen supply is low.

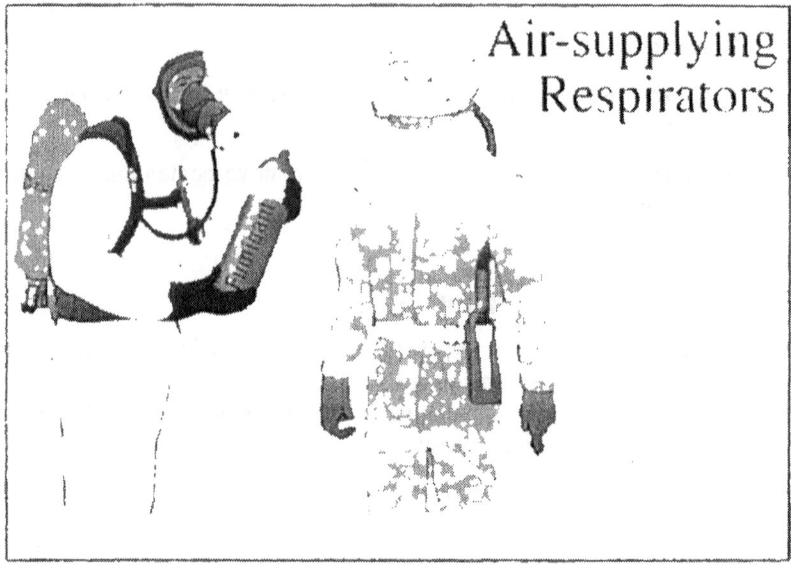


**Figure 6.13** Air-supplying respirators, airline and self-contained breathing apparatus.

*a. Functions of Air-Purifying Respirators*

Air-purifying respirators remove contaminants from the air in two ways:

- by filtering dusts, mists, and particles
- by removing gases and vapors.

Sometimes you will need only a respirator that filters dusts and mists from the air; at other times, one that removes gases and vapors as well. Figures 6.14 and 6.15 show the various APR body styles available. APRs are available in both full-face and half-mask styles.Wear a dust/mist-filtering respirator if the pesticide labeling tells you to or if you will be exposed to pesticide dusts, powders, mists, or sprays in your breathing zone. Wear a respirator that also removes vapors if the pesticide labeling tells you to or if you will be exposed to gases or vapors in your breathing zone.

*b. Styles of Air-Purifying Respirators*

Air-purifying respirators are of three basic styles:

- Dust/mist masks, which usually are shaped filters that cover the nose and mouth to filter out dusts, mists, and particles.

- Devices consisting of a body and one or more cartridges that contain air-purifying materials.
- Devices consisting of a body and a canister that contains air-purifying materials.

**Cartridges** may contain either dust/mist-filtering material or vapor-removing material. For pesticide handling tasks where vapor removal is needed, a prefilter must be used with the vapor-removing cartridge. The prefilter removes dusts, mists, and other particles before the air passes through the vapor-removing cartridge. A few vapor-removing cartridges have an attached prefilter, but most are sold separately. Separate prefilters are preferred for use with pesticides, because they often need to be replaced before the vapor-removing cartridge is used up.

Some cartridge-type respirators are one-piece units with cartridges permanently attached to the facepiece. After use, the entire unit is discarded. Other cartridge respirators are two-piece units with removable cartridges and a body that can be cleaned and reused. The dust/mist filtering or vapor-removing cartridges and the prefilters can be replaced when they lose their effectiveness. Figure 6.14 shows a worker using an air-purifying respirator-half mask. Worker is also wearing goggles with indirect air vents and a large brimmed hat for protection.

A **canister** contains both dust/mist-filtering and vapor-removing material. Canisters contain more air-purifying material than cartridges. They last much longer and may protect better in situations where the concentration of gas or vapor in the air is high. They are also much heavier and more uncomfortable to wear.

Canister-type respirators are often called gas masks. They usually have the canister connected directly to the facepiece or worn on a belt and connected to the facepiece by a flexible hose. The body is designed to be cleaned and reused. The canisters can be replaced when necessary.

### a. *Selecting and Using Dust/Mist Filtering Devices*

Dust/mist filtering masks and cartridges are approved by NIOSH and MSHA. You must wear one that has their stamp of approval. Nonapproved filters are not as protective and are not acceptable.

- Pesticide handlers must wear dust/mist-filtering masks or cartridges with NIOSH/MSHA approval number prefix TC21C.

Look for a dust/mist mask that is held in place by two straps. One-strap styles are not approved by NIOSH and MSHA, because they do not keep the respirator adequately sealed against the face.

**Figure 6.14** Air-purifying respirator-half mask. Worker is also wearing goggles with indirect air vents and a large brimmed hat for protection (courtesy of University of Wisconsin-Extension).

When wearing a dust/mist filter mask, cartridge, or prefilter, you will have more trouble breathing as more dusts, mists, and other particles become trapped in the filter material. When breathing becomes too difficult, replace the filter. Eight hours of use is usually the limit for these filters. During continual use, you may need to change filters twice a day, or even more often in dusty or dirty conditions. Do not use a dust/mist mask when the pesticide will completely soak the mask and be held close to the skin and breathing passages. Replace the mask if it gets soaked or loses its shape. Figure 6.15 illustrates the assortment of APRs available in the market place.

*b. Selecting and Using Vapor-Removing Devices*

Vapor-removing devices are rated by NIOSH for the types of gases and vapors they will remove. For pesticide-handling tasks where vapor protection is needed, NIOSH requires that an organic vapor-removing material and a pesticide prefilter be used.

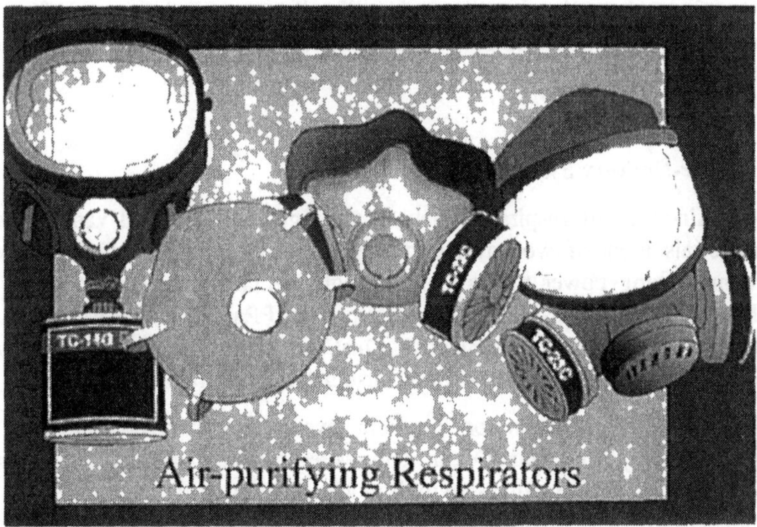


**Figure 6.15** Assorted air-purifying respirators with organic vapor-filtering cartridges (University of Wisconsin, 40-hour OSHA Training, 1991).

Pesticide handlers must use either:

- Cartridge approved for organic vapor removal plus a prefilter approved for pesticides (NIOSH/MSHA approval number prefix for both is TC23C).
- Canister approved for pesticides (NIOSH/MSHA approval number prefix is 14G).

When you wear a vapor-removing respirator, remember that vapor-removing materials gradually lose their ability to hold more gases and vapors. Their useful life can vary greatly depending on:

- the amount of particles in the air
- the concentration of vapor being filtered
- the amount of absorbent material they contain
- the breathing rate of the wearer
- the temperature and humidity
- the length of time they have been stored before use and between uses.

If you notice an odor, taste, irritation, or dizziness, that is a signal that you are no longer being protected. Some vapor-removing materials have a "service life indicator" to tell when the material is nearly depleted. The instructions on some other materials will tell you to replace them after a specific number of hours of use. If there are no instructions about replacement, change the cartridge or canister after about 8 hours of use.

### c. *Air-Delivery Systems*

Air-purifying respirators draw air through the filters and vapor-removing materials in one of two ways. Ordinary air-purifying respirators depend on the wearer's lung power to draw air through the purifying material with each breath. Powered air-purifying respirators (PAPRs) assist the wearer by pulling the air through mechanically. Dust/mist masks and most cartridge and canister respirators are nonpowered air-purifying respirators.

If you have a respiratory problem, even a temporary problem such as a cold or allergy, you cannot wear nonpowered cartridge and canister respirators. You need strong lung pressure to draw the air through the cartridges into your lungs. Even persons with normal lung capacity cannot wear these respirators for long periods of time because they tend to be hot, uncomfortable, and exhausting.

Before using these respirators, have a medical examination to make sure that you do not have a medical condition that would prevent you from using such devices. If you have trouble breathing while wearing a respirator even though you have used and cared for it correctly, see your physician to find out whether you have a health problem.

Powered air-purifying respirators use a blower to draw air to the user. PAPRs should not be confused with air-supplying respirators, because they do not supply clean air. The air is cleaned by cartridges or canisters, as it is with other air-purifying respirators. These respirators are available as lightweight backpacks, or they may be mounted on or in application equipment where the power is supplied by the vehicle's electrical systems.

### d. *Fitting Air-Purifying Respirators*

Respirators fit wearers in one of two ways. Most must seal tightly to the face; others are loose-fitting.

**Face-sealing respirators** must form a tight seal against your face to be effective. Otherwise, pesticides can leak in around the edges. People with beards cannot wear this style of respirator, because a tight seal cannot be formed through the hair. These respirators must be fitted to each wearer and are not interchangeable among handlers.

Dust/mist masks are face-sealing respirators. They fit over your nose and mouth and have a clip that you press around the bridge of your nose to help form a seal. Most cartridge and canister respirators are also face-sealing respirators. Full-face styles form and keep a tight seal better than half-face styles.

Many pesticide handlers are not adequately protected while wearing face-sealing cartridge and canister respirators, because they often break the seal by pulling the respirator away from their face to get temporary relief from the heat, sweat, itching, or difficult breathing. Once the seal is broken in the exposure area, the respirator's ability to protect you is greatly reduced. Face-sealing cartridge and canister respirators are most useful for short-term tasks.

Your face-sealing respirator should be tested before wearing it in a situation where you may inhale pesticides. There are two types of fit tests: qualitative and quantitative. They ensure that the respirator is operating correctly and that you are being protected.

Have a **fit test** before using your cartridge or canister respirator the first time, and then be retested periodically. Get the fit test through a program approved by NIOSH and OSHA, the agencies that regulate respirator fit testing. The Cooperative Extension Service in your locality may be able to tell you where to find an approved fit testing program. Figure 6.16 shows a qualitative fit test.

The two main types of fit tests are:

- qualitative testing—whether the wearer can detect a test substance by irritation, odor, or taste
- quantitative testing—measuring the actual amount of a test substance that gets inside the facepiece.

A **fit check** is an on-the-spot check that you should do to make sure the respirator is still working correctly. Do a fit check each time you wear a face-sealing respirator.

There are two methods for checking the seal of the facepiece against your face. To check by the first method:

- Close off the inlet of the canister or cartridge (cover it with your palm, replace the caps, or squeeze the breathing tube so that it does not allow air to pass).
- Inhale gently so that the facepiece collapses slightly.
- Hold your breath for about 10 seconds.

If the facepiece remains slightly collapsed and no inward leakage is detected, the respirator probably fits tightly enough and will work correctly. This method does not work for dust/mist masks.

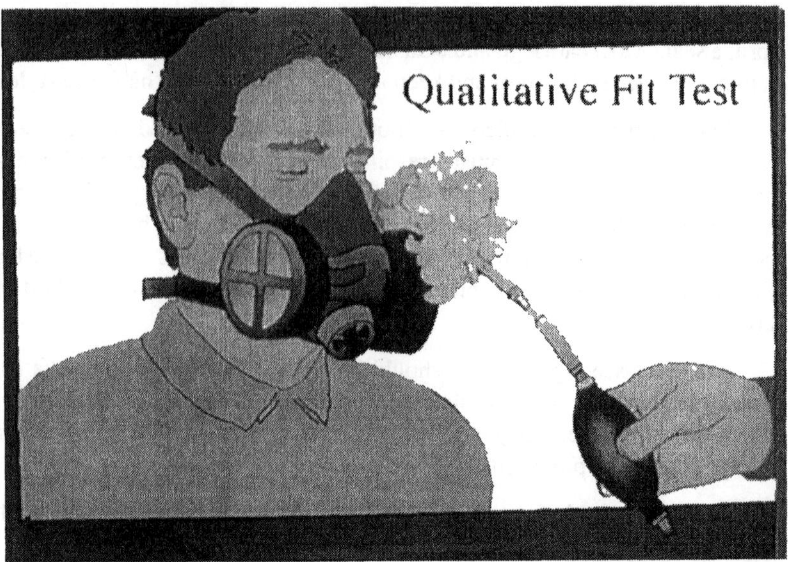


**Figure 6.16** Qualitative fit testing is performed every time a mask is repaired or new mask is used (University of Wisconsin, 40-hour OSHA training, 1991).

The second method for testing the facepiece seal is to close the exhalation valve with your palm and exhale gently into the facepiece. If slight pressure builds up inside the facepiece without any evidence of outward leakage, the respirator probably fits tightly enough and will work correctly. This method is not appropriate for respirators with an exhalation valve cover that would have to be removed first. Figure 6.17 illustrates a negative pressure fit check.

**Loose-fitting respirators** are powered air-purifying respirators that constantly pump air through a cartridge or canister into a loosefitting helmetlike or hoodlike head covering. The positive outward pressure caused by the steady outflow of air prevents contaminants from entering the headpiece. The purified air circulates over the user's head, face, and neck and provides some cooling.

Not all loose-fitting respirators move the air at the same rate. Most pesticide handling tasks require a minimum airflow rate of 4 cubic feet per minute. If you are doing physically strenuous work, use a respirator with an airflow rate of at least 6 cubic feet per minute.

Loose-fitting respirators do not have to form a seal on your face, so people with facial hair can use them safely. They do not require extra lung power and are not nearly as tiring or as hot as face-sealing respirators.

Loose-fitting respirators are much more expensive than face-sealing respirators. In some situations, however, they are the only safe option. For example, you might have to use one if you have facial hair that prevents an adequate seal with the respirator facepiece.

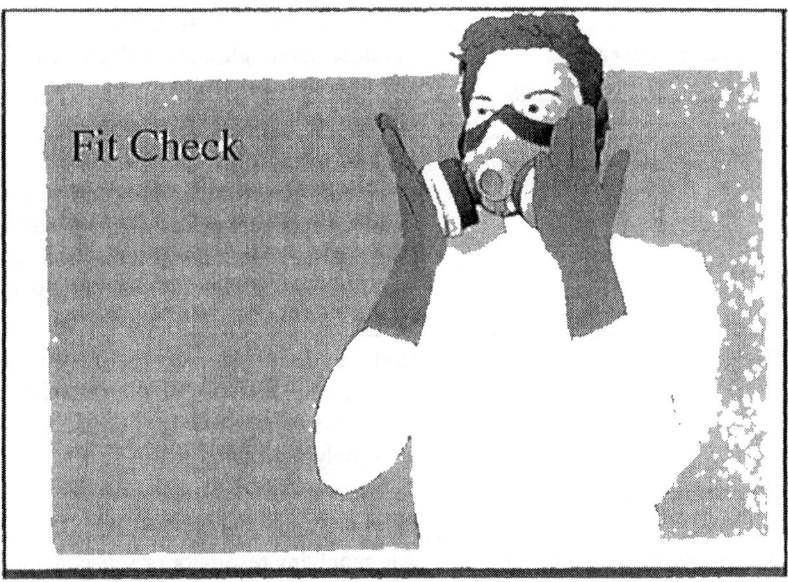


**Figure 6.17** Fit check should be performed every time you use your respirator. If leak is detected, a further examination should be performed (University of Wisconsin, 40-hour OSHA training, 1991).

In many situations, loose-fitting respirators are a good choice. For example, you might choose to use one:

- to avoid the need for fit tests and fit checks
- if you will be exposed to pesticides for several hours at a time
- if you are working in situations where heat stress is a concern.

Table 6.11 lists the acceptable PPE that should be used for each labeling statement on the pesticide container.

**Table 6.11** Interpreting labeling PPE statements.

| Labeling Statement | Acceptable PPE |
|---|---|
| Chemical-resistant hood or wide-brimmed hat | Rubber- or plastic-coated safari-style hat, rubber- or plastic-coated firefighter-style hat, plastic- or other barrier-coated hood, rubber or plastic hood, or full hood or helmet that is part of some respirators |
| Protective eyewear | Shielded safety glasses, or shield, goggles, or full-face style respirator |
| Goggles | Goggles or full-face style respirator |
| Dust/mist filtering respirator | Dust/mist respirator, respirator with dust/mist filtering cartridge, respirator with organic vapor-removing cartridge and pesticide prefilter, respirator with canister approved for pesticides, or air-supplying respirator |
| Cartridge respirator | Respirator with organic vapor-removing cartridge and pesticide prefilter, respirator with canister approved for pesticides, or air-supplying respirator |
| Canister respirator (gas mask) | Respirator with canister approved for pesticides or air-supplying respirator |
| Air-supplying respirator or self-contained breathing apparatus (SCBA) | Air-supplying respirator or self-contained breathing apparatus (SCBA) |

## IV. DISPOSABLES AND REUSABLES

Personal protective equipment items either should be disposable or should be easy to clean and sturdy enough for repeated use.

### A. DISPOSABLES

Disposable personal protective equipment items are not designed to be cleaned and reused. Discard them when they become contaminated with pesticides.

Chemical-resistant gloves, footwear, and aprons that are labeled as disposable are designed to be worn only once and then thrown away. These items often are made of thin vinyl, latex, or polyethylene. These inexpensive disposables may be a good choice for brief pesticide handling activities that require flexibility and will not tear the thin plastic.

For example, you might use disposable gloves, shoe covers, and apron while pouring pesticides into a hopper or tank, cleaning or adjusting a nozzle, or making minor equipment adjustments.

Nonwoven (including coated nonwoven) coveralls and hoods usually are designed to be disposed of after use. Most are intended to be worn for only one workday's exposure period. The instructions with some coated nonwoven suits and hoods may permit you to wear them more than once if each period of use is short and if they are not contaminated or torn. Be especially alert when reusing these items, and be ready to change them whenever there are signs that pesticides could be getting through the material or that the inside surface is contaminated. Figure 6.18 shows two different types of Tyvek® coveralls that are disposable.

**Figure 6.18** Disposable coveralls made of Tyvek®. Tyvek® is a DuPont registered trademark for its brand of spunbonded olefin. Only DuPont makes Tyvek®.

Dust/mist masks, prefilters, canisters, filtering and vapor-removing cartridges, and a few cartridge respirators are disposables. They cannot be cleaned, and they should be replaced often.

## B. REUSABLES

Some personal protective equipment may be designed to be cleaned and reused several times. However, do not make the mistake of reusing these items when they are no longer protecting you.

Rubber, rubber-like, and plastic suits, gloves, boots, aprons, capes, and headgear often are designed to be cleaned and reused, but even these reusables should be replaced often. Wash them thoroughly between uses. Before you put them on, inspect reused items carefully for signs of wear or abrasion. If they show any sign of wear, throw them out. Even tiny holes or thin places can allow large quantities of pesticides to move to the inside surface and transfer to your skin. Check for rips and leaks during cleaning by holding the items up to the light.

Even if you can see no signs of wear, replace reusable chemical-resistant items regularly. The ability of a chemical-resistant material to resist the pesticide decreases each time the items are worn and after repeated exposure to pesticides. Even though you do not see any changes in the material, the pesticide may be moving through the material and getting on your skin. The pesticide moves through the material in the same way air leaks through the surface of a balloon—slowly, but steadily.

A good rule of thumb is to throw out gloves that have been worn for about 5 to 7 days of work. Extra-heavy-duty gloves, such as those made of butyl or nitrile rubber, may last as long as 10 to 14 days. Because hand protection is the most important concern for pesticide handlers, make glove replacement a high priority. The cost of frequently replacing gloves is a prudent investment. Footwear, aprons, headgear, and protective suits may last longer than gloves, because they generally receive less exposure to the pesticides and less abrasion from rough surfaces. However, they should be replaced regularly and at any sign of wear.

Fabric coveralls are designed to be cleaned after each day's use and reused. However, absorbent materials such as cotton, polyester, cotton blends, denim, and canvas cannot be cleaned adequately after they are drenched or thoroughly contaminated with concentrated pesticides labeled with the signal word "DANGER" or "WARNING." Always discard any such clothing or footwear. They cannot be safely reused.

Most protective eyewear and respirator bodies, facepieces, and helmets are designed to be cleaned and reused. These items may last many years if they are good quality and are maintained correctly.

## C. MAINTAINING PERSONAL PROTECTIVE EQUIPMENT

When you finish an activity where you are handling pesticides or are exposed to them, remove your personal protective equipment right away. Wash the outside of your gloves with detergent and water before you remove them. Consider washing the outside of other chemical-resistant items as well before removing them. This helps avoid contact with the contaminated part of the items while you are removing them and helps keep the inside surface uncontaminated. If any other clothes have pesticides on them, change them also. Determine whether the items should be disposed of or cleaned for reuse.

Place reusable items in a plastic bag or hamper away from your other personal clothes and away from the family laundry. Place disposables in a separate plastic bag or container. The pesticides remaining on your personal protective equipment, work clothing, and other work items could injure persons who touch them. Do not allow children or pets near them. Do not allow contaminated gloves, boots, respirators, or other equipment to be washed in streams, ponds, or other bodies of water.

Clean all reusable personal protective equipment items between uses. Even if they were worn for only a brief period of exposure to pesticides during that day, wash them before you wear them again. Pesticide residues that remain on the personal protective equipment are likely to continue to move slowly through the personal protective equipment material, even chemical-resistant material. If you wear the personal protective equipment again, pesticide may already be on the inside next to your skin. Also, personal protective equipment that is worn several times between laundering may build up pesticide residues. The residues can reach a level that can harm you, even if you are handling pesticides that are not highly toxic.

### 1. Washing Personal Protective Equipment

Wash pesticide-contaminated items separately from uncontaminated clothing and laundry. Otherwise, the pesticide residues can be transferred to the other clothing or laundry and can harm you or your family.

#### a. *Alert the Persons Who Do the Washing*

Be sure that the people who clean and maintain your personal protective equipment and other work clothes know that they can be harmed by touching the pesticide that remains on the contaminated items. Tell them that they should:

- Wear gloves and an apron, especially if handling contaminated items regularly or handling items contaminated with highly toxic pesticides.
- Work in a well-ventilated area, if possible, and avoid inhaling steam from the washer or dryer.

*b. Washing Procedure*

Follow the manufacturer's instructions for cleaning chemical-resistant items. If the manufacturer instructs you to wash the item but gives no detailed instructions, or offers no cleaning instructions at all, follow the procedure below. Some chemical-resistant items that are not flat, such as gloves, footwear, and coveralls, must be washed twice—once to thoroughly clean the outside of the item and a second time after turning the item inside out. Some chemical-resistant items, such as heavy-duty boots and rigid hats or helmets, can be washed by hand using hot water and a heavy-duty liquid detergent. They should be dried and aired as directed below.

The best procedure for washing nonchemical-resistant items, such as cotton, cotton/polyester, denim, canvas, and other absorbent materials, and most chemical-resistant items is:

1. Rinse in a washing machine or by hand.
2. Wash only a few items at a time so there will be plenty of agitation and water for dilution.
3. Wash in a washing machine, using a heavy-duty liquid detergent and hot water for the wash cycle.
4. Rinse twice using two entire rinse cycles and warm water.
5. Use two entire machine cycles to wash items that are moderately to heavily contaminated.
6. Run the washer through at least one additional cycle without clothing, using detergent and hot water, to clean the machine after each batch of pesticide contaminated items and before any other laundry is washed.

*c. Drying Procedure*

Hang the items to dry, if possible. It is best to let them hang for at least 24 hours in an area with plenty of fresh air. Even after thorough washing, some items still may contain pesticides. When the items are exposed to clean air, remaining pesticide residues move to the surface and evaporate. You may wish to buy two or more sets of equipment at a time so you can leave one set airing in a clean place while using the other set. Do not hang items in enclosed living areas, because pesticides that remain in the items may evaporate and expose people or animals in the area.

Using a clothes dryer is acceptable for fabric items, if it is not possible to hang them to dry. However, over a period of time, the dryer may become contaminated with pesticide residues.

## 2. Maintaining Eyewear and Respirators

Wash goggles, face shields, shielded safety glasses, and respirator bodies and facepieces after each day of use. Use a detergent and hot water to wash them thoroughly. Sanitize them by soaking for at least 2-10 minutes in a mixture of two tablespoons of chlorine bleach in a gallon of hot water. Rinse thoroughly to remove the detergent and bleach. Dry thoroughly or hang them in a clean area to dry.

Pay particular attention to the headbands. Headbands made of absorbent materials should be replaced with chemical-resistant headbands. After each day of use, inspect all headbands for signs of wear or deterioration and replace as needed.

Store respirators and eyewear in an area where they are protected from dust, sunlight, extreme temperatures, excessive moisture, and pesticides or other chemicals. A zip-closable sturdy plastic bag works well for storage.

Respirator maintenance is especially important. Inspect your respirator before each use. Repair or replace it whenever any part shows sign of wear or deterioration. Maintain an inventory of replacement parts for the respirators you own, and do not try to use makeshift substitutes or incompatible brands. If you keep a respirator for standby or emergency use, inspect it at least monthly and before use.

If you remove your respirator between handling activities:

- Wipe the respirator body and facepiece with a clean cloth.
- Replace caps, if available, cover cartridges, canisters, and prefilters.
- Seal the entire respirator in a sturdy, airtight container, such as a zip-closable plastic bag. If you do not seal the respirator immediately after each use, the disposable parts will have to be replaced more often. Cartridges, canisters, prefilters, and filters will continue to collect impurities as long as they are exposed to the air.

At the end of any work day when you wore a reusable respirator:

- Remove the filter or prefilter. Most filters should be discarded. A few are designed to be washed and reused.
- Take off the cartridges or canisters. Discard them or, if still usable, replace their caps and seal them in an airtight container, such as a zip-closable plastic bag.
- Clean and store respirator as directed above. Discard disposable respirators according to manufacturer's instructions. Do not try to clean them.

# V. PESTICIDE STORAGE, DISPOSAL, AND SPILL MANAGEMENT

## A. PESTICIDE STORAGE

Many pesticide formulators and handlers use existing buildings or areas within existing buildings for pesticide storage. However, if large amounts of pesticides will be stored, it is best to build a special storage building just for pesticide needs. Figure 6.19 shows two storage facilities specially designed for the storage of hazardous chemicals and pesticides.

### 1. Selecting and Designing a Storage Site

A correctly designed and maintained pesticide storage site is essential. A suitable storage site:

- protects people and animals from accidental exposure
- protects the environment from accidental contamination
- prevents damage to pesticides from temperature extremes and excess moisture
- protects the pesticides from theft, vandalism, and unauthorized use
- reduces the likelihood of liability.

#### *a. Security at the Site*

Keeping out unauthorized people is an important function of the storage site. Whether the storage site is as small as a cabinet or closet or as large as an entire room or building, keep it secure. Post signs on doors and windows to alert people that pesticides are stored there. Post "No smoking" warnings.

*b. Prevent Water Damage*

Choose a storage site where water damage is unlikely to occur. Water from burst pipes, spills, overflows, excess rain or irrigation, or flooding streams can damage pesticide containers and pesticides. Water or excess moisture can cause:

- metal containers to rust
- paper and cardboard containers to split or crumble
- pesticide labeling to peel, smear, run, or otherwise become unreadable
- dry pesticides to clump, degrade, or dissolve
- slow-release products to release their pesticide
- pesticides to move from the storage site into other areas.

If the storage site is not protected from the weather or if it tends to be damp, consider placing metal, cardboard, and paper containers in sturdy plastic bags or cans for protection. Large metal containers, which may rust when damp, often can be placed on pallets within the storage site.

*c. Control the Temperature*

The storage site should be indoors, whenever possible. Choose a cool, well-ventilated room or building that is insulated or temperature-controlled to prevent freezing or overheating. The pesticide labeling may tell at what temperatures the product should be stored. Freezing temperatures can cause glass, metal, and plastic containers to break. Excessive heat can cause plastic containers to melt, some glass containers to explode, and some pesticides to volatilize and drift away from the storage site. Temperature extremes can destroy the potency of some pesticides.

*d. Provide Adequate Lighting*

The storage site should be well lighted. Pesticide handlers using the facility must be able to see well enough to:

- read pesticide container labeling
- notice whether containers are leaking, corroding, or otherwise disintegrating
- clean up spills or leaks completely.

**Figure 6.19** Hazardous chemical storage facilities. (University of Wisconsin, 40-hour OSHA training, 1991.)

### *e. Use Nonporous Materials*

The floor of the storage site should be made of sealed cement, glazed ceramic tile, no-wax sheet flooring, or another easily cleaned material. Carpeting, wood, soil, and other absorbent floors are difficult or impossible to decontaminate in case of a leak or spill. For ease of cleanup, shelving and pallets should be made of nonabsorbent materials such as plastic or metal. If wood or fiberboard materials are used, they should be coated or covered with plastic or polyurethane or epoxy paint.

### *f. Prevent Runoff*

Inspect the storage site to determine the likely path of pesticides in case of spills, leaks, drainage of equipment wash water, and heavy pesticide runoff from firefighting or floods. Pesticide movement away from the storage site could contaminate sensitive areas, including surface water or groundwater. If your storage site contains large amounts of pesticides, you may need to use a collection pad to contain pesticide runoff. Figure 6.20 shows a typical collection pad.

**Figure 6.20** Pesticide storage containment. Chemicals are surrounded by a cement curb to contain any spilled material (courtesy of University of Wisconsin-Extension).

*g. Provide Clean Water*

Each storage site must have an immediate supply of clean water. Potable running water is ideal. If running water is not practical, use a carboy or other large, sealable container with clean water. Changing the water in a container at least once each week will ensure that it remains safe for use on skin and eyes. Keep an eyewash dispenser immediately available for emergencies.

## 2. Maintain the Storage Site

*a. Prevent Contamination*

Store only pesticides, pesticide containers, pesticide equipment, and a spill cleanup kit at the storage site. Do not keep food, drinks, tobacco, feed, medical or veterinary supplies or medication, seeds, clothing, or personal protective equipment (other than personal protective equipment necessary for emergency response) at the site. These could be contaminated by vapors, dusts, or spills and cause accidental exposure to people or animals.

*b. Keep Labels Legible*

Store pesticide containers with the label in plain sight. Costly errors can result if the wrong pesticide is used by mistake. Labels should always be legible. They may be damaged or destroyed by exposure to moisture, dripping pesticide, diluents, or dirt. You can use transparent tape or a coating of lacquer or polyurethane to protect the label. If the label is destroyed or damaged, request a replacement from the pesticide dealer or the pesticide formulator immediately.

*c. Keep Containers Closed*

Keep pesticide containers securely closed whenever they are being stored. Tightly closed containers help protect against:

- a spill
- cross-contamination with other stored products
- evaporation of liquid pesticides or the solvent
- clumping or caking of dry pesticides in humid conditions
- dust, dirt, and other contaminants getting into the pesticide, causing it to be unusable.

**Figure 6.21** Shelf storage of pesticides. Packages labeled and labels visible.

*d. Use Original Containers*

Store pesticides in their original containers. Never put pesticides in containers that might cause children and other people to mistake them for food or drink. You are legally responsible if someone or something is injured by pesticides you have placed in unlabeled or unsuitable containers.

*e. Watch For Damage*

Inspect containers regularly for tears, splits, breaks, leaks, rust, or corrosion. When a container is damaged, put on appropriate personal protective equipment and take immediate action. If the damaged container is an aerosol can or fumigant tank that contains pesticides under pressure, use special care to avoid accidentally releasing the pesticide into the air. When a container is damaged:

- Use the pesticide immediately at a site and rate allowed by the label.
- Transfer the pesticide into another pesticide container that originally held the same pesticide and has the same label still intact.
- Transfer the contents to a sturdy container that can be tightly closed. If possible, remove the label from the damaged container and use it on the new container. Otherwise, temporarily mark the new container with the name and EPA registration number of the pesticide,

and get a copy of the label from the pesticide dealer or formulator (whose telephone number is usually on the label) as soon as possible.

- Place the entire damaged container and its contents into a suitable larger container. Consider this option carefully, however. Many times the label on the leaking container becomes illegible. The pesticide is useless and becomes a disposal problem unless you know the name and registration number and can get a copy of the label.

*f. Store Volatile Products Separately*

Volatile pesticides, such as some types of 2,4-D, should be stored apart from other types of pesticides and other chemicals. A separate room is ideal. Vapors from opened containers of these pesticide can move into other nearby pesticides and chemicals and make them useless. The labeling of volatile herbicides usually will direct you to store them separately from seeds, fertilizers, and other types of pesticides.

*g. Isolate Waste Products*

If you have pesticides and pesticide containers that are being held for disposal, store them in a special section of the storage site. Accidental use of pesticides meant for disposal can be a costly mistake. Clearly mark containers that have been triple rinsed or cleaned by an equivalent method, because they are more easily disposed of than unrinsed containers.

*h. Know Your Inventory*

Keep an up-to-date inventory of the pesticides you have in storage. Each time a pesticide is added to or removed from the storage site, update the inventory list. The list will help keep track of your stock and will be essential in a fire or flood emergency. The inventory list also will aid in insurance settlements and in estimating future pesticide needs.

Do not store unnecessarily large quantities of pesticides for long periods of time. Buy only as much as you will need for a year at most. Pests, pesticides, or pesticide registrations may change by the next year and make the pesticides useless. Some pesticides have a relatively short shelf life and cannot be carried over from year to year.

*i. Consider Shelf Life*

Mark each pesticide container with the date of purchase before it is stored. Use older materials first. If the product has a shelf life listed in the labeling, the purchase date will indicate whether it is still usable. Excessive clumping, poor suspension, layering, or abnormal coloration may be indications that the pesticide has broken down. However, sometimes pesticide deterioration from age or poor storage conditions becomes obvious only after application. Poor pest control or damage to the treated surface can occur. If you have doubts about the shelf life of a pesticide, call the dealer or manufacturer for advice.

## 3. Prevent Pesticide Fires

Some pesticides are highly flammable; others do not catch fire easily. The labeling of pesticides that require extra precautions often will contain a warning statement in either the "Physical/Chemical Hazards" section or the "Storage and Disposal" section. Pesticides that contain oils or petroleum-based solvents are the ones most likely to contain these warning statements. Some dry products also present fire and explosion hazards.

Store flammable and combustible pesticides away from open flames and other heat sources, such as steam lines, heating systems, kerosene heaters or other space heaters, gas powered equipment, or incinerators. Do not store glass containers in sunlight where they can focus the heat rays and possibly explode or ignite. Install fire detection systems in large storage sites, and equip each storage site with a working fire extinguisher that is approved for all types of fires, including chemical fires.

If you store highly toxic pesticides or large amounts of any pesticides, inform your local fire department, hospital, public health officials, and police of the location of your pesticide storage building before a fire emergency occurs. Tell fire department officials what types of pesticides are regularly stored at the site, give them a floor plan, and work with them to develop an emergency response plan.

## 4. Labeling Statements About Storage

Typical pesticide labeling instructions about storage include:

- Store at temperatures above 32°F.
- Do not contaminate feed, foodstuffs, or drinking water during storage.
- Store in original container only.
- In outside storage areas, store drums on sides to avoid accumulation of rain water in top or bottom of recessed areas.

- Do not store near ignition sources such as electrical sparks, flames, or heated surfaces.
- Do not use, pour, spill, or store near heat or open flame. Do not cut or weld container.

## REGULATORY COMPLIANCE

Some pesticide applicators, application businesses, and dealers may be affected by Title III of the Superfund Amendments and Reauthorization Act of 1986 (SARA Title III), administered by the Environmental Protection Agency. SARA Title III has three sections that relate to the storage of pesticides:

### Emergency Planning and Notification

Under certain conditions, the law requires you to notify State and local officials about the location and amount of hazardous chemicals at your site. EPA has assigned a Threshold Planning Quantity (TPQ) for a number of active ingredients (not total weight of formulated product). When the product in storage is at or above the TPQ, you must notify the State Emergency Response Commission (SERC) in writing. Each facility must designate a coordinator to work with the Local Emergency Planning Committee (LEPC). The State will notify the LEPC that your operation is covered under SARA. This is a one-time notification.

### Material Safety Data Sheet (MSDS) Reporting

Employers are required to obtain and keep material safety data sheets. They must submit copies of each MSDS (or a listing of MSDSs that must be maintained) to their local fire department, the LEPC, and the SERC. There is one exclusion: if a chemical is used solely for household, consumer, or agricultural purposes, notification is not required. However, under OSHA regulations, pesticide users (except homeowners) must have an MSDS for each pesticide they handle.

### Annual Inventory Reporting

All regulated facilities must submit an annual chemical inventory to the local fire department, LEPC, and SERC. This inventory must include:

- all hazardous chemicals stored at the facility in quantities of 10,000 pounds or more
- all extremely hazardous chemicals stored in quantities of 500 pounds (or 55 gallons) or more, or in a quantity that exceeds the TPQ, whichever is less.

Agricultural producers are exempt from this section.

## B. PESTICIDE DISPOSAL

Pesticide users are responsible for properly dealing with empty pesticide containers, excess usable pesticides, and waste materials that contain pesticides or their residues. There is growing concern about the serious harm to humans and the environment that improper disposal of pesticide wastes can cause. For information on disposal options available in your local area, contact your State or tribal pesticide authority.

### 1. Excess Pesticides

The best solution to the problem of what to do with excess pesticides is to take steps to avoid having them:

- Buy only the amount needed for a year or a season.
- Calculate carefully how much diluted pesticide is needed for a job and mix only that amount.
- Use all the mixed pesticide in accordance with labeling instructions.

If you have excess pesticides that are usable, first try to find a way to use them as directed on the label. The best option is to apply the pesticide on a site listed in the use directions on the pesticide labeling, under the following conditions:

- The total amount of pesticide active ingredient applied to the site, including all previous applications, must not exceed the rate and frequency allowed on the labeling.
- You must comply with other application instructions specified on the labeling.

If you have pesticide products in their original containers that you cannot use, you may be able to find another pesticide handler who can. Or you may be able to return them to a dealer, formulator, or manufacturer.

Most container rinsates should not become excess pesticides, because they can be added into the tank during the mixing process. You also may be able to add some rinsates from equipment cleaning, spill cleanup, and other activities to a tank mixture that contains the same pesticide, as long as doing so will not violate labeling instructions. Some rinsates will contain dirt, cleaning agents, or other substances that will make them unusable, however.

**Figure 6.22** Bulk storage facilities (courtesy of Growmark).

## 2. Pesticide Wastes

Excess pesticides and rinsates that cannot be used must be disposed of as wastes. Other pesticide wastes include such things as contaminated spill cleanup material and personal protective equipment items that cannot be cleaned and reused. Whenever possible, avoid creating pesticide wastes that require disposal.

Sometimes pesticide wastes can be disposed of in a landfill operating under EPA, state, tribal, or local permit for hazardous wastes. Most sanitary landfills are not suitable. Some regions have pesticide incinerators for disposing of pesticide wastes. Never burn, bury, or dump excess pesticides, and never dispose of them in a way that will contaminate public or private groundwater or surface water or sewage treatment facilities.

Pesticide wastes that cannot be disposed of right away should be marked to indicate the contents and then stored safely and correctly until disposal is possible.

**Figure 6.23** Improper disposal of empty pesticide containers and wastes.

*a. Labeling Statements About Waste Disposal*

Typical pesticide labeling instructions about disposal of pesticide wastes include:

- Do not contaminate water by disposal of wastes.
- Pesticide wastes are toxic. Improper disposal of excess pesticide is a violation of federal law. If these wastes cannot be disposed of by use according to label instructions, contact your State Pesticide or Environmental

Control Agency, or the Hazardous Waste representative at the nearest EPA Regional Office for guidance.

## 3. Containers

Try to avoid the need to dispose of pesticide containers as wastes. For example, you may be able to:

- use containers that are designed to be refilled by the pesticide dealer or the chemical company
- arrange to have the empty containers recycled or reconditioned
- use soluble packaging.

If you have containers that must be disposed of, be sure to rinse them, if possible. Rinsed containers are easier to dispose of than unrinsed containers.

### a. *Refillable Containers*

Some types of containers are designed to be refilled with pesticide repeatedly during their lifetime, which may be many years. They usually are not designed to be triple rinsed or pressure rinsed by the pesticide user. When necessary, they are cleaned by the pesticide dealer or chemical company before refilling. Common types of refillable containers include minibulks and small-volume returnables.

### b. *Recyclable and Reconditionable Containers*

You may be able to take your rinsed metal or plastic containers to a facility that can recycle them. Some 55- and 30-gallon drums can be returned to the dealer, manufacturer, or formulator to be reconditioned and reused.

### c. *Soluble Containers*

Soluble containers are designed to be placed, unopened, into the mixing tank. The container dissolves in the solvent (usually water) in the tank. Only the overpackaging remains, and it may be disposed of as nonhazardous waste in a sanitary landfill.

### d. *Triple-Rinsed or Pressure-Rinsed Containers*

Containers that have been correctly triple-rinsed or pressure-rinsed usually may be disposed of as regular trash in a sanitary landfill, unless prohibited by the pesticide labeling or by state, tribal, or local authorities. Mark the containers to show that they have been rinsed.

*e. Unrinsed Containers*

To dispose of unrinsed containers, you may take them to an incinerator or landfill operating under EPA, state, or tribal permit for hazardous waste disposal. If this is not possible, check with your state, tribal, or local authorities to find out what to do. Otherwise, you may have to pay a disposal company to accept them for proper disposal, or you may need to store the containers until you have a way to dispose of them.

*f. Burnable Containers*

The labeling of some paper, cardboard, and plastic containers may list "burning, if allowed by state and local authorities" as a disposal option for pesticide containers. However, open burning of pesticide containers and waste pesticides is a questionable practice and may be in violation of federal regulations that could take precedence over the instructions on the pesticide labeling. Because of possible air pollution hazard and the risks of liability, your best option is to use another disposal method for these containers.

*g. Labeling Statements about Container Disposal*

Typical pesticide labeling instructions about disposal of pesticide containers include:

- Do not reuse empty containers.
- Offer for recycling or reconditioning, or puncture and dispose of in a sanitary landfill or incinerate.
- Dispose of bag in a sanitary landfill or by incineration.

**REGULATORY COMPLIANCE**

The EPA regulates wastes under the Resource Conservation and Recovery Act (RCRA). EPA issues a list of materials that are considered hazardous. However, RCRA applies to certain flammable, corrosive, reactive, or toxic wastes, even if they are not on the list. Therefore, some other pesticides could be "regulated hazardous wastes" under RCRA. States and tribes often have their own hazardous waste laws, which may be more stringent than RCRA. Contact your state or tribal authority for applicable requirements.

"Wastes" include rinsed containers, excess pesticides, pesticide dilutions and rinse and wash water that contain a listed chemical and cannot be used. Triple-rinsed pesticide containers are not considered hazardous waste under RCRA and can be disposed of in sanitary landfills.

RCRA regulates pesticide users who accumulate wastes of acutely toxic pesticides totaling 2.2 pounds or more a month or wastes of any RCRA-regulated pesticides totaling 2,200 pounds per month. Such users must register as a generator of hazardous waste and obtain an ID number from EPA, state, or tribe and follow certain disposal requirements. To find out if a pesticide is listed in RCRA, call the EPA RCRA Hotline 1-800-424-9346.

## C. SPILL MANAGEMENT

A spill is any accidental release of a pesticide. As careful as people try to be, pesticide spills can and do occur. The spill may be minor, involving only a dribble from a container, or it may be major, involving large amounts of pesticide or pesticide-containing materials such as wash water, soil, and absorbents.

You must know how to respond correctly when a spill occurs. Stopping large leaks or spills is often not simple. If you cannot manage a spill by yourself, get help. Even a spill that appears to be minor can endanger you, other people, and the environment if not handled correctly. Never leave a spill unattended. When in doubt, get assistance.

You can get help from Chemtrec (Chemical Transportation Emergency Center) by calling 1-800-424-9300. Use this number only for emergencies.

The faster you can contain, clean-up, and dispose of a spill, the less chance there is that it will cause harm. Clean up most spills immediately. Even minor dribbles or spills should be cleaned up before the end of the work day to keep unprotected persons or animals from being exposed.

A good way to remember the steps for a spill emergency is the "three Cs": Control, Contain, Clean-up.

### 1. Control the Spill Situation

#### a. *Protect Yourself*

Put on appropriate personal protective equipment before contacting the spill or breathing its fumes. If you do not know how toxic the pesticide is or what type of personal protective equipment to wear, don't take a chance. Wear a chemical suit or apron, footwear, and gloves, eye protection, and a respirator.

*b. Stop the Source*

If a small container is leaking, place it into a larger chemical-resistant container, such as a plastic drum or bag. If a spray tank is overflowing, stop the inflow and try to cap off the tank. If a tank, hopper, or container has burst or has tipped over and is too heavy to be righted, you will not be able to stop the source.

*c. Protect Others*

Isolate the spill site by keeping children, other unprotected people, and animals well back. Rope off the site if necessary. If you suspect the spill contains a highly volatile or explosive pesticide, you may need to keep people back even farther. Warn people to keep out of reach of any drift or fumes. Do not use road flares or allow anyone to smoke if you suspect the leaking material is flammable.

*d. Stay At The Site*

Do not leave the spill site until another knowledgeable and correctly protected person arrives. Someone should be at the spill site at all times until the spill is cleaned up.

**2. Contain the Spill**

*a. Confine the Spill*

As soon as the source of the leak is under control, move quickly to keep the spill in as small an area as possible. Do everything you can to keep it from spreading or getting worse. For small spills, use containment snakes to surround the spill and keep it confined. For larger spills, use a shovel, a rake, or other tool or equipment to make a dike of soil, sod, or absorbent material.

*b. Protect Water Sources*

Keep the spill out of any body of water or any pathway that will lead to water, such as a ditch, floor drain, well, or sinkhole. If the spilled pesticide is flowing towards such an area, block it or redirect it.

**Figure 6.24** Containing the spill with absorbent materials (EPA, *Protect Yourself from Pesticides*, 1993).

*c. Absorb Liquids*

Liquid pesticide spills can be further contained by covering the entire spill with absorbent materials, such as spill pads, blankets or pillows, absorbent clay, vermiculite, sawdust, corn cobs, expanded silica, or other acceptable material.

*d. Spill-Absorbent Materials*

Many different products are in the marketplace. Most of these materials have been in use for many years, and many will continue to be of value in the remediation of spilled materials. Since the purpose of this guide is to give the user enough information to respond quickly, effectively, and safely to most common incidents, we will briefly discuss only a few of the many different materials that will serve the above purposes. For more information, the user can refer to the reference listings.

1. Diatomite (diatomaceous earth)

This is a natural inorganic material composed of the silica skeletons of microorganisms that lived millions of years ago. This material is mined in the same manner as surface coal and is commonly used to remediate spills of aqueous and nonaqueous or oil-based chemicals.

Diatomite is incompatible with and therefore should not be used with the following:

- inorganic acids
- bases
- cyanates and isocyanates
- inorganic halides
- heavy metals.

The sorption capacity of diatomite (the ratio of lbs of sorbent required to clean up lbs of spilled liquid) ranges from 1.1 for aliphatic hydrocarbons such as isoprene to 2.5 for aliphatic halogenated hydrocarbons such as trichloroethylene. The sorption capacity of diatomite is rather poor. However, this material is relatively inexpensive and can be purchased for approximately $1.50/lb, depending upon quantity ordered.

As a particulate, the material is extremely dusty, and a dust respirator and eye protection should be worn during its application.

2. Clay (sorbent clay)

This is also a natural substance composed of both inorganic and organic materials. This material has also been used for many years to remediate spills of aqueous and nonaqueous or oil-based chemicals.

Clay is incompatible with the following:

- inorganic acids
- cyanates and isocyanates
- inorganic halides
- heavy metals

The sorption capacity of clay ranges from 1.0 for alcohols and basic compounds such as allyl alcohol and ammonium hydroxide, respectively, to 1.6 for aliphatic halogenated hydrocarbons such as trichloroethylene. Therefore, the sorption capacity of clay is very low when compared to the other materials available. However, this material is one of the most inexpensive materials and can be purchased for approximately $0.04/lb to $0.12/lb, depending upon supplier and quantity ordered.

As a particulate, the material is extremely dusty, and a dust respirator and eye protection should be worn during its application.

3. Expanded Mineral

The materials termed "expanded minerals" describe micafil or perlite and vermiculite. These are naturally occurring materials composed of inorganic components. These materials have also been used for many years to remediate spills of aqueous and nonaqueous or oil-based chemicals.

These materials are incompatible with the following:

- hydrazines and hydrazides
- heavy metals

The sorption capacities range from 3.0 for amines, such as triethylamine, to 7.1 and 8.1 for inorganic acids and inorganic halides, such as nitric acid and phosphorus trichloride, respectively. Therefore, the sorption capacity of expanded minerals is far superior to that of clay and diatomite. However, this material is more expensive than clay but cheaper than diatomite. Micafil/perlite can be purchased for approximately $0.12/lb to $0.31/lb and vermiculite for approximately $0.14/lb, depending upon supplier and quantity ordered.

As a particulate, the material is not dusty, and a dust respirator is not mandatory but still recommended. However, eye protection should be worn during its application.

4. Foamed Glass (amorphous silica)

This is a natural inorganic material composed of off-white, free-flowing granules. This material is very light and can adsorb large quantities of spilled liquid. This material has also been used in recent years to remediate spills of aqueous and nonaqueous or oil-based chemicals.

Foamed glass is incompatible with:

- hydrofluorosilic acid
- hydrazines and hydrazides
- heavy metals

The sorption capacity of foamed glass products ranges from 6.7 for ketones such as methyl isobutyl ketone to 15.2 for nitric acid. Therefore the sorption capacity of foamed glass is very good, and the foamed glass is one of the best materials for spill remediation. It is also available in several forms besides a particulate, such as pillows, dikes, and booms. However, this material is more expensive, approximately $3.00/lb for the particulate form and $3.20-4.40/lb for pillows, depending upon supplier and quantity ordered.

*e. Cover Dry Materials*

Prevent dry, dusty pesticide spills, such as dusts, powders, or granules, from becoming airborne by covering them with a sweeping compound or a plastic covering or by very lightly misting the material with water. Do not mist too much, because water may release the pesticidal action or may cause the pesticide to form clumps and be unusable.

| Warning: Certain pesticides are oxidizers, such as calcium hypochloride (a common sanitizer), and some herbicides and desiccants contain chlorites, which should not be contained with sawdust or other combustible material. These absorbent materials can combined with the oxidizer to create a fire hazard and could burst into flame. |
|---|

## 3. Clean Up

After containing the spill, you must pick up the spilled material and decontaminate the spill site and any contaminated items or equipment.

*a. Clean Up the Spill*

For spilled liquid pesticides, sweep up the absorbent material containing the pesticide and place it into a heavy-duty plastic drum or bag. Keep adding the absorbent material until the spilled liquid is soaked up and removed. Do not use more absorbent than is necessary to clean up the spill. It also must be disposed of properly.

Spills of dry pesticides should be swept up for reuse if possible. Avoid contaminating the spilled materials with soil or other debris so it can be used in the usual application equipment and will not clog the nozzles or hopper openings. However, if the dry spill has become wet or full of debris, it must be swept up and placed in a heavy-duty plastic drum or bag for disposal.

*b. Decontaminate the Spill Site*

Once you have collected as much of the spilled material as possible, decontaminate the spill site as well as possible. Do not hose down the site with water, unless the spill is on a containment tray or pad.

If the surface on which the pesticide has spilled is nonporous, such as sealed concrete, glazed ceramic tile, or no-wax sheet flooring, use water (or the chemical listed on the label to dilute the pesticide) and a strong detergent to remove the residues of the spill from the surface. Do not allow any of the wash solution to run off the site being cleaned. Place fresh absorbent material over the wash solution until it is all soaked up. Then sweep up the absorbent material and place it in a plastic drum or bag for disposal as an excess pesticide.

**Figure 6.25** Pesticide handler cleaning up a small spill.

If the surface upon which the pesticide has spilled is porous, such as soil, unsealed wood, or carpet, you may have to remove the contaminated surface and dispose of it as an excess pesticide. Depending on the size of the spill and the toxicity of the pesticide, however, sometimes the site can be successfully neutralized.

*c. Neutralize the Spill and Site*

The labeling of a few pesticides will instruct you to neutralize a spill of that pesticide. Sometimes an authority, such as the pesticide manufacturer or Chemtrec, will also instruct you to neutralize the spill site. Follow the instructions carefully.

Neutralizing a spill often consists of mixing full-strength bleach with hydrated lime and working this mixture into the spill site with a coarse broom. Fresh absorbent material is then spread over the spill site to soak up the neutralizing liquid. This material is swept up and placed in a plastic drum or bag for disposal. You may be instructed to repeat the process several times to make sure that the site is thoroughly neutralized.

Soil is sometimes neutralized by removing and disposing of the top 2 to 3 inches and then neutralizing the remaining soil. You may be instructed to mix activated charcoal into the soil or to cover the spill site with two or more inches of lime and cover the lime with fresh topsoil.

Sometimes you may be instructed to cover minor spills with activated charcoal. The activated charcoal can adsorb or tie up enough pesticide to avoid adverse effects to plants and animals that contact the soil in the future. However, activated charcoal is not effective for large spills.

*d. Decontaminate Equipment*

Clean any vehicles, equipment, and personal protective equipment that were contaminated by the spill or during the containment and cleanup process. Use a strong mixture of chlorine bleach, dishwasher detergent, and water to clean the vehicles and equipment. Wash personal protective equipment thoroughly, following manufacturers' instructions and the guidelines in the personal protective equipment section of this text. Remember particularly that porous materials, such as brooms, leather shoes, and clothing, cannot be cleaned effectively if they are thoroughly saturated with pesticide. They should be disposed of along with the other spent cleanup materials. Figure 6.26 shows a pesticide worker decontaminating a truck after a spill.

*e. Decontaminate Yourself and Other Personnel*

As soon as you are finished with the spill and equipment cleanup, the personnel should wash thoroughly with detergent and water. Wash any part of skin that might have been exposed, and always wash your face, neck, hands, and forearms.

## 4. Spill Assistance

Chemtrec, the Chemical Transportation Emergency Center, is a public service of the Chemical Manufacturing Association. Located in Washington, D.C., Chemtrec is staffed 24 hours a day by competent, trained personnel who are able to advise you to manage chemical emergencies.

When you request help from Chemtrec or any other source, have the product label on hand.

**Figure 6.26** Decontamination of truck after spill.

Many pesticide labels list an emergency telephone number that gives you direct access to the manufacturer and people who know how to manage emergencies for that product.

If the spill occurs on a highway, call the highway patrol or highway department right away. If the spill occurs on a county road or city street, call the county sheriff, city police, or fire department. These authorities are trained for such emergencies and will be able to assist in the cleanup. Many local and state authorities require that you notify them of a pesticide spill.

If you suspect that a large spill is flammable, call the fire department for assistance. However, do not let them hose down the spill unless an authority directs them to do so.

If the spill may expose the public to pesticides or pesticide residues, contact public health officials. If anyone is poisoned by contacting the spill or if you suspect that an exposure may lead to poisoning, call the hospital emergency room and provide them with the brand name, active ingredients, and any other labeling information about human health hazards, signs and symptoms of poisoning, and antidotes.

## 5. Labeling Statements About Spill Management

Typical pesticide labeling instructions about spill procedures include:

- If container is broken or contents have spilled, clean it up immediately. Before cleaning up, put on full-length trousers, long-sleeved shirt, protective gloves, and goggles or face shield. Soak up the spill with absorbent media such as clay, saw dust, corn cods, or other suitable material, and dispose of waste at an approved waste disposal facility.
- If the container is leaking or material is spilled, carefully sweep the material into a pile. Refer to Precautionary Statements on the label for hazards associated with the handling of this material. Do not walk through spilled material. Keep unauthorized people away.
- Contact the emergency response team or your response contractor for decontamination procedures.

## 6. Spill Kit

Keep a spill cleanup kit immediately available whenever you handle pesticides or their containers. If a spill occurs, you will not have the time or the opportunity to find all of the items.

The kit should consist of:

- telephone numbers for emergency assistance
- chemical-resistant gloves, footwear, and apron, suit, etc.
- protective eyewear

- an appropriate respirator, if any of the pesticides require the use of one during handling activities or for spill cleanup
- containment "snakes" to confine the leak or spill to a small area
- absorbent materials, such as spill pads, pillows, absorbent clay, expanded silica, activated charcoal, vermiculite, sawdust, or corn cobs to soak up liquid spills
- sweeping compound to keep dry spills from drifting or wafting during cleanup
- a shovel, broom, and dustpan (foldable brooms and shovels are handy, because they can be carried easily)
- heavy-duty detergent
- a fire extinguisher rated for all types of fires
- any other spill cleanup items specified on the labeling of any products you use regularly
- a sturdy plastic container that will hold the quantity of pesticide from the largest pesticide container being handled and that can be tightly closed.

All of these items can be stored in the plastic container and kept clean and in working order until a spill occurs.

## REGULATORY COMPLIANCE

If you are involved in a pesticide spill, you may need to comply with the provisions of two different laws administered by the EPA.

In the event of a spill, Title III of SARA requires you to report any accidental release of a extremely hazardous substance. Reporting is required if all the following occur:

- the pesticide was spilled
- the pesticide is covered under SARA Title III
- the spill quantity was greater than the reportable quantity (RQ) specified in the law
- the spill created offsite exposure

If such an accident occurs, you must:

- notify the State Emergency Response Commission (SERC)

- notify the Local Emergency Planning Committee (LEPC)
- report the release to the National Response Center (1-800-424-8802)

In addition, any spill that has the potential to contaminate ground or surface waters must be reported to the EPA under the authority of the Clean Water Act.

If you do not know whether the spill is large enough to be a "reportable quantity" under SARA Title III or whether the spill might enter ground or surface waters, call your local, state, or tribal pesticide agency or the EPA regional office for help.

## VI. FIRE HAZARDS AND PREVENTION

### A. NFPA-HAZARD RATING

The NFPA rating provides a simple system for the rapid recognition of the hazards associated with the many materials found at facilities that produce, ship, store, sell, and/or utilize chemicals and pesticides. The rating is concerned with the health, fire, reactivity (or instability), and other hazards created by short-term exposure as might be encountered under fire, spill, or other related emergency conditions.

This system recognizes the hazards of a material in terms of three principal categories, namely "health," "flammability," and "reactivity" and ranks the severity numerically ranging from "four (4)," indicating a severe hazard, to "zero (0)," indicating no special hazard. The information is displayed in the NFPA 704 diamond, which is shown in Figure 6.27. For more information, you should consult the NFPA Fire Protection Guide on Hazardous Materials.

### B. NFPA CLASSIFICATION OF FIRES

Class A: Fires involving ordinary combustible materials (such as wood, cloth, paper, rubber, and many plastics) requiring the heat-absorbing (cooling) effects of water or water solutions or the coating effects of certain dry chemicals that retard combustion.

Class B: Fires involving flammable or combustible liquids, flammable gases, greases, and similar materials where extinguishment is most readily secured by excluding air (oxygen), inhibiting the release of combustible vapors, or interrupting the combustion chain reaction.

Class C: Fires involving energized electrical equipment where safety to the operator requires the use of electrically nonconductive extinguishing agents (after the electric equipment has been de-energized, the use of Class A or B extinguishers may be indicated).

Class D: Fires involving certain combustible metals such as magnesium, titanium, zirconium, sodium, potassium, and others that require a heat-absorbing extinguishing medium not reactive with burning metals.

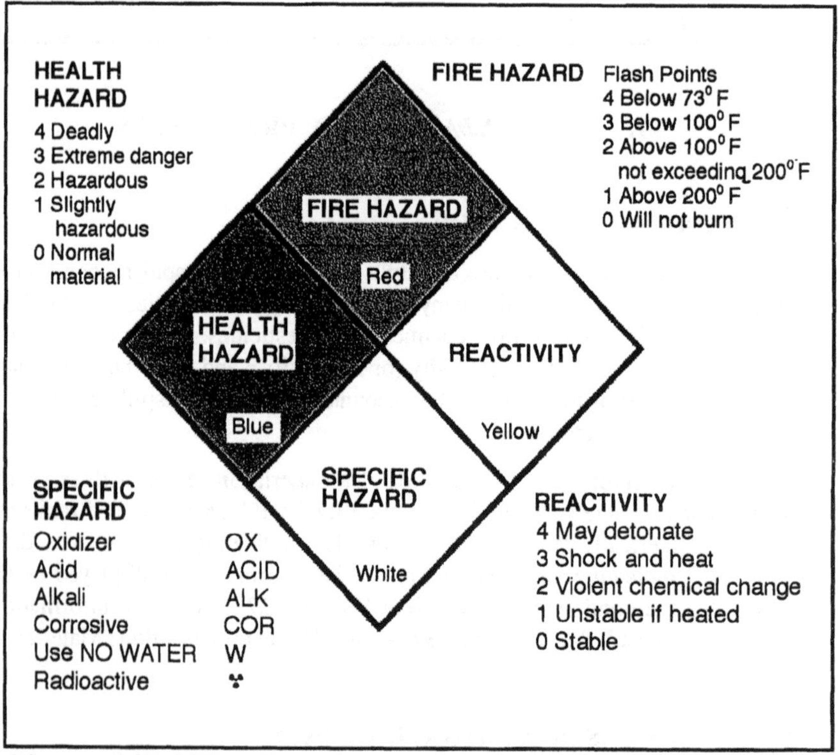


**Figure 6.27** NFPA-704M diamond (National Fire Protection Association, *Fire Protection Guide on Hazardous Materials*, 9th ed., 1986).

## REFERENCES

American National Standards Institute, *Practice for Occupational and Educational Eye and Face Protection*, ANSI, Inc., 1430 Broadway, NY, 1989.

American Society for Testing and Materials, *Standard Test Method F903-87-Resistance of Protective Clothing Materials to Penetration by Liquids*, American Society for Testing and Materials, Philadelphia, 1987.

Coletta, G. C., Schwope, A. D., Arons I. J., King, J. W. and A. Sivak, *Development of Performance Criteria for Protective Clothing Used Against Carcinogenic Liquids*, National Institute for Occupational Safety and Health, Pub. No. 79-106, Cincinnati, 1978.

Ehntholt, D. J., Bodek, I., Doerfler, T. E., Schwope, A. D., Stolki, T. J. and J. Valentine, *Permeation Resistance of Polymer Glove Materials to Various Strengths of Active Pesticide Ingredients and Carrier Solvents*, Draft Internal Report, EPA Office of Research and Development Contract No. 68-03-3293, Cincinnati, 1989.

Fraser, A. J. and V. B. Keeble, *Factors Influencing Design of Protective Clothing for Pesticide Application, Performance of Protective Clothing*, Second Symposium, ASTM STP 989, Mansdorf, S. Z., Sager, R., and A. P. Nielsen, Eds., American Society for Testing and Materials, Philadelphia, pp. 565-572, 1988.

Freed, V. H., Davies, J. E., Peters, L. J. and F. Parveen, *Minimizing Occupational Exposure to Pesticides: Repellency and Penetrability of Treated Textiles to Pesticide Sprays*, Pesticide Reviews, 75, pp. 159-167, 1980.

Lloyd, G. A., *Efficiency of Protective Clothing for Pesticide Spraying Performance of Protective Clothing*, ASTM STP 900, Barker, R. L., and G. C. Coletta, Eds., American Society for Testing and Materials, Philadelphia, pp. 121-135, 1986.

Mickelsen, R. L. and R. C. Hall, "A Breakthrough Time Comparison of Nitrile and Neoprene Glove Materials Produced by Different Glove Manufacturers," *Am. Ind. Hyg. Assoc. J.*, 48, 11, pp. 941-947, 1987.

Mickelsen, R. L., Roder, M. M. and S. P. Berardinelli, *"Permeation of Chemical Protective Clothing by Three Binary Solvent Mixtures," Am. Ind. Hyg. Assoc. J.*, 47, pp. 236-240, 1986.

National Fire Protection Association, *Fire Protection Guide on Hazardous Materials,* 9th ed., Quincy, MA, 1986.

Sansone, E.B. and Y.B. Tewari, "Differences in the Extent of Solvent Penetration through Natural Rubber and Nitrile Gloves from Various Manufacturers," *Am. Ind. Hygiene Assoc. J.*, 41, pp. 527-528, 1980.

Schwope, A. D., Costas, P. P., Jackson, J. O., Stull, J. O. and D. J. Weitzman, *Guidelines for the Selection of Chemical Protective Clothing*, 3rd ed., National Technical Information Service No. AD-A179 164, Springfield, VA, 1987.

Schwope, A. D., Goydan, R., Reid, R. C., and S. Krishnamurthy, "State of the Art Review of Permeation Testing and the Interpretation of Its Results," *Am. Ind. Hyg. Assoc. J.*, 49, pp. 557-565, 1988.

Slocum, A. C., Nolan, R. J., Shern, L. C., Gay, S. L., and A. J. Turgeon, *Development and Testing of Protective Clothing for Lawn Care Specialists, Performance of Protective Clothing*, Second Symposium, ASTM STP 989, Mansdorf, S. Z., Sager, R., and A. P. Nielsen, Eds., American Society for Testing and Materials, Philadelphia, pp. 557-564, 1988.

U.S. Environmental Protection Agency, *Applying Pesticides Correctly: A Guide for Private and Commercial Applicators*, revised 1991.

U.S. Environmental Protection Agency, *Guidance Manual for Selecting Protective Clothing for Agricultural Pesticide Operations*, EPA 736-B-94-001, 1993.

Waxman, M.F., *Hazardous Waste Site Operations: A Training Manual for Site Professionals*, John Wiley & Sons, New York, 1996.

Waxman, M.F. and D. W. Kammel, *A Guidebook for the Safe Use of Hazardous Agricultural Farm Chemicals and Pesticides*, North Central Regional Publication 402, University of Wisconsin-Extension, 1991.

# CHAPTER 7

# FIRST AID AND THE HARMFUL EFFECTS OF PESTICIDES

## I. HARMFUL EFFECTS OF PESTICIDES

Most pesticides are designed to harm or kill pests. Because some pests have systems similar to the human system, some pesticides also can harm or kill humans. Fortunately, humans usually can avoid harmful effects by avoiding exposure to pesticides.

Humans may be harmed by pesticides in two ways; they may be poisoned or injured. Pesticide poisoning is caused by pesticides that harm internal organs or other systems inside the body. Pesticide-related injuries usually are caused by pesticides that are external irritants.

Pesticides that are chemically similar to one another cause the same type of harmful effects to humans. These effects may be mild or severe, depending on the pesticide involved and the amount of exposure, but the pattern of illness or injury caused by each chemical group is usually the same. Some pesticide chemical families can cause both external irritation injuries and internal poisoning illnesses.

Some pesticides are highly toxic to humans; only a few drops in the mouth or on the skin can cause extremely harmful effects. Other pesticides are less toxic, but too much exposure to them will also cause harmful effects. A good equation to remember is:

**Hazard = Toxicity x Exposure**

**Hazard** is the risk of harmful effects from pesticides. Hazard depends on both the **toxicity** of the pesticide and the **exposure** received in any situation.

### A. EXPOSURE

When a pesticide comes into contact with a surface or an organism, that contact is called a pesticide exposure. For humans, a pesticide exposure means getting pesticides in or on the body. The toxic effect of a pesticide depends on how much pesticide is involved and how long it remains in contact.

## 1. Types of Exposures

Pesticides contact your body in four main ways:

- oral exposure (when you swallow a pesticide)
- inhalation exposure (when you inhale a pesticide)
- ocular exposure (when you get a pesticide in your eyes)
- dermal exposure (when you get a pesticide on your skin)

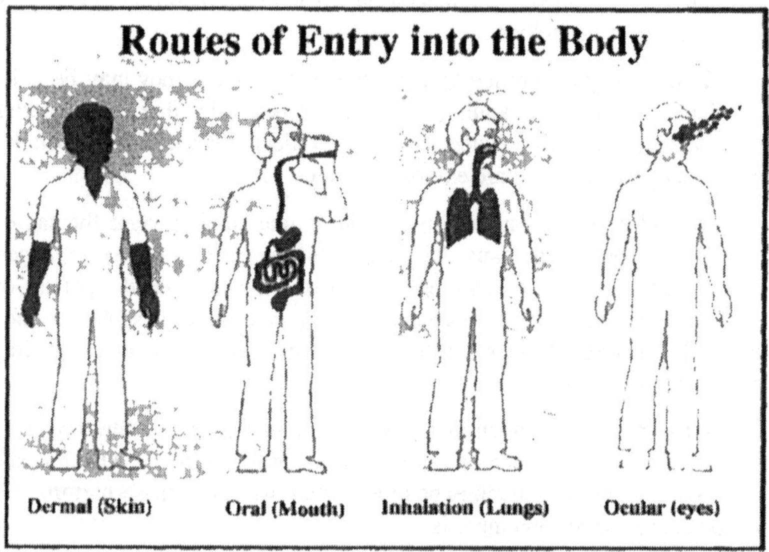


**Figure 7.1** The four possible routes of entry of pesticides into the body (Waxman, M.F., *Pesticide Safety*, University of Wisconsin Pesticide Safety Training, 1991).

## 2. Avoiding Exposure

Avoiding and reducing exposure to pesticides will reduce the harmful effects from pesticides. You can avoid exposure by using safety systems, such as closed systems and enclosed cabs, and you can reduce exposure by wearing appropriate personal protective equipment, by washing exposed areas often, and by keeping your personal protective equipment clean and in good operating condition.

**Figure 7.2** The improper and proper attire (left and right photos, respectively) to apply pesticides using a high pressure sprayer. (Both figures are courtesy of the EPA.)

In most pesticide handling situations, the skin is the part of the body that is most likely to receive exposure. Evidence indicates that about 97 percent of all body exposure that happens during pesticide spraying is by contact with the skin. The only time that inhalation is a greater hazard than skin contact is when you are working in a poorly ventilated enclosed space and are using a fumigant or other pesticide that is highly toxic by the inhalation route.

The amount of pesticide that is absorbed through your skin (and eyes) and into your body depends on:

- the pesticide itself and the material used to dilute the pesticide. Emulsifiable concentrates, oil-based liquid pesticides, and oil-based diluents (such as xylene) are, in general, absorbed most readily. Water-based pesticides and dilutions (such as wettable and soluble powders and dry flowables) usually are absorbed less readily than the oil-based liquid formulations but more readily than dry formulations. Dusts, granules, and other dry formulations are not absorbed as readily as liquids.

- the area of the body exposed. The genital area tends to be the most absorptive. The scalp, ear canal, and forehead are also highly absorptive.

- the condition of the skin exposed. Cuts, abrasions, and skin rashes allow absorption more readily than intact skin. Hot, sweaty skin will absorb more pesticide than dry, cool skin. Figures 7.2, 7.3A, and 7.3B illustrate both improper and proper attire for applying and mixing pesticides.

### 3. Causes of Exposure

One of the best ways to avoid pesticide exposure is to avoid situations and practices where exposures commonly occur.

**Oral exposures** often are caused by:

- not washing hands before eating, drinking, smoking, or chewing
- mistaking the pesticide for food or drink
- accidentally applying pesticides to food
- splashing pesticide into the mouth through carelessness or accident.

**Inhalation exposures** often are caused by:

- prolonged contact with pesticides in closed or poorly ventilated spaces
- breathing vapors from fumigants and other toxic pesticides
- breathing vapors, dust, or mist while handling pesticides without appropriate protective equipment
- inhaling vapors present immediately after a pesticide is applied, for example, from drift or from reentering the area too soon
- using a respirator that fits poorly or using an old or inadequate filter, cartridge, or canister.

**Dermal exposures** often are caused by:

- not washing hands after handling pesticides or their containers
- splashing or spraying pesticides on unprotected skin or eyes
- wearing pesticide-contaminated clothing (including boots and gloves)
- applying pesticides (or flagging) in windy weather
- wearing inadequate personal protective equipment while handling pesticides
- touching pesticide-treated surfaces.

**Figure 7.3A** The improper method to fill a hand-operated duster. The worker is not taking precautions to protect himself and the environment.
**Figure 7.3B** The proper method to fill equipment. The worker is wearing a long sleeved shirt, gloves, respirator, goggles, and a hat and also the equipment is placed on newspaper to avoid contamination of ground while the equipment is filled slowly with a scoop. (Both figures are courtesy of the EPA.)

**Eye exposures** often are caused by:

- splashing or spraying pesticides in eyes
- applying pesticides in windy weather without eye protection
- rubbing eyes or forehead with contaminated gloves or hands
- pouring dust, granule, or powder formulations without eye protection.

## B. TOXICITY

Toxicity is a measure of the ability of a pesticide to cause harmful effects. Toxicity depends on:

- type and amount of active ingredient(s)
- type and amount of carrier or solvent ingredient(s)
- type and amount of inert ingredient(s)
- type of formulation, such as dust, granule, powder, or emulsifiable concentrate.

The toxicity of a particular pesticide is measured by subjecting laboratory animals (usually rats, mice, rabbits, and dogs) or tissue cultures to different dosages of the active ingredient and of the formulated product over various time periods. However, some people react more severely or more mildly than estimated. Be alert to your body's reaction to the pesticides you are handling. Some people seem to be especially sensitive to individual pesticides or to groups of similar pesticides.

You may have a choice of pesticides for a particular pest problem. One of the factors you should consider is how toxic each possible choice is to persons who will use it or be exposed to it.

## C. HARMFUL EFFECTS

Pesticides can cause three types of harmful effects: acute effects, delayed or chronic effects, and allergic effects.

### 1. Acute Effects

Acute effects are illnesses or injuries that may appear immediately after exposure to a pesticide (usually within 24 hours). Studying a pesticide's relative capability of causing acute effects has been the main way to assess and compare how toxic pesticides are. Acute effects can be measured more accurately than delayed effects, and they are more easily diagnosed than effects that do not appear until long after the exposure. Acute effects usually are obvious and often are reversible if appropriate medical care is given promptly.

Pesticides cause four types of acute effects:

- acute oral effects
- acute inhalation effects
- acute dermal effects
- acute eye effects.

**Figure 7.4** Collapsed worker suffering acute effects of pesticide exposure (EPA, source unknown).

*a. Acute Oral Effects*

Your mouth, throat, and stomach can be burned severely by some pesticides. Other pesticides that you swallow will not burn your digestive system, but will be absorbed and carried in your blood throughout your body and may cause you harm in various ways. These chemicals that are transported throughout the body are termed systemic pesticides. For some pesticides, swallowing even a few drops from a splash or wiping your mouth with a contaminated glove can make you very ill or make it difficult to eat and drink and get nourishment.

*b. Acute Inhalation Effects*

Your entire respiratory system can be burned by some pesticides, making it difficult to breathe. Other pesticides inhaled may not harm your respiratory system, but are carried quickly in your blood throughout the whole body where they can cause harm in various ways.

*c. Acute Dermal and Skin Irritation Effects*

Contact with some pesticides will harm your skin. These pesticides may cause your skin to itch, blister, crack, or change color. Other pesticides can pass through your skin and eyes and get into your body. Once inside your body, these pesticides are carried throughout your system where they can cause harm in various ways.

*d. Acute Eye Effects*

Some pesticides that get into your eyes can cause temporary or permanent blindness or severe irritation. Other pesticides may not irritate your eyes, but pass through your eyes and into your body. These pesticides can travel throughout your body, causing harm in various ways. Table 7.1 lists the typical precautionary statements found in pesticide labeling.

## 2. Delayed or Chronic Effects

Delayed effects are illnesses or injuries that do not appear immediately (within 24 hours) after exposure to a pesticide or combination of pesticides. Often the term "chronic effects" is used to describe delayed effects, but this term is applicable only to certain types of delayed effects.

Delayed effects may be caused by:

- **Repeated exposure** to a pesticide, a pesticide group, or a combination of pesticides over a long period of time.
- A **single exposure** to a pesticide (or combination of pesticides) that causes a harmful reaction that does not become apparent until much later.

Some pesticides cause delayed effects only with **repeated exposure** over a period of days, months, or even years. For example, if a rat eats a large amount of the pesticide cryolite at one time, the pesticide passes through the rat's system quickly and is eliminated without harmful effects. However, if the rat regularly eats small amounts of cryolite, it soon becomes ill and dies. Cryolite does not readily dissolve in water. The small amount of pesticide that is absorbed into the rat's system from a one-time exposure is not enough to cause illness. But if that same small amount is absorbed day after day, enough poison will be absorbed into the rat's system to cause illness and death.

**Table 7.1** Typical precautionary statements on pesticide labeling.

| | Highly Toxic | Moderately Toxic | Slightly Toxic |
|---|---|---|---|
| **Acute Oral** | "Fatal if swallowed," or "Can kill you if swallowed." | "Harmful or fatal if swallowed," or "May be fatal if swallowed." | "Harmful if swallowed," or "May be harmful if swallowed." |
| **Acute Inhalation** | "Poisonous if inhaled," or "Can kill you if breathed," combined with the statement "Do not breathe dust vapors or spray mist." | "Harmful or fatal if inhaled," or "May be fatal if breathed," followed by a statement such as "Do not breathe dusts, vapors. or spray mist." | "Harmful if inhaled," or "May be harmful if breathed," combined with the statement "Avoid breathing dusts, vapors or spray mist." |
| **Acute Dermal** | "Fatal if absorbed through the skin," or "Can kill you by skin contact," combined with the statement "Do not get on skin or clothing." | "Harmful or fatal if absorbed through the skin," or "May be fatal by skin contact," followed by a statement such as "Do not get on skin or clothing." | "Harmful or fatal if absorbed through the skin," or "May be harmful by skin contact," combined with the statement "Avoid contact with skin or clothing." |
| **Skin Irritation** | "Corrosive—causes severe skin burns," combined with the statement "Do not get on skin." | "Causes skin irritation," or "Causes skin burns," followed by a statement such as "Do not get on skin." | "May irritate skin," combined with the statement "Avoid contact with skin." |
| **Eye Irritation** | "Corrosive—causes irreversible eye damage," or "Causes severe eye burns or blindness," combined with the statement "Do not get in eyes." | "Causes eye irritation," or "Causes eye burns," followed by a statement such as "Do not get in eyes." | "May irritate eyes," combined with the statement "Avoid contact with eyes." |

Sometimes repeated exposures to a pesticide or family of pesticides will result in a delayed effect, but a larger exposure will cause an acute effect. Organophosphate and carbamate pesticides inhibit a chemical, called cholinesterase, in the nervous system of humans. A large exposure causes immediate acute illness. Smaller exposures cause no apparent problem at first. They inhibit the cholinesterase, but not enough to cause immediate illness. Small, repeated exposures to these pesticides over several days or weeks may greatly reduce cholinesterase levels in the body. At that point, even a small exposure to a pesticide with relatively low cholinesterase-inhibiting properties may trigger severe illness.

A person who is repeatedly exposed to two or more specific chemicals may become ill even though any one of the chemicals alone would have had no harmful health impact. Some organophosphate pesticides have been shown to have this effect when they are used in combination.

In some cases, a **single exposure** to a pesticide (or combination of pesticides) could adversely affect the exposed person's health after a period of time. For example, large exposures to paraquat, a herbicide, may cause severe or fatal lung injury that does not appear for 3 to 14 days after the initial exposure. After an exposure, paraquat slowly builds up in the lungs and destroys lung cells.

Some kinds of harmful effects may not occur unless a certain set of circumstances is present. These effects can occur after the first exposure, but the likelihood is small. Continuous or frequent exposures over a long period of time make it more likely that all the necessary factors will be present. Some genetic changes that result in the development of cancer or other delayed effects are in this category.

Types of delayed effects include:

- chronic effects
- developmental and reproductive effects
- systemic effects.

### *a. Chronic Effects*

Chronic effects are illnesses or injuries that appear a long time, usually several years, after exposure to a pesticide. Some delayed effects that are suspected to result from pesticides' chronic toxicity include:

- production of tumors (oncogenic effect)
- production of malignancy or cancer (carcinogenic effect)
- changes in the genes or chromosomes (mutagenic effect).

Typical precautionary statements on pesticide labeling include:

- *Cancer Hazard Warning Statement: This product contains an ingredient which has been determined to cause tumors in laboratory animals.*
- *NOTE: This product has been shown to cause cancer in laboratory animals.*
- *The use of this product may be hazardous to your health. This product contains an ingredient which has been determined to cause tumors in laboratory animals.*

*b. Developmental and Reproductive Effects*

A **developmental effect** is an injury or illness that occurs to a fetus in the womb of a woman who has been exposed to a pesticide. These effects include:

- birth defects (teratogenic effect)
- illness or death (miscarriage or stillbirth) of a fetus (fetotoxic effect).

A **reproductive effect** is an injury to the reproductive system of exposed men or women. These effects include:

- infertility or sterility in men or women
- impotence in men.

Some developmental or reproductive effects are thought to occur immediately after exposure to a pesticide or combination of pesticides, but they may not be apparent for some time after the exposure. For example, a birth defect may be seen only after the birth of a child, which may be several months after the exposure.

Other developmental or reproductive effects are thought to result from repeated exposures to a pesticide or combination of pesticides over a period of time.

A typical precautionary statement on pesticide labeling is:

- *This product may be hazardous to your health. This product has been determined to cause birth defects in laboratory animals.*

*c. Systemic Effects*

A delayed systemic effect is an illness or injury to a system in the body that does not appear immediately (within 24 hours) after exposure to a pesticide or combination of pesticides. Such effects include:

- blood disorders (hemotoxic effects), such as anemia or an inability to coagulate
- nerve or brain disorders (neurotoxic effects), such as paralysis, nervous excitation, behavioral changes, tremor, blindness, and brain damage
- skin disorders, such as rash, irritation, discoloration, and ulceration
- lung and respiratory disorders, such as emphysema and asthma
- liver and kidney disorders, such as jaundice and kidney failure.

Typical precautionary statements on pesticide labeling include:

- *May produce kidney and liver damage upon prolonged exposure.*
- *Inhalation may cause delayed lung, nerve, or brain injury.*
- *Liquid or vapor may cause serious skin or eye injury which may have a delayed onset.*

*d. Determining Delayed Effects*

Because of the time delay between the exposure and the observable effect, and because many other types of exposures may have occurred during the delay, it is sometimes hard to identify the cause of a delayed effect. Although some pesticides may cause delayed effects in laboratory animals, further studies are needed to determine whether these pesticides will affect humans the same way.

When there is clear evidence that a pesticide may cause chronic, developmental, reproductive, or systemic effects in humans, the Environmental Protection Agency will determine what steps are appropriate to reduce or eliminate the risk. Such actions include:

- removing the pesticide from use
- requiring label warning statements about the possible effect
- requiring specific personal protective equipment or safety systems be used during handling of the pesticide
- requiring changes in dosages, method or frequency of application, and waiting times before entry or harvest/slaughter/grazing
- restricting the use to certified applicators.

*e. Avoiding Delayed Effects*

Scientists, pesticide manufacturers, and the Environmental Protection Agency cannot yet be sure what the delayed effects of too much exposure to individual pesticides or combinations of pesticides may be. It may be years before there are clear answers on the effects of all the pesticides and combinations of pesticides in use today. Meanwhile, it makes good sense to reduce your exposure to all pesticides as much as possible.

## 3. Allergic Effects

**Allergic effects** are harmful effects that some people develop in reaction to substances that do not cause the same reaction in most other people. Allergic reactions are not thought to occur during a person's first exposure to a substance. The first exposure causes the body to develop repelling response chemicals to that substance. A later (second, third, or more) exposure results in the allergic response. This process is called **sensitization,** and substances that cause people to become allergic to them are known as **sensitizers.**

Certain substances cause many people to develop an allergic reaction. Poison ivy, for example, causes a severe skin rash in many people. Other substances cause allergic reactions in only a few people. Turfgrass, for example, causes a severe skin rash in relatively few people.

### *a. Types of Allergic Effects*

Some people are sensitized to certain pesticides. After being exposed once or a few times without effect, they develop a severe allergy-like response upon later exposures. These allergic effects include:

- systemic effects, such as asthma or even life-threatening shock
- skin irritation, such as rash, blisters, or open sores
- eye and nose irritation, such as itchy, watery eyes and sneezing.

Unfortunately, there is no way to tell which people may develop allergies to which pesticides. However, certain people seem to be more chemically sensitive than others. They develop an allergic response to many types of chemicals in their environment. These persons may be more likely to develop allergies to pesticides.

Typical precautionary statements on pesticide labeling include:

- *This product may produce temporary allergic side effects characterized by redness of the eyes, mild bronchial irritation, and redness or rash on exposed skin areas. Persons having an allergic reaction should contact a physician.*
- *May be a skin sensitizer.*
- *The active ingredient may cause skin sensitization reactions in certain individuals.*

*b. Avoiding Allergic Effects*

Depending on how severe the allergic reaction is, persons with allergies to certain pesticides may have to stop handling or working with those pesticides. They may be unable to tolerate even slight exposures. Sometimes persons with allergies to certain pesticides can continue to work in situations where those pesticides are present by reducing their exposure to them.

## D. SIGNS AND SYMPTOMS OF HARMFUL EFFECTS

Watch for two kinds of clues to pesticide-related illness or injury. Some clues are symptoms/reactions that only the person who has been poisoned can notice, such as nausea or headache. Others clues, like vomiting or fainting, can be noticed by someone else. You should know:

- what your own symptoms might mean
- what signs of poisoning to look for in your co-workers and others who may have been exposed.

Many of the signs and symptoms of pesticide poisoning are similar to signs and symptoms of other illnesses you might experience, such as the flu or even a hangover. If you have been working with pesticides and then develop suspicious signs and symptoms, call your physician or poison control center. Only a physician can diagnose pesticide poisoning injuries.

**External irritants** cause:

- redness, blisters, rash, and/or burns on skin
- swelling, a stinging sensation, and/or burns in eyes, nose, mouth, and throat.

**Pesticide poisoning** may cause:

- nausea, vomiting, diarrhea, and/ or stomach cramps
- headache, dizziness, weakness, and/or confusion
- excessive sweating, chills, and/ or thirst
- chest pains
- difficult breathing
- cramps in muscles or aches all over your body.

### 1. Telltale Signs or Symptoms

Poisoning by some pesticide chemical families results in distinctive signs that help others to recognize the cause of the poisoning. Organophosphate and n-methyl carbamate poisoning, for example, is often identified by the presence of very small (pinpoint) pupils in the victim's eyes. Poisoning by pesticides containing arsenic or phosphorus is often identified by a garlic odor on the victim's breath.

You should know the kinds of harmful effects most likely to be caused by the pesticides you use. The Pesticide Guides in Part Two of this text and the appendix entitled "Effects of Pesticides on the Human Body" (EPA, *Applying Pesticides Correctly*, 1991) both contain guides to help judge how the products you are using might be expected to affect you. The guides and tables list the major groups of pesticides. For each group, it tells:

- the action of the poison on the human system
- acute poisoning (systemic) effects

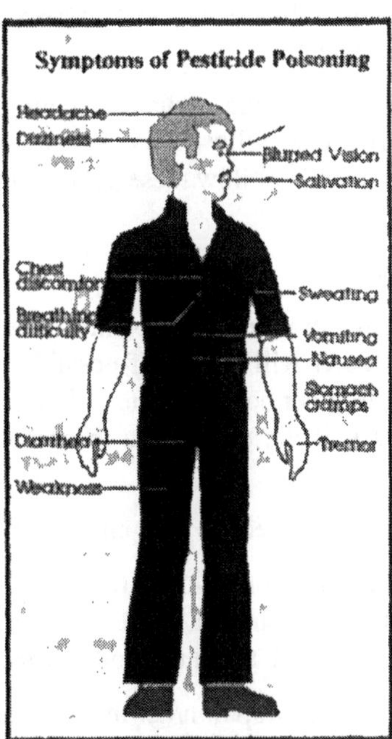


**Figure 7.5** Common symptoms of pesticide poisoning (Waxman, M.F., *Pesticide Safety*, University of Wisconsin Pesticide Safety Training, 1991).

- acute irritation effects
- delayed or allergic effects
- type of pesticide.

## E. RESPONDING TO A POISONING EMERGENCY

Get medical advice quickly if you or any of your fellow workers have unusual or unexplained symptoms starting at work or later the same day. Do not let yourself or anyone else get dangerously sick before calling your physician or going to a hospital. It is better to be too cautious than too late. Take the pesticide container (or the labeling) to the physician. Do not carry the pesticide container in the passenger space of a car or truck.

### 1. First Aid for Pesticide Poisoning

The best first aid in pesticide emergencies is to stop the source of pesticide exposure as quickly as possible. First aid is the initial effort to help a victim while medical help is on the way. If you are alone with the victim, make sure the victim is breathing and is not being further exposed to the pesticide before you call for emergency help. Apply artificial respiration if the victim is not breathing. Do not become exposed to the pesticide yourself while you are trying to help.

In an emergency, look at the pesticide labeling, if possible. If it gives specific first aid instructions, follow those instructions carefully. If labeling instructions are not available, follow these general guidelines for first aid:

#### Pesticide on Skin:

- Drench skin and clothing with plenty of water. Any source of relatively clean water will serve. If possible, immerse the person in a pond, creek, or other body of water. Even water in ditches or irrigation systems will do, unless you think they may have pesticides in them.
- Remove personal protective equipment and contaminated clothing.
- Wash skin and hair thoroughly with a mild liquid detergent and water. If one is available, a shower is the best way to completely and thoroughly wash and rinse the entire body surface.
- Dry the victim and wrap him/her in a blanket or any clean clothing at hand. Do not allow the victim to become chilled or overheated.

- If skin is burned or otherwise injured, cover the skin immediately with loose, clean, dry, soft cloth or bandage.
- Do not apply ointments, greases, powders, or other drugs in first aid treatment of burns or injured skin.

**Pesticide in Eye:**

- Wash eye quickly but gently.
- Use an eyewash dispenser, if available. Otherwise, hold the eyelid open and wash with a gentle drip of clean running water positioned so that it flows across the eye rather than directly into the eye.
- Rinse the eye for 15 minutes or more.
- Do not use chemicals or drugs in the rinse water. They may increase the injury.

**Inhaled Pesticide:**

- Get victim to fresh air immediately.
- If other people are in or near the area, warn them of the danger.
- Loosen tight clothing on the victim that would constrict breathing.
- Apply artificial respiration if breathing has stopped or if the victim's skin is blue. If pesticide or vomit is on the victim's mouth or face, avoid direct contact and use a shaped airway tube, if available, for mouth-to-mouth resuscitation.

**Pesticide in Mouth or Swallowed:**

- Rinse mouth with plenty of water.
- Give the victim large amounts (up to 1 quart) of milk or water to drink.
- Induce vomiting only if instructions to do so are on the labeling.

**Procedure for Inducing Vomiting:**

- Position the victim face down or kneeling forward. Do not allow the victim to lie on his back, because vomit could enter the lungs and cause additional damage.
- Put a finger or the blunt end of a spoon at the back of victim's throat or give the victim syrup of ipecac.

- Do not use salt solutions to induce vomiting.

**Do not induce vomiting:**

- If the victim is unconscious or is having convulsions.
- If the victim has swallowed a corrosive poison. A corrosive poison is a strong acid or alkali. It will burn the throat and mouth as severely coming up as it did going down. It may get into the lungs and burn there also.
- If the victim has swallowed an emulsifiable concentrate or oil solution. Emulsifiable concentrates and oil solutions may cause death if inhaled during vomiting.

## II. HEAT STRESS

A worker whose body temperature rises above a safe level may be affected by heat stress—a general term that refers to conditions ranging from heat cramps through the more serious heat exhaustion and the potentially life threatening heat stroke. Heat stress is marked by a variety of symptoms ranging from muscle spasms to dizziness, nausea, and coma. This section describes the different heat illnesses and the recommended treatments for each.

### A. HEAT ILLNESSES AND FIRST AID MEASURES

#### 1. Early Heat Stress

- Signs and Symptoms—mild dizziness, fatigue, or irritability; decreased concentration and/or impaired judgment.
- Cause and Problems—reduced flow of blood to the brain which may lead to heat exhaustion or heat stroke.
- Treatment—loosen or remove clothing, rest in shade 30 minutes or more, and drink plenty of water.

#### 2. Heat Rash (Prickly Heat)

- Signs and Symptoms—Tiny, blister-like red spots on the skin; prickly sensations. Commonly found on clothed areas of the body.
- Cause and Problems—Sweat glands become plugged and inflamed from unrelieved exposure of skin to heat, humidity, and sweat.

- Treatment—Clean skin, apply mild drying lotion or cornstarch. Wear loose clothing. Preventable by regular bathing and drying the skin and by periodic relief from humid conditions of work. See physician if rash persists.

3. **Heat Cramps**

- Signs and Symptoms— Painful spasms of leg, arm, or abdominal muscles. Heavy sweating and thirst which occurs during or after hard work.
- Cause and Problems— Loss of body salt in sweat may be totally disabling.
- Treatment—Loosen clothing, drink isotonic beverages, and rest.

4. **Heat Exhaustion**

- Signs and Symptoms— Fatigue, headache, dizziness, muscle weakness, loss of coordination, fatigue, collapse. Profuse sweating; pale, moist, cool skin; excessive thirst, dry mouth; dark yellow urine. Fast pulse, if conscious. Low or normal oral temperature, rectal temperature usually 99.5-101°F. May also have heat cramps, nausea, urge to defecate, rapid breathing, chills, tingling of the hands or feet, confusion, giddiness, slurred speech, and/or irritability.
- Cause and Problems—Dehydration, lack of acclimatization; reduction of blood in circulation, strain on respiratory system, reduced flow of blood to the brain. May lead to heat stroke.
- Treatment—Removal to cooler, shaded area as quickly as possible. Rest lying down. If conscious, have worker drink as much water as possible. Do not give salt. If unconscious or if heat stroke is also suspected, treat for heat stroke. Loosen or remove clothing. Splash cold water on the body. Massage legs and arms. If worker collapsed, get an evaluation by physician, nurse, or EMT before worker leaves for the day. Shower in cold water and rest for the balance of the day and overnight.

5. **Heat Stroke**

- Signs and Symptoms—Fatigue, headache, dizziness, muscle weakness, loss of coordination, fatigue, collapse. Profuse sweating; pale, moist, cool skin; excessive thirst, dry mouth; dark yellow urine. Fast pulse, if conscious. Low or normal oral temperature, rectal temperature usually 99.5-101°F. May also have heat cramps, nausea, urge to defecate, rapid breathing, chills, tingling of the hands or feet, confusion, giddiness, slurred speech, and/or irritability.

**Figure 7.6** Worker exhibiting the early stages of heat-stress.

- Cause and Problems—Sustained exertion in heat, lack of acclimatization, dehydration, individual risk factors such as high blood pressure, etc.; reduced flow of blood to the brain and other vital organs. Body's temperature-regulating system fails, body cannot cool itself. Risk of damage to vital organs.

- Treatment—Move to shaded area. Remove outer clothing. Immediately wrap in wet sheet, pour water on and fan vigorously, avoid over-cooling. Treat shock if present, once temperature is lowered. If worker vomits, make sure all vomit is cleared from the mouth and nose to prevent choking. Transport to nearest medical facility at once. While waiting for transport, elevate legs, and continue pouring on water and fanning. If conscious, have worker drink as much water as possible. Do not give salt.

The potential for heat stress is influenced by many factors, including ambient temperature, relative humidity, amount of sunshine, wind speed, amount and type of PPE, and the workload.

Whether a worker actually develops heat stress symptoms also depends on the worker, including his/her level of hydration, fitness, weight, height, and degree of acclimatization. As a general rule, workers who have worked two hours or more in hot environments or in PPE that restricts heat loss from the body for seven or more days in a row are considered to be *acclimatized.*

Unfortunately, PPE used for protection from pesticides can increase the potential for heat stress. The PPE features that keep pesticides out also serve to prevent body heat from escaping. People cannot work as long or as hard when the PPE they are wearing reduces the body's ability to maintain its thermal balance. Under some environmental conditions, certain activities cannot be performed without jeopardizing the health of the worker unless auxiliary cooling is provided. The risk that PPE may aggravate heat stress must be balanced with the PPE's potential worker protection benefits.

One way to reduce the risk of heat stress is to modify the work/rest cycle, i.e., reduce the work time and increase the rest time during each hour on the job.

## B. MANAGING WORK ACTIVITIES: APPROACHES FOR SETTING WORK/REST PERIODS

Employers should include the following approaches for managing work in their heat stress control program:

- setting up rest breaks
- rotating tasks
- shifting times for doing heavy work and work requiring protective gear
- reducing workloads
- postponing non-essential tasks.

These measures contradict some traditional notions about work and productivity, but as part of an overall heat stress control program, these measures help protect employee health and help maintain worker efficiency and safety. Heat stress control programs are now standard in many industries and in the armed forces.

## 1 Setting Up Rest Breaks

Workers recover from the heat more effectively with shorter, more frequent rest breaks than they do with longer, less frequent breaks. Longer, frequent rest breaks are necessary for heavier work and for work in higher temperatures and humidities. Rest breaks should be promoted among workers as a time for drinking water. Whenever possible, rest breaks should be taken in a shaded or air-conditioned area. It is hard to rest effectively in a hot environment. Rest breaks help workers recover from heat because:

- the heart rate slows
- the body cools down
- body fluids lost from sweating are replaced with drinking water.

Under mild conditions, workers wearing protective gear might take a ten-minute break every hour. When heat stress conditions increase, a five-minute break every half hour would be better. When working in chemical-resistant suits and/or in higher temperatures and humidities, the length of rest breaks and the amount of water consumed must increase sharply (see Tables 7.2 and 7.3).

Work/rest cycles should be flexible. Even among acclimatized workers, there are large individual differences in work capacity and tolerance to heat, and this tolerance can vary from day to day. Many persons are able to work under hot conditions for longer periods than those given in Tables 6.18 and 6.19; for others, these work periods may be too long. It is more important that supervisors understand the trends and underlying principles of these tables than that they follow work/rest times exactly.

The suggested work/rest periods in these tables are not a guarantee of protection against heat illness and should not be used as a substitute for good judgment and experience

People naturally want to persevere and finish a task at hand. In agriculture in particular, the demands of crop production and pest control during warm and hot weather create pressure to get as much done in as short a period of time as possible. In addition, workers and employers often face strong economic pressure not to interrupt work.

Do not allow workers to push themselves or to be pushed to keep on working when they begin to feel ill from the heat. Supervisors should insist that workers take the time necessary to cool down. Under extreme circumstances, it may be necessary to stop work altogether, unless additional cooling garments are used.

"Toughing it out" is dangerous. Workers are in danger of becoming seriously ill or dying when either they or their supervisors fail to recognize the need for breaks to cool off under heat stress conditions. With heat exhaustion and heat stroke, the flow of blood to the brain decreases and mental function becomes impaired. A worker will become less aware of his or her condition and may become aggressive, try to work harder, and resist being told to stop work. There have been fatalities where workers pushed themselves harder when they should have been forcibly stopped from continuing to work.

### 2. Rotating Tasks

When possible, rotate heavier tasks among workers in the best physical condition, and alternate heavy work with light and medium work.

### 3. Shifting Times for Certain Activities

When possible, schedule heavy tasks and work requiring protective gear for cooler hours of the day, such as in the early morning or at night, but be aware that cooler, early morning temperatures can mislead some workers to think that heat stress will not occur under these conditions. The relative humidity is usually highest at this time. Very high humidity limits evaporation of sweat, even at cooler temperatures. This can prevent adequate cooling of the body and lead to serious heat illness. When the relative humidity is above 70%, heavy work can be particularly risky. Workers should still be alert to heat strain, take rest breaks to cool down, and drink enough water to replace body fluid lost from perspiration.

### 4. Postponing Non-Essential Tasks During Heat Spells

With unusually hot weather lasting longer than two days, heat can build up in the environment both at work and at home, and the body can become progressively dehydrated. Chronic dehydration can occur without any signs of thirst; in its early stages it may be indicated by a lack of alertness. During heat spells, increase water consumption both on and off the job, and postpone nonessential tasks likely to cause severe heat strain until after the heat spell is over. For tasks which cannot be postponed, special measures should be taken, such as:

- establishing sharply limited work periods
- restricting overtime work
- monitoring workers closely for heat stress
- wearing cooling garments such as ice vests (vests with ice or ice packs)
- assigning heavy tasks only to workers who are in good condition and are fully acclimatized
- conducting safety meetings to emphasize special heat spell procedures.

## 5. Recommended Work/Rest Periods

Tables 7.2 and 7.3 list the recommended work/rests periods for different work loads being performed at various temperatures and the recommendations on water intake. The recommendations are taken in part from the EPA document *A Guide to Heat Stress in Agriculture,* 1994.

**Table 7.2** Approach for setting work/rest periods and the amount of drinking water for workers wearing normal work clothes (EPA, *A Guide to Heat Stress in Agriculture*, 1994).

| Air Temperature °F | Light Work | Moderate Work | Heavy Work | Minimum Water to Drink** |
|---|---|---|---|---|
| 90 | Normal | Normal | Normal | 1/2 pint every 30 minutes |
| 91 | Normal | Normal | Normal | same as 90°F |
| 92 | Normal | Normal | Normal | same as 90°F |
| 93 | Normal | Normal | Normal | same as 90°F |
| 94 | Normal | Normal | Normal | same as 90°F |
| 95 | Normal | Normal | 45/15†† | same as 90°F |
| 96 | Normal | Normal | 45/15 | same as 90°F |
| 97 | Normal | Normal | 40/20 | same as 90°F |
| 98 | Normal | Normal | 35/25 | same as 90°F |
| 99 | Normal | Normal | 35/25 | same as 90°F |
| 100 | Normal | 45/15†† | 30/30 | same as 90°F |
| 101 | Normal | 40/20 | 30/30 | same as 90°F |

**Table 7.2** continued.

| Air Temperature °F | Light Work | Moderate Work | Heavy Work | Minimum Water to Drink** |
|---|---|---|---|---|
| 102 | Normal | 35/25 | 25/35 | same as 90°F |
| 103 | Normal | 30/30 | 20/40 | 1/2 pint every 15 minutes |
| 104 | Normal | 30/30 | 20/40 | same as 103°F |
| 105 | Normal | 25/35 | 15/45 | same as 103°F |
| 106 | 45/15++ | 20/40 | Caution+ | same as 103°F |
| 107 | 40/20 | 15/45 | Caution+ | 1/2 pint every 10 minutes |
| 108 | 35/25 | Caution+ | Caution+ | same as 107°F |
| 109 | 30/30 | Caution+ | Caution+ | same as 107°F |
| 110 | 15/45 | Caution+ | Caution+ | same as 107°F |
| 111 | Caution+ | Caution+ | Caution+ | same as 107°F |
| 112 | Caution+ | Caution+ | Caution+ | same as 107°F |

** Varies from person to person and increases under heavier work and hotter conditions.
+Indicates very high levels of heat stress.
++ 45/15 = 45 minutes work and 15 minutes rest during each hour.

Table 7.3 lists the recommended work/rests periods for different work loads being performed at various temperatures and the recommendations on water intake in various sunlight conditions where radiant heating may have profound effects. The recommendations are taken in part from the EPA document *A Guide to Heat Stress in Agriculture,* 1994.

**Table 7.3** Approach for setting work/rest periods and amount of drinking water for workers wearing chemical-resistant suits* (EPA, *A Guide to Heat Stress in Agriculture*, 1994).

| Air Temperature | Work/Rest Periods | | | | | | | | | Minimum Water to Drink** |
|---|---|---|---|---|---|---|---|---|---|---|
| | Light Work | | | Moderate Work | | | Heavy Work | | | |
| | Full Sun | Partly Cloudy | No Sun | Full Sun | Partly Cloudy | No Sun | Full Sun | Partly Cloudy | No Sun | |
| 75°F | Normal schedule | Normal schedule | Normal schedule | Normal schedule | Normal schedule | Normal schedule | 35/25†† | Normal schedule | Normal schedule | One half pint every 30 minutes |
| 80°F | 30/30 | Normal schedule | Normal schedule | 20/40 | Normal schedule | Normal schedule | 10/50 | 40/20 | Normal schedule | One to 1.5 pints every 30 minutes |
| 85°F | 15/45 | 40/20 | Normal schedule | 10/50 | 25/35 | Normal schedule | Caution† | 15/45 | 40/20 | One pint or more every 15 minutes |
| 90°F | Caution† | 15/45 | 40/20 | Caution† | Caution† | 25/35 | Stop work | Caution† | 15/45 | Same as above |
| 95°F | Stop work | Stop work | 15/45 | Stop work | Stop work | Stop work | Stop work | Stop work | Stop work | Same as above |

* This table is taken in part from *A Guide to Heat Stress in Agriculture*, EPA, 1994.
** Varies from person to person and increases under heavier work and hotter conditions.
†Indicates very high levels of heat stress.
†† 45/15 = 45 minutes work and 15 minutes rest during each hour.

## REFERENCES

Briggs, S. A., *Basic Guide to Pesticides: Their Characteristics and Hazards*, Hemisphere Publishing Corp., New York, 1992.

Brown, V. K., *Acute Toxicity in Theory and Practice*, John Wiley & Sons, New York, 1980.

Cheremisinoff, N. P. and J. A. King, *Toxic Properties of Pesticides*, Marcel Dekker, Inc., New York, 1994.

Hayes, W. J. Jr. and E. R. Laws Jr., *Handbook of Pesticide Toxicology*, Academic Press, 1990.

Klaassen, C. D., Amdur M. O. and J. Doull (Eds.), *Casarett and Doull's Toxicology, The Basic Science of Poisons*, 3rd ed., Macmillan Publishing Co., New York, 1986.

Matsumura, F., *Toxicology of Insecticides*, 2nd ed., Plenum Press, New York, 1985.

U.S. Environmental Protection Agency, *A Guide to Heat Stress in Agriculture*, HW77, 1994.

U.S. Environmental Protection Agency, *Applying Pesticides Correctly: A Guide for Private and Commercial Applicators*, revised 1991.

U.S. Environmental Protection Agency, *Controlling Heat Stress Made Simple*, EPA 750-h-93-001, 1995.

U.S. Environmental Protection Agency, *Guidance Manual for Selecting Protective Clothing for Agricultural Pesticide Operations*, EPA 736-B-94-001, 1993.

U.S. Environmental Protection Agency, *Recognition and Management of Pesticide Poisonings*, 4th. ed., 1989.

U.S. National Institute of Safety and Health, *Occupational Exposure to Hot Environments: Revised Criteria 1986*, U.S. Department of Health and Human Services, 1986.

U.S. National Institute of Safety and Health, *Working in Hot Environments*, revised 1986.

Waxman, M.F., *Hazardous Waste Site Operations: A Training Manual for Site Professionals*, John Wiley & Sons, New York, 1996.

# CHAPTER 8

# APPLICATION EQUIPMENT

The application process is getting the pesticide to the target crop. This process usually involves a carrier, liquid, solid, or gas, which transports the pesticide to the intended surface or target. The application may be the simple act of spraying an aerosol repellent on our skin or applying a fumigant with a high-pressure applicator.

The pesticide application equipment used is important to the success of your pest control job. You must first select the right kind of application equipment, then you must use it correctly and take good care of it. Here are some things you should know about choosing, using, and caring for equipment.

## I. SPRAYERS

Sprayers are the most common pesticide application equipment. They are standard equipment for nearly every professional pesticide applicator and are used in every type of pest control operation. Sprayers range in size and complexity from simple, hand-held models to intricate machines weighing several tons.

### A. HAND SPRAYERS

Hand sprayers are often used to apply small quantities of pesticides. They can be used both in structures and outside for spot treatments and hard-to-reach areas. Most operate on compressed air, which is supplied by a manually operated pump. These sprayers are usually used in the structural pest control industry and are common for home gardens.

**Advantages:**

- economical
- simple to operate
- easy to clean and store.

**Limitations:**

- pressure and output rate fluctuate
- often provide too little agitation to keep wettable powders in suspension
- must be shaken frequently.

### 1. Pressurized Cans (Aerosols)

This type of sprayer consists of a sealed container of compressed gas and pesticides. The pesticide is driven through an aerosol-producing nozzle when the valve is activated. Pressurized cans usually have a capacity of less than one quart (one liter) and are not reusable. Larger reusable cylinders are available for structural pest control and for some greenhouse uses. Figure 8.1 depicts an aerosol can containing an insecticide under pressure.

**Figure 8.1** Aerosol can containing an insecticide under pressure, used for small, indoor projects.

## 2. Trigger Pump Sprayer

With trigger pump sprayers, the pesticide is not packaged under pressure. Instead, the pesticide and diluent are forced through the nozzle by pressure created when the pump is squeezed. Their capacity ranges from one pint to one gallon. Figure 8.2 illustrates a trigger pump sprayer used for household pest control.

**Figure 8.2** Trigger pump sprayers are useful for mixing or diluting a limited quantity of pesticide for a small job.

## 3. Hose-End Sprayer

This device causes a fixed rate of pesticide to mix with the water flowing through the hose to which it is attached. The mixture is expelled through a high-volume nozzle. These sprayers usually hold no more than one quart (one liter) of concentrated pesticide, but because the concentrate mixes with the water, they may deliver 20 gallons or more of finished spray solution before refilling. Figure 8.3 depicts a hose-end sprayer used for small pest control projects.

## 4. Push-Pull Hand Pump Sprayer

This type of sprayer depends on a hand-operated plunger which forces air out of a cylinder, creating a vacuum at the top of a siphon tube. The suction draws pesticide from the small tank and forces it out with the air flow. Capacity is usually one quart (one liter) or less (see Figure 8.4).

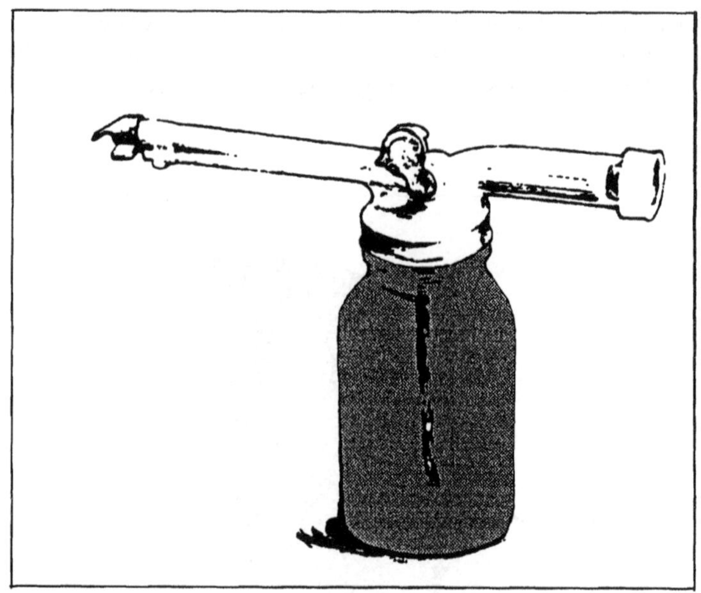

**Figure 8.3** Hose-end sprayers are used for small pest control projects.

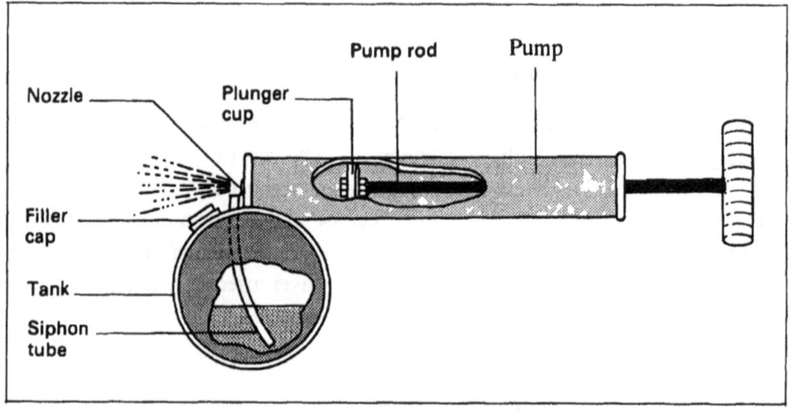


**Figure 8.4** Push-pull hand pump sprayers are also used for small pest control projects.

### 5. Compressed Air Sprayer

A hand-carried sprayer which operates under pressure, created by a self-contained manual pump. Capacity usually 1-3 gallons. Commonly used by homeowners and pest control operators. Figure 8.5 shows an illustration of hand-carried sprayer, and Figure 8.6 shows the application of insecticide also using a hand-carried sprayer.

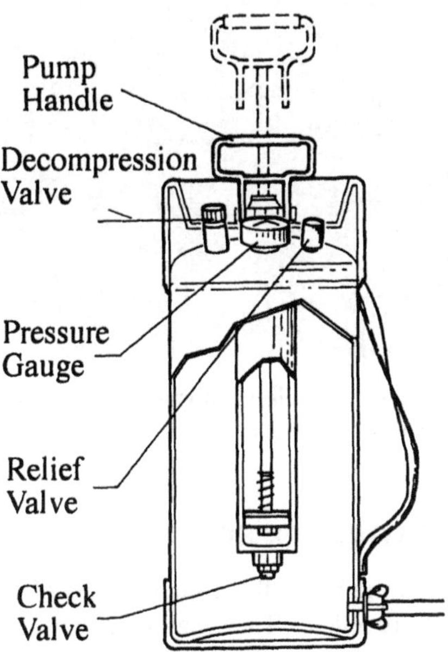


**Figure 8.5** Diagram of a compressed air sprayer showing working parts.

### 6. Backpack Sprayer

A backpack sprayer is similar to a push-pull sprayer, except that it is a self-contained unit (tank and pump) and is carried on the operator's back. A mechanical agitator plate may be attached to the pump plunger. Capacity of these sprayers is usually less than five gallons. Figure 8.7 shows a pest control worker using a backpack sprayer in a greenhouse.

### 7. Bucket Or Trombone Sprayer

These sprayers involve a double-action hydraulic pump which is operated with a push-pull motion. The pesticide is sucked into the cylinder and pushed out through the hose and nozzle with the stroke. Pressures up to 150 psi can be achieved. The separate tank often consists of a bucket with a capacity of five gallons or less. Figure 8.8 shows an applicator using a trombone sprayer.

**Figure 8.6** Applying insecticide to mosquito producing stagnant water using a hand-carried sprayer.

## B. SMALL MOTORIZED SPRAYERS

Some small sprayers have all the components of larger field sprayers but are usually not self-propelled. They may be mounted on wheels so they can be pulled manually, mounted on a small trailer for pulling behind a small tractor, or skid-mounted for carrying on a small truck. They may be low-pressure or high-pressure, according to the pump and other components with which they are equipped.

Standard equipment includes a hose and an adjustable nozzle on a handgun. Some models have multi-nozzle booms. These sprayers are suitable for relatively small outdoor areas, especially small orchards, ornamental and nursery plantings, and golf course greens.

**Advantages:**

- larger capacity than hand sprayers
- low- and high-pressure capability
- built-in hydraulic agitation

- small enough for limited spaces.

**Figure 8.7** Worker using a backpack sprayer in a greenhouse.

**Limitations:**

- not suitable for general field use
- cost is relatively high.

### 1. Estate Sprayers

These sprayers are mounted on a two-wheel cart with handles for pushing. Trailer hitches are available for towing the units. Spray material is hydraulically agitated. Some models have 15- to 30-gallon tanks. Pumps deliver 1.5 to 3 gallons per minute at pressures up to 250 psi.

Larger models have 50-gallon tanks and pumps that deliver 3 to 4 gallons per minute at pressures up to 400 psi. Power is supplied by an air-cooled engine of up to five horsepower.

**Figure 8.8** Trombone or bucket sprayers used for small pest control projects.

## 2. Power Wheelbarrow Sprayer

This sprayer is simply a powered version of the manually operated wheelbarrow sprayer described above. It may deliver up to three gallons per minute and can develop pressures up to 250 psi. The 1.5 to 3 horsepower engine is usually air-cooled. The tank size ranges from 12 to 18 gallons. The spray mixture may be either mechanically or hydraulically agitated.

## C. LARGER POWER-DRIVEN SPRAYERS

### 1. Low-Pressure Sprayers

These sprayers are designed to distribute diluted liquid pesticides over large areas. They are most often used in agricultural, ornamental, turf, forestry, aquatic, regulatory, and right-of-way pest control operations. They deliver a low to moderate volume of spray—usually 10 to 60 gallons per acre—at working pressures ranging from 10 to 80 psi.

These sprayers are usually mounted on tractors, trucks, or boats, but some are self-propelled. Roller and centrifugal pumps are most often used and provide outputs from 5 to more than 20 gallons per acre. Tank sizes range from 50 gallons or less to 1,000 gallons. The spray material is usually hydraulically agitated, but mechanical agitation may be used.

**Figure 8.9** Low-pressure sprayer commonly used by lawn services to apply pesticides and fertilizers. These sprayers are used to apply chemicals to several lawns before refilling is necessary (EPA, *Applying Pesticides Correctly*).

**Advantages:**

- medium to large tanks
- lower cost than high-pressure sprayers
- versatility.

**Limitation:**

- low pressure limits pesticide penetration and reach.

*a. Boom Sprayers*

Low-pressure sprayers are often equipped with sprayer booms ranging from 10 to 60 feet in length. The most common booms are between 20 and 35 feet long and contain many nozzles. The height of the sprayer boom must be easily adjustable to meet the needs of the job. Boom supports should allow the boom to be set at any height from 12 to 72 inches above the surface being sprayed. Many nozzle arrangements are possible, and special-purpose booms are available. Figures 8.10A-E show different types of boom sprayers.

*b. Boomless Sprayers*

Low-pressure sprayers which are not equipped with booms generally have a central nozzle cluster that produces a horizontal spray pattern. The resulting swath is similar to the pattern made by a boom sprayer. These sprayers are useful in irregularly-shaped areas, because they can move through narrow places and avoid trees and other obstacles. Some low-pressure sprayers are equipped with a hose and handgun nozzle for applications in small or hard-to-reach areas.

## 2. High-Pressure Sprayers

These sprayers are used to spray through dense foliage, thick animal hair, to the tops of tall trees, and into other areas where high-pressure sprays are necessary for adequate penetration. Commercially, they are used in agricultural, livestock, ornamental, turf, forestry, right-of-way, and some structural pest control operations. Often called "hydraulic" sprayers, they are equipped to deliver large volumes of spray—usually 20 to 500 gallons per acre—under pressures ranging from 150 to 400 psi or more.

These sprayers are usually mounted on tractors, trailers, trucks, or boats, or are self-propelled. Piston pumps are used and provide outputs of up to 60 gallons or more per minute. Because the application rate is usually 100 gallons per acre or more, large tanks (500 to 1,000 gallons) are used.

Mechanical agitators are usually standard equipment, but hydraulic agitators may be used. When fitted with correct pressure unloaders, these sprayers can be used at low pressures. All hoses, valves, nozzles, and other components must be designed for high-pressure applications. High-pressure sprayers may be equipped with a hose and single handgun nozzle for use in spraying trees and animals. These sprayers are also fitted with a boom for agricultural, ornamental, nursery, and aquatic applications. Figure 8.11 shows several pictures of workers with high-pressure sprayers.

**Advantages:**

- provide good penetration and coverage of plant surfaces
- usually well built and long lasting if properly cared for.

**Limitations:**

- high cost
- large amounts of water, power, and fuel needed
- high pressure may produce fine droplets which drift easily.

### 3. Air Blast Sprayers

Air blast sprayers use a combination of air and liquid rather than liquid alone to deliver the pesticide to the surface being treated. They are used in agricultural, ornamental and turf, biting fly, forestry, livestock, and right-of-way pest control operations.

These sprayers usually include the same components as low-pressure or high-pressure sprayers, plus a high-speed fan. Nozzles operating under low pressure deliver spray droplets directly into the high-velocity airstream. The air blast shatters the drops of pesticide into fine droplets and transports them to the target. The air blast is directed to one or both sides as the sprayer moves forward, or it may be delivered through a movable nozzle.

Most air blast sprayers are trailer-mounted, but tractor-mounted models are available. Tank capacity ranges from 100 to 1,000 gallons. Most of these sprayers can be adapted to apply either high or low volumes of spray material as well as concentrates. Mechanical agitation of the spray mixture is common. An air blast sprayer may cover a swath up to 90 feet wide and reach trees up to 70 feet tall.

**Figures 8.10A-E** Boom sprayers (8.10A, courtesy of University of Wisconsin Extension, 8.10B courtesy of Monsanto, and 8.10 C,D,E courtesy of American Cyanamid Company).

**Advantages:**

- good coverage and penetration
- mechanical agitation
- high capacity
- can spray high or low volumes
- low pump pressures.

**Limitations:**

- high cost of equipment
- drift hazards
- use of concentrated pesticides may increase chance of dosage errors
- not suitable for windy conditions
- hard to confine discharge to limited target area
- difficult to use in small areas
- high power requirement and fuel use.

## D. OTHER SPRAYERS

### 1. Ultralow Volume (ULV) Sprayers

These are sprayers that use special pesticide concentrates. They may be used in agricultural, ornamental, turf, forestry, right-of-way, biting fly, and some structural pest control operations. ULV sprayers may be hand-held or mounted on either ground equipment or aircraft.

**Advantages:**

- no water is needed, so less time and labor are involved
- equal control with less pesticide.

**Limitations:**

- does not provide for thorough coverage
- hazards of using high concentrates
- chance of overdosage
- small number of pesticides registered for ULV use.

**Figure 8.11** High-pressure sprayers equipped with a hose and single hand-gun nozzle for use in difficult to reach spots.

## 2. Spinning Disc Sprayers

These sprayers use a special type of nozzle which spins at a high speed and breaks the liquid into uniformly sized droplets by centrifugal force. The droplets may be carried to the target by gravity or by an airstream created by a fan. Power to spin the nozzles is provided by small electric or hydraulic motors. Sizes range from a small hand-held type to large tractor-mounted and trailer-mounted units. Figure 8.11 shows a spinning disc herbicide applicator.

**Advantages:**

- low volume of water required
- produces narrower range of droplet sizes than conventional nozzles, thus reducing drift

- droplet size can be adjusted by speed of rotation
- low-pressure pump and components.

**Limitations:**

- relatively high cost
- foliar penetration may be limited on gravity type
- not suitable for use in windy conditions.

**Figure 8.12** An air blast sprayer (EPA, *Applying Pesticides Correctly*).

### 3. Recirculating Sprayers

These devices usually are used to apply contact herbicides to weeds which are taller than the crop in which they are growing. Solid streams of highly concentrated herbicides are directed across rows above the crop. Spray material which is not intercepted by the weeds is caught in a box or sump on the opposite side of the row and is recirculated.

**Advantages:**

- uses small quantities of pesticide
- less pesticide moves off target and into environment
- permits treatment of weeds which have escaped other control measures
- protects susceptible nontarget plants from injury.

**Limitations:**

- use limited to special situations
- relatively high cost.

### 4. Electrostatic Sprayers

Electrostatic sprayer systems are designed to reduce drift and apply less pesticide per acre. The pesticide is charged with a positive electric charge as it leaves the nozzles. Plants have a natural negative charge, so the pesticides are attracted to the plants. The spray is directed horizontally through or above the crop (depending on the pesticide being applied).

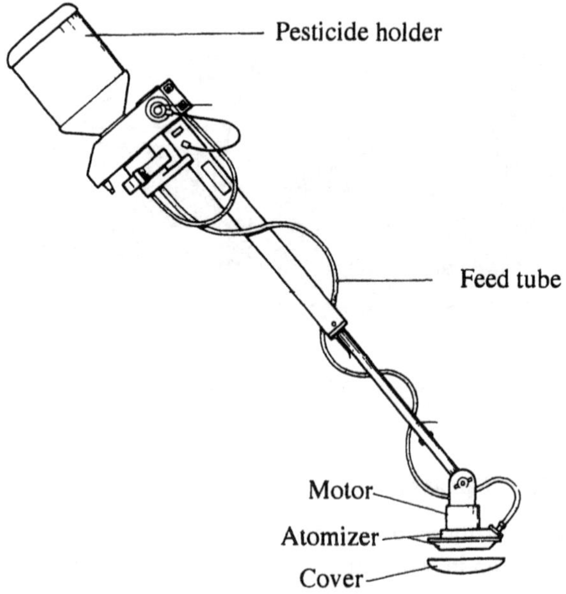


**Figure 8.13** Spinning disc herbicide applicator (Waxman, 1991).

## E. SPRAYER PARTS

### 1. Tanks

Tanks should have large openings for easy filling and cleaning. They should allow straining during filling and have provision for mechanical or hydraulic agitation. The tank should be made of corrosion-resistant material such as stainless steel or fiberglass. If made of mild steel, it should have a protective lining or coating.

The tank should have a large drain, and other outlets should be sized to the pump capacity. If you use dual tanks, make sure the plumbing allows for agitation and adequate withdrawal rates in both tanks. All tanks should have a gauge to show the liquid level. External gauges should be protected to prevent breakage. All tanks should have a shut-off valve for storing liquid pesticide temporarily while other sprayer parts are being serviced. Figures 8.9A-E show boom sprayers with large tanks.

## 2. Pumps

The pump must have sufficient pumping capacity to supply the needed volume to the nozzles and to the hydraulic agitator (if necessary) and to maintain the desired pressure. The pump parts should be resistant to corrosion and abrasion if abrasive materials such as wettable powders are to be used. Select gaskets, plunger caps, and impellers that are resistant to the swelling and chemical breakdown caused by many liquid pesticides. Consult your dealer for available options.

Never operate a sprayer pump at speeds or pressures above those recommended by the manufacturer. Pumps will be damaged if run dry or with restricted inlet or outlet. Pumps depend on the spray liquid for lubrication and removal of the heat of friction.

### *a. Roller Pumps*

Roller pumps are among the least expensive and most widely used of all sprayer pumps. They provide moderate volumes (8 to 30 gpm) at low to moderate pressure (10 to 300 psi). Often used on low-pressure sprayers, roller pumps are positive-displacement, self-priming pumps. The rollers, made of nylon, teflon, or rubber, wear rapidly in wettable powders but are replaceable. A pump that will be subjected to such wear should have a capacity at least 50 percent greater than that needed to supply the nozzles and agitator. This reserve capacity will extend the life of the pump. The pump case is usually cast iron or a nickel alloy. Roller pumps are best for emulsifiable concentrates, soluble powders, and other nonabrasive pesticide formulations.

### *b. Gear Pumps*

Gear pumps are used on sprayers with low operating pressures. They provide low to moderate volume (5 to 65 gpm) at low to moderate pressures (20 to 100 psi). Often used on special-purpose sprayers, gear pumps are positive-displacement, self-priming pumps. The self-priming ability is rapidly lost as the pump wears. These pumps are designed for oil solution formulations and wear rapidly when suspensions of wettable powders are used. The parts are generally not replaceable. The pump is not affected by solvents, since all parts are metal. The case may be bronze with stainless steel impellers, or it may be made entirely of bronze. Figure 8.14A illustrates the gear pump.

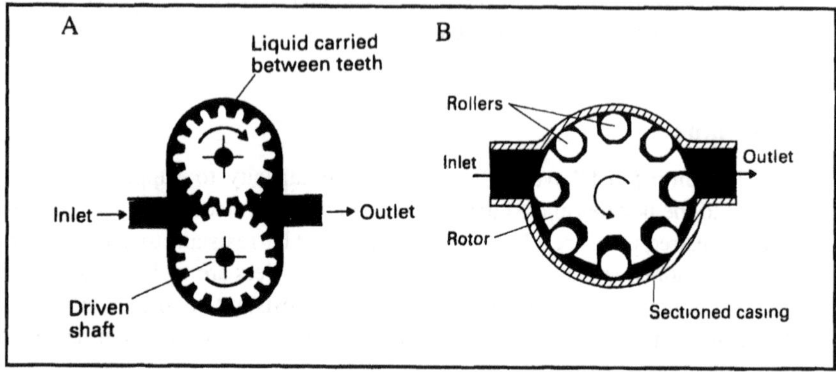


**Figure 8.14A and B** Figure A illustrates the gear pump and B the roller pump.

### c. *Diaphragm Pumps*

Diaphragm pumps deliver low volume (3 to 10 gpm) at low to moderate pressure (10 to 100 psi). They withstand abrasion from wettable powder mixtures much better than the gear or roller pumps because the spray mixture does not contact any moving metal parts except the valves. Diaphragm pumps are positive-displacement, self-priming pumps. The rubber or neoprene diaphragm may be damaged by some solvents. The pump case is usually iron. Figure 8.15A diagrams the internal structure of a diaphragm pump.

### d. *Centrifugal Pumps*

Centrifugal pumps are relatively inexpensive pumps adaptable to a wide variety of spray applications. Generally, they deliver high volume (up to 200 gpm) at low pressures (5 to 70 psi); however, two-stage pumps develop high pressures (up to 200 psi). Used on agricultural sprayers, commercial spray-dip machines, and other equipment, these are not positive-displacement pumps, so pressure regulators and relief valves are not necessary. They are not self-priming and must be mounted below the tank outlet or with a built-in priming system. Centrifugal pumps are well adapted for spraying abrasive materials because the impeller does not contact the pump housing. Slany models are easily repairable. The pump case is usually iron; the impeller is iron or bronze. Figure 8.15B diagrams the movement of fluid in centrifugal pumps.

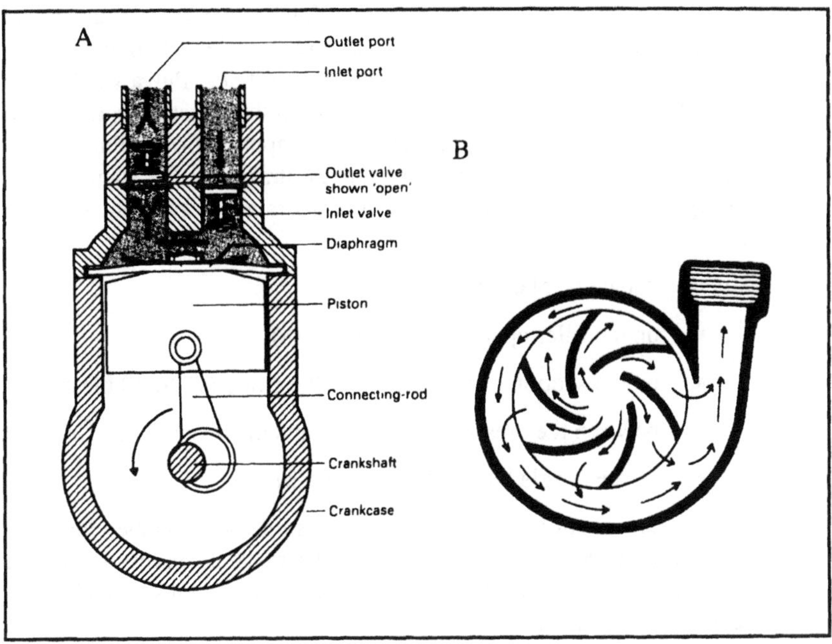


**Figures 8.15 A and B** Figure A illustrates the diaphragm pump and B the centrifugal pump.

*e. Piston Pumps*

Piston pumps are the most expensive of the commonly used sprayer pumps. They deliver low to medium volumes (2 to 60 gal) at low to high pressures (20 to 800 psi). Used for high-pressure sprayers or when both low and high pressures are needed, piston pumps are positive-displacement, self-priming pumps. They have replaceable piston cups made of leather, neoprene, or nylon fabric which make the pump abrasion-resistant and capable of handling wettable powders for many years. The cylinders are iron, stainless steel, or porcelain-lined. The pump casing is usually iron.

## 3. Strainers (Filters)

Pesticide mixtures should be filtered to remove dirt, rust flakes, and other foreign materials from the tank mixture. Proper filtering protects the working parts of the sprayer from undue wear and avoids time loss and uneven application caused by clogged nozzle tips.

Filtering should be progressive, with the largest mesh screens in the filler opening and in the suction line between the tank and the pump. They should be keyed to the size of the nozzle opening. Total screen area should be large enough to prevent pump starvation. This requires at least two square inches of screen area for each gpm of flow in the suction line. Put a smaller mesh strainer in the pressure line between the pump and the pressure regulator, with at least one square inch of screen area for each gpm of flow. Put the finest mesh strainer nearest the nozzles. Do not use a strainer in the suction line of a centrifugal pump, but be sure the tank has a strainer to take out large trash particles.

Strainers should be placed:

- on the filler opening (12 to 25 mesh)
- on the suction or supply line to the pump (15 to 40 mesh)
- between the pressure relief valve and the boom (25 to 100 mesh)
- on the nozzle body (50 to 100 mesh).

Clean strainers after each use, or during use if they become clogged. A shut-off valve between the tank and suction strainer is necessary to allow cleaning the strainer without draining the contents of the tank. Replace damaged or deteriorated strainers.

Strainers are the best defense against nozzle plugging and pump wear. Nozzle screens should be as large as nozzle size permits; however, the screen opening should be less than the nozzle opening. Nozzle catalogs specify the proper screen size for each nozzle. Figure 8.16 illustrates strainers, filters, or screens.

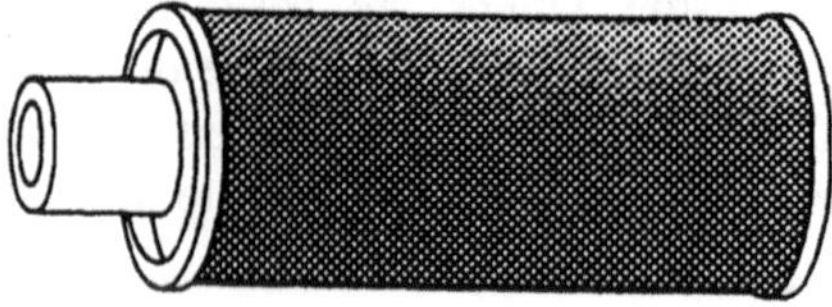

Pump intake strainer

Nozzel tip strainer

**Figure 8.16** Strainers, filters, or screens. Pump intake strainer and nozzle tip strainer.

## 4. Hoses

Select neoprene, rubber, or plastic hoses that:

- have burst strength greater than the peak operating pressures
- have a working pressure at least equal to the maximum operating pressure
- resist oil and solvents present in pesticides
- are weather resistant.

Suction hoses should be reinforced to resist collapse. They should be larger than pressure hoses, with an inside diameter equal to or larger than the inlet part of the pump. All fittings on suction lines should be as large as or larger than the inlet part of the pump.

Keep hoses from kinking or being rubbed. Rinse them often, inside and outside, to prolong life. During the off season, store the unit out of the sun. Replace hoses at the first sign of surface deterioration (cracking or checking).

## 5. Pressure Gauges

Pressure gauges monitor the function of the spraying system. They must be accurate and have the range needed for your work. For example, a zero to 100 psi gauge with 2-pound gradations would be adequate for most low-pressure sprayers.

Check frequently for accuracy against an accurate gauge. Excess pressure will destroy a gauge. If your gauge cannot be zeroed, replace it. Use gauge protectors to guard against corrosive pesticides and pressure surges.

## 6. Pressure Regulators

The pressure regulator controls the pressure and, indirectly, the quantity of spray material delivered by the nozzles. It protects pump seals, hoses, and other sprayer parts from damage due to excessive pressure.

The bypass line from the pressure regulator to the tank should be kept fully open and unrestricted and should be large enough to carry the total pump output without excess pressure buildup. The pressure range and flow capacity of the regulator must match the pressure range you plan to use and the capacity of the pump. Agitation devices should never be attached to the bypass line discharge. Pressure regulators are usually one of three types:

**Throttling valves** simply restrict pump output, depending on how much the valve is open. These valves are used with centrifugal pumps, whose output is very sensitive to the amount of restriction in the output line.

**Spring-loaded bypass valves** (with or without a diaphragm) open or close in response to changes in pressure, diverting more or less liquid back to the tank to keep pressure constant. These valves are used with roller, diaphragm, gear, and small piston pumps.

**Unloader valves** work like a spring-loaded bypass valve when the sprayer is operating. However, when the nozzles are shut down, they reduce strain on the pump by moving the overflow back into the tank at low pressure. These valves should be used on larger positive-displacement pumps (piston and diaphragm) to avoid damage to the pump or other system components when the nozzles are cut off.

## 7. Agitators

Every sprayer must have agitation to keep the spray material uniformly mixed. If there is too little agitation, the pesticide will be applied unevenly. If there is too much agitation, some pesticides may foam and interfere with pump and nozzle operation. The type of agitation necessary depends on the pesticide formulation used.

### *a. Bypass Agitators*

Soluble powders and liquid formulations such as solutions and emulsifiable concentrates require little agitation. Bypass agitation is sufficient for these formulations. Bypass agitation uses the returning liquid from the pressure relief valve to agitate the tank. The return must extend to the bottom of the tank to prevent excessive foaming.

Bypass agitation is not sufficient for wettable powders or in tanks larger than 55 gallons unless a centrifugal pump is used. Centrifugal pumps usually have outputs large enough to make bypass agitation adequate even for wettable powders in tanks less than 100 gallons.

### *b. Hydraulic (Jet Action) Agitators*

Hydraulic agitation is required for wettable powder and flowable formulations in small tanks and for liquid formulations in 100-gallon or larger tanks with gear, roller, piston, or diaphragm pumps. Hydraulic agitation is provided by the high-pressure flow of surplus spray material from the pump. The jet or jets are located at the bottom of the tank. The agitator is connected to the pressure side of the pump. Jet agitator nozzles should never be placed in the bypass line. The pump and tank capacity and operating pressure determine the minimum jet number and size:

- 55 gallon = 1 or more jets
- 100 to 150 gallon = 3 or more jets
- 200 gallon and larger = 5 or more jets.

### *b. Mechanical Agitation*

Wettable powder formulations are best mixed and kept in suspension with mechanical agitation. The mechanical agitator usually consists of flat blades or propellers mounted on a shaft which is placed lengthwise along the bottom of the tank. The paddles or propellers are rotated by the engine to keep the material well mixed. Mechanical agitators are usually found only on large, high-pressure hydraulic sprayers.

## 8. Control Valves

Quick-acting cutoff valves should be located between the pressure regulator and the nozzles to provide positive on-off action. These control valves should be rated for the pressures you intend to use and should be large enough not to restrict flow when open. Cutoff valves to stop all flow or partial flow to any section of the spraying system should be within easy reach of the sprayer operator.

There are many kinds of control valves. Mechanical valves must be accessible to the operator's hand; electrically operated valves permit remote control of flow. For tractors or self-propelled sprayers with enclosed cabs, the remote-controlled valves permit all hoses carrying pesticides to be kept safely outside the cab.

## 9. Nozzles

Nozzles are made up of four major parts: the nozzle body, the cap, the strainer (screen), and the tip or orifice plate. They may also include a separate spinner plate. Successful spraying depends on the correct selection, assembly, and maintenance of the nozzles. Figure 8.17 illustrates a typical cone pattern nozzle with its parts.

The nozzle body holds the strainer and tip in proper position. Several types of tips that produce a variety of spray patterns may be interchanged on a single nozzle body made by the same manufacturer.

The cap is used to secure the strainer and the tip to the body. The cap should not be overtightened.

The nozzle strainer is placed in the nozzle body to screen out debris which may clog the nozzle opening. The type of nozzle strainer needed depends on the size of the nozzle opening and the type of chemical being sprayed.

Special nozzle screens fitted with a check valve help prevent nozzle dripping. Check valves should be used in situations where a sprayer must be stopped and started frequently, such as in small target areas, near sensitive crops or areas, indoors, or for right-of-way treatments. The operator must check these spring-loaded ball valves frequently to assure proper operation.

Nozzle tips break the liquid pesticide into droplets. They also distribute the spray in a predetermined pattern and are the principal element that controls the rate of application. Nozzle performance depends on:

- nozzle design or type
- operating pressure
- size of the opening
- discharge angle
- distance of nozzle from the target.

*a. Nozzle Patterns*

Nozzle patterns are of three basic types: solid stream, fan, and cone. Some special-purpose nozzle tips or devices produce special patterns. These include "raindrops," "flooding," and others that produce wide angle fan or cone-shaped patterns.

Solid stream nozzles—These nozzles are used in handgun sprayers to spray a distant or specific target such as livestock, nursery, or tree pests, and for crack and crevice treatment in and around buildings. They also may be attached to booms to apply pesticides in a narrow band or inject them into the soil.

Fan pattern nozzles—At least three types of nozzle tips have fan patterns. They are used mostly for uniform spray coverage of surfaces, for example, applying herbicides or fertilizers to soil.

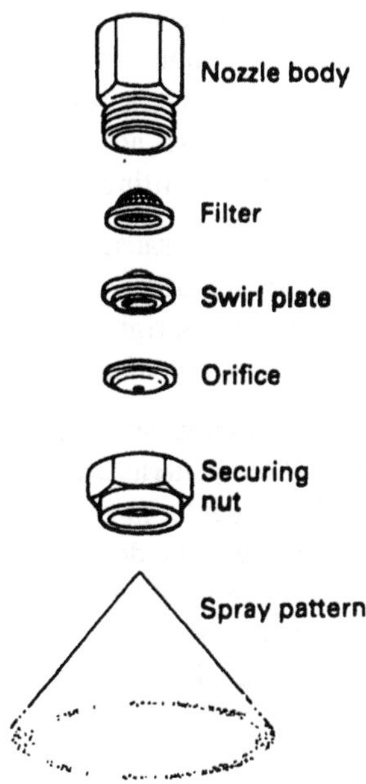


**Figure 8.17** Typical cone pattern nozzle with its parts.

The regular flat fan nozzle tip makes a narrow oval pattern with tapered ends. It is used for broadcast herbicide and insecticide spraying at 15 to 60 psi. The pattern is designed to be used on a boom and to be overlapped 30 to 50 percent for even distribution. Spacing on the boom, spray angle, and boom height determine proper overlap and should be carefully controlled. Tips are available in brass, plastic, stainless steel, and hardened stainless steel.

The even net fan nozzle makes a narrow oval pattern. Spray delivery is uniform across its width. It is used for band spraying and for treating walls and other surfaces. It is not useful for broadcast applications. Boom height and nozzle spray angle determine the width of the band sprayed. These tips are available in brass, plastic, stainless steel, and hardened stainless steel.

The flooding (net fan) nozzle delivers a wide-angle net spray pattern. It operates at very low pressure and produces large spray droplets. Its pattern is fairly uniform across its width but not as even as the regular net fan nozzle pattern. If used for broadcast spraying, it should be overlapped to provide double coverage. It is frequently used for applying liquid fertilizers, fertilizer-pesticide mixtures, or for directing herbicide sprays up under plant canopies. These tips are available in plastic, brass, or stainless steel.

Cluster nozzles are used either without a boom or at the end of booms to extend the effective swath width. One type is simply a large flooding deflector nozzle which will spread spray droplets over a swath up to 70 feet wide from a single nozzle tip. Cluster nozzles are a combination of a center-discharge and two or more off-center-discharge fan nozzles. The spray droplets vary in size from very small to very large, so drifting is a problem.

Coverage may be variable, because the spray pattern is not uniform. Since no boom is required, these nozzles are particularly well suited for spraying roadside hedgerows, fence rows, rights-of-way, and other inaccessible locations where uniform coverage is not critical.

Cone pattern nozzles—Hollow and solid cone patterns are produced by several types of nozzles. These patterns are used where penetration and coverage of plant foliage or other irregular targets are desired. They are most often used to apply fungicides and insecticides to foliage, although some types are used for broadcast soil applications of herbicides or fertilizers or combinations of the two. When cone pattern nozzles are used for air blast sprayer broadcast application, they should be angled to spray between 15° and 30° from the horizontal and should be spaced to overlap up to 100 percent at the top of the manifold. These tips are available in stainless steel, hardened stainless steel, and tungsten carbide. Figure 8.17 shows a typical cone pattern nozzle.

The side-entry hollow cone or "whirl-chamber" nozzle produces a very wide angle hollow cone spray pattern at very low pressures. It has a large opening and resists clogging. Because of the wide spray angle, the boom can be operated in a low position, reducing drift. Spacing for double coverage and angling 15° to 45° to the rear is recommended for uniform application. These nozzles may be used in place of net fan nozzle tips in broadcast applications.

Core-insert cone nozzles produce either a solid or hollow cone spray pattern. They operate at moderate pressures and give a finely atomized spray. They should not be used for wettable powders because of small passages which tend to clog and wear rapidly due to abrasion.

Disc-core nozzles produce a cone-shaped spray pattern which may be hollow or solid. The spray angle depends on the combination of disc and core used and also, to some extent, on the pressure. Discs made of very hard materials resist abrasion well, so these nozzles are recommended for spraying wettable powders at high pressures.

Adjustable cone nozzles change their spray angle from a wide cone pattern to a solid stream when the nozzle collar is turned. Many manual sprayers are equipped with this type of nozzle. Handguns for power sprayers have adjustable nozzles which usually use an internal core to vary the spray angle.

*b. Nozzle Materials*

Most nozzle parts are available in several materials. Here are the main features of each kind:

**Brass:**

- moderately expensive
- resists corrosion from most pesticides
- wears quickly from abrasion
- probably the best material for general use
- liquid fertilizers will corrode.

**Plastic:**

- moderately expensive
- will not corrode
- resists abrasion better than brass
- may swell when exposed to organic solvents.

**Stainless Steel:**

- moderately expensive,
- resists abrasion, especially if hardened
- good corrosion resistance
- suited for high pressures, especially with wattle powders.

**Aluminum:**

- inexpensive
- resists some corrosive materials
- is easily corroded by some fertilizers.

**Tungsten Carbide and Ceramic:**

- very expensive
- highly resistant to abrasion and corrosion

- best materials for high pressures and wettable powders.

## F. SPRAYER SELECTION, USE, AND CARE

Choosing the correct sprayer for each job is important. Your sprayer should be:

- designed to do the job
- durable
- convenient to fill, operate, and clean.

Always read and follow the operator's manuals for all spray equipment. They will tell you exactly how to use and care for it. After each use, rinse out the entire system. Check for leaks in lines valves, seals, and tank. Remove and clean nozzles, nozzle screens, and strainers.

Be alert for nozzle clogging and changes in nozzle patterns. If nozzles clog or other trouble occurs in the field, be careful not to contaminate yourself while correcting the problem. Shut off the sprayer and move it to the edge of the field before dismounting. Wear protective clothing while making repairs. Clean plugged nozzles only with a toothbrush or wooden toothpick or similar instrument. Never use your mouth.

To prepare spray equipment for storage, rinse and clean the system. Then fill the tank almost full with clean water. Add a small amount of new light oil to the tank. Coat the system by pumping this mixture out through the nozzles or handgun. Drain the pump and plug its openings or fill the pump with light oil or antifreeze. Remove nozzles and nozzle screens and store in light oil or diesel fuel.

# II. AEROSOL GENERATORS AND FOGGERS

Aerosol generators and foggers convert special formulations into very small, fine droplets. Single droplets cannot be seen, but large numbers of droplets are visible as a fog or mist. Aerosol generators and foggers are usually used to completely fill a space with a pesticidal fog. Some insects in the treated area are killed when they come in contact with the poison. Other insects are simply repelled by the mist and return quickly after it has settled.

Thermal foggers, also called thermal generators, use heat to vaporize a special oil formulation of a pesticide. As the pesticide vapor is released into the cooler air, it condenses into very fine droplets, producing a fog. Figure 8.18 shows a vehicle-mounted fogger.

Other aerosol generators (cold foggers) break the pesticide into aerosols by using mechanical methods such as:

- rapidly spinning discs
- extremely fine nozzles and high pressure (atomizing nozzles)
- strong blasts of air.

This specialized equipment is most often used in greenhouses, barns, warehouses, and other structures and for biting fly and mosquito control in outdoor recreation areas.

**Advantages:**

- no unsightly residue
- safe reentry immediately after ventilating the area
- penetration in dense foliage
- penetration of cracks and crevices and furniture
- some indoor devices are automatic and do not require presence of applicator.

**Limitations:**

- aerosols and fogs drift easily from target area —use outdoors is limited
- no residual control—pests may return to the area as soon as fog dissipates
- risk of explosion in enclosed areas.

## A. SELECTION, USE, AND CARE

Choose an aerosol generator according to where you will use it—indoors or outdoors. Aerosol and fog generators are manufactured for many special uses. There are truck- and trailer-mounted machines for use outdoors. Most hand-operated or permanently mounted automatic machines are for use indoors.

In general, use and care for an aerosol generator as you would a sprayer. They do require special precautions. Be sure that the pesticides used in them are registered for such use. Keep them on the target. Because of the effects of weather conditions during application, follow special use instructions. The operator, other humans, and animals must be kept out of the fog or smoke cloud.

# III. SOIL FUMIGATION EQUIPMENT

The equipment needed for applying soil fumigants depends on the kind of fumigant being used. There are two types of fumigants:

- low-pressure (low volatility) liquid fumigants
- highly volatile fumigants which remain as liquids only when placed under pressure.

## A. LOW-PRESSURE LIQUID FUMIGATORS

Equipment for applying low-pressure fumigants is widely varied but uses two basic designs for delivering the amount of fumigant to be metered out. These delivery systems are either pressure (pump)-fed or gravity-fed.

**Pressure-fed applicators** have a pump and metering device and deliver fumigant at pressure to the nozzle openings (orifices) as with a low-pressure sprayer.

**Figure 8.18** Fogger mounted on a vehicle.

**Gravity-fed applicators** use the size of the nozzle orifice and the pressure created by gravity to regulate the output of fumigant. Constant speed is necessary to maintain a uniform delivery rate. In most applicators, a constant head gravity flow device keeps the pressure at the orifice(s) constant as the tank or container of fumigant empties. Needle valves, orifice plates or discs, and capillary tubes are used to adjust the flow rate.

Low-pressure fumigators usually use the soil itself or water to keep the fumigant from vaporizing and moving off target too quickly. Some of the methods used are:

- soil injection
- soil incorporation
- drenching or flooding.

## 1. Soil Injection

Soil injectors use a variety of mechanisms to insert the fumigant into the soil (usually to a depth of 6 inches or more) and then cover the area with soil again to seal in the fumigant. Figure 8.19 shows an operator using a soil injection chisel.

## 2. Soil Incorporation

Soil incorporators are used when applying low-volatility fumigants. The fumigant usually is sprayed onto the soil surface. The area is immediately cultivated, usually to a depth of 5 inches or less, and then compacted with a drag, float, or cultipacker. Power-driven rotary cultivators are also used.

## 3. Drenching or Flooding

This method uses water as a sealant. The fumigant may be applied in the water as a drench.

Equipment used depends on the size and timing of the application. It may be applied with a sprinkling can, sprinkler system, or irrigation equipment, or the fumigant may be applied by spraying the soil surface and immediately flooding the area. The depth of the water seal (usually 1 to 4 inches of wetted soil) depends on the volatility of the fumigant.

The principal mechanisms include:

- chisel cultivators, blades, or shovels
- sweep cultivator shovels

- planter shoes
- plows.

## B. HIGH-PRESSURE FUMIGATORS

Effective application of highly volatile fumigants depends on tightly sealing the soil with tarps, plastic film, or similar covers. Figure 8.20 shows workers setting tarps for fumigation of a field.

**Figure 8.19** Hand-operated soil injection chisel.

There are two major methods of using vapor-proof tarps:

- Tarp supported off the ground and sealed around the edges: fumigant introduced under the tarp.
- Tarp applied to the soil by the injection chisel applicator immediately after the fumigant is injected.

Highly volatile fumigants must be handled in closed pressurized containers or tanks. The equipment is similar to gravity-flow, low-pressure fumigators. The pressure in the tank maintains the pressure at the nozzle orifices. The tank is either precharged with sufficient pressure to empty its contents, or an inert pressurized gas is fed into the tank during application to displace the fumigant. A gas pressure regulator maintains uniform pressure in the system. To insure accurate application, the fumigant must be under enough pressure to maintain a liquid state in the tank, pressure lines, manifold, and metering devices.

## 1. Selection

Pumps, tanks, fittings, nozzles or metering orifices, and lines must be corrosion-resistant. Soil injection knives should be designed to shed trash and allow the soil to seal over the fumigant.

Select high-pressure fumigators designed to handle the pressure created by the fumigant and the corrosive action of the product you plan to use.

**Figure 8.20** Workers setting tarps for fumigation of a field.

# IV. DUSTERS AND GRANULAR APPLICATORS

## A. DUSTERS

Dusters are used mostly by home gardeners and by pest control operators in structures.

### 1. Hand Dusters

Hand dusters may consist of a squeeze bulb, bellows, tube, shaker, sliding tube, or a fan powered by a hand crank.

**Advantages:**

- lightweight—do not require water
- the pesticide is ready to apply without mixing
- good penetration in confined spaces.

**Limitations:**

- high cost for pesticide
- hard to get good foliar adherence
- dust is difficult to direct and is subject to drifting.

### 2. Power Dusters

Power dusters use a powered fan or blower to propel the dust to the target. They include knapsack or backpack types, units mounted on or pulled by tractors, and specialized equipment for treating seeds. Their capacity in area treated per hour compares favorably with some sprayers.

**Advantages:**

- lightweight—no water required
- simply built
- easy to maintain
- low in cost.

**Limitations:**

- drift hazards
- high cost of pesticide

- application may be less uniform than with sprays
- difficult to get foliar adherence.

## B. GRANULAR APPLICATORS

Granular applicators are used mainly in agricultural, ornamental, turf, forestry, and aquatic pest control. They distribute granular pesticides by several different methods, including:

- forced air
- spinning or whirling discs (fertilizer spreaders)
- multiple gravity-feed outlets (lawn spreaders, grain drills)
- soil injectors (furrow treatments)
- ram-air (agricultural aircraft).

Granular applicators may be designed to apply the pesticides:

- broadcast—even distribution over the entire area
- to specific areas—banding, in-furrow, side-dress
- by drilling—soil incorporation or soil injection.

**Advantages:**

- inexpensive
- simple in design
- eliminates mixing—no water needed
- minimal drift hazard
- less exposure hazard to applicator.

**Limitations:**

- high cost for pesticides
- limited use against some pests because granules will not adhere to most foliage
- need to calibrate for each different granular formulation
- spinning disc types may give poor lateral distribution, especially on side slopes.

## C. SELECTION, USE, AND CARE

Look for a power duster that is easy to clean. It should give a uniform application rate as the hopper is emptied. Look for both hand and power dusters that direct the dust cloud away from the user.

Choose a granular applicator that is easy to fill and clean. It should have mechanical agitation over the outlet holes. This prevents clogging and helps keep the flow rate constant. Application should stop when drive stops, even if outlets are still open.

Both dusters and granular applicators are speed sensitive, so maintain uniform speed. Do not travel too fast for ground conditions. Bouncing equipment will cause the application rate to vary. Stay out of any dust created by action of the equipment.

Watch band applicators to see that the band width stays the same. Small height changes due to changing soil conditions may cause rapid changes in band width.

Clean equipment as directed by the operator's manual.

**Figure 8.21** Spinning or whirling disc granular applicator (EPA, *Applying Pesticides Correctly*).

## V. SEED TREATERS

Seed treaters are used to coat seeds with a pesticide. The amount of pesticide the seeds receive is important. Too much can injure the seed; too little will not control the pests. The three basic types of commercial seed treaters are:

- dust treaters
- slurry treaters
- liquid treaters.

### A. DUST TREATERS

These treaters mix seeds with a pesticide dust in a mechanical mixing chamber until every seed is thoroughly covered. The amount of pesticide to be added depends on the weight of seeds in the mixing chamber. Seed flow is controlled by adjusting the gate opening on the feed hopper. The gate opening is correct when the required weight of seeds is dumped into the mixing chamber with each batch. The amount of pesticide added is controlled by a vibrating feeder which is adjusted to achieve the desired dosage.

### B. SLURRY TREATERS

These treaters coat seeds with wettable powder pesticide formulations in the form of a slurry. Only a small amount of water is used with the pesticide so that seed germination or deterioration will not be triggered. As with dust treaters, a specific amount of pesticide is added to a specific weight of seeds in a mechanical mixing chamber. Slurry tanks have 15- to 35-gallon capacities, depending on the size of the treater. Agitators keep the material mixed during the treating operation. Wettable powders will rapidly separate from the water if not continuously agitated. Be sure to mix the pesticide and water thoroughly before starting the treater and before resuming treatment after any stoppage. Sediment on the bottom of the individual slurry cups must be stirred into the liquid after any stoppages.

### C. LIQUID OR DIRECT TREATERS

These treaters are designed to apply a small amount of pesticide solution to a large quantity of seeds. The pesticides suitable for this treatment are moderately volatile liquids which need not cover each seed entirely to achieve good pest control. Some of these treaters have dual tanks which allow treatment with more than one pesticide at a time.

### 1. Panogen Liquid Seed Treater

The Panogen treater meters one treatment cup of pesticide per dump of seed into a revolving mixing drum. The pesticide flows into the drum from a tube and is distributed over the seed as the seeds rub against the walls of the drum. The correct dosage is achieved by selecting the appropriate size or treatment cup for the chosen seed dump weight.

### 2. Mist-O-Matic Seed Treater

The Mist-O-Matic treater applies the pesticide as a mist directly onto the seed. The treater delivers one treatment cup of pesticide per dump of seed. The pesticide flows onto a rapidly whirling disc which breaks the liquid pesticide into a fine mist. The seeds fall onto a large cone which spreads the seeds out so they are evenly coated with the pesticide spray mist. The desired dosage is obtained by selecting the appropriate treatment cup size and adjusting the seed dump weight.

## VI. ANIMAL APPLICATION EQUIPMENT

### A. DIPPING VATS

Dipping vats are large tanks (vats) of liquid pesticide mixture used to treat livestock for external parasites. They are used in farm, ranch, and regulatory pest control operations. Portable dipping vats are usually trailer-mounted tanks with a set of folding ramps and railings. The animals are driven up the ramp onto a platform and forced into the tank so they are completely immersed. The animal's head may have to be pushed under the surface.

Maintaining the proper concentration of pesticide in the vat is very important. The bath volume should never be allowed to fall below the 7/8 level. Replenishment ratios are usually based on a knowledge of the amount of liquid used from the vat.

### B. SPRAY-DIP MACHINES

Spray-dip machines are used to treat livestock for external parasites. They are used in farm, ranch, regulatory, and other livestock pest control operations.

A spray-dip machine usually consists of a trailer-mounted chute with solid walls and gates at either end. The chute is located above a shallow tank and is equipped with several rows of large nozzles mounted in a manner that directs the spray mixture to thoroughly cover each animal. A large centrifugal pump supplies the pesticide to the nozzles. Surplus and runoff spray falls back into the tank where it is filtered and recycled to the nozzles. Figure 8.22 shows a diagram of a spray-dip machine application.

## C. FACE AND BACK RUBBERS

Face and back rubbers are bags or other containers of dry or liquid pesticide formulation used to control external parasites of livestock. They are used in farm, ranch, and other livestock confinement operations. The face and back rubbers are hung or mounted in areas adjacent to high livestock traffic, such as feeding troughs, watering troughs, and narrow gate entrances. When the animal rubs against the device, the pesticide is transferred to the animal's face, back, sides, or legs.

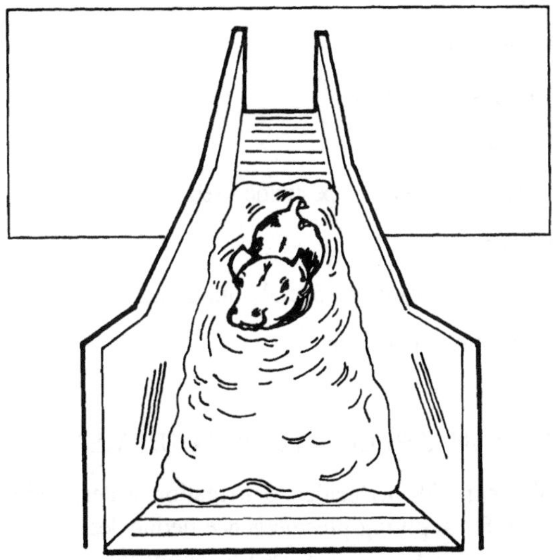

**Figure 8.22** Spray-dip machines are used to treat livestock for external parasites (EPA, *Applying Pesticides Correctly*).

## VII. SPECIALIZED APPLICATION EQUIPMENT

Some other types of application equipment do not fit into the common categories. They include wiper applicators, irrigation application equipment, and wax bars.

### A. WIPER APPLICATORS

These devices are used to apply contact or translocated (systemic) herbicides selectively to weeds in crop areas. Wicks made of rope, rollers made of carpet or other material, or absorbent pads made of sponges or fabric are kept wet with a concentrated mixture of contact herbicide and water and brought into direct contact with weeds. The herbicide is "wiped" onto the weeds but does not come in contact with the crop. Application may be to tall weeds growing above the crop or to lower weeds between rows, depending on the way the wiper elements are designed. Pumps, control devices, and nozzles are minimal or are eliminated altogether, and tanks are quite small because of the small amount of liquid applied.

**Advantages:**

- low cost
- simple to operate
- no drift
- reduces amount of pesticide used.

**Limitations:**

- applicable only in special situations
- difficult to calibrate.

### B. IRRIGATION APPLICATION EQUIPMENT

This equipment adds herbicides to irrigation water. This has become a common method for applying preemergence herbicides in most irrigated areas of the United States, and it is also used to apply insecticides, fungicides, and other pesticides. The pesticide is injected into the main irrigation stream with a positive-displacement pump. Accuracy of calibration and distribution is achieved by metering a large volume of dilute pesticide into the irrigation system. Antisiphon and switch valves prevent contamination of the irrigation water source or overflow into the slurry feed tank.

**Advantages:**

- inexpensive
- convenient
- field access unnecessary.

**Disadvantages:**

- constant agitation needed in slurry tank.
- application of more water per acre than recommended on label will result in leaching the chemical from the effective soil zone.
- sprinkler distribution must have appropriate overlap pattern to avoid underdosing or overdosing.
- injection of pesticides into flood and furrow irrigation systems may result in uneven concentrations of pesticides throughout the field, depending on soil permeability and field contours.

### C. WAX BARS

Herbicides are sometimes applied to turfgrass and vineyards with wax bars. The wax bars are impregnated with herbicides, then dragged slowly over the area to be protected.

**Advantages:**

- no drift
- no calibration.

**Disadvantages:**

- highly specialized, not readily available.

## III. CONCLUSION

Many factors affect the ability to place a pesticide on the target in the manner and amount for most effective results, with the least undesirable side effects, and at the lowest possible cost. Certainly the selection and use of equipment are of utmost importance and deserve major emphasis when considering pesticide application. However, without proper consideration to calibration, formulation, adjuvants, compatibility, and use records, successful application is not very probable. Successful application must also involve principles of proper timing and drift control.

## REFERENCES

Baker, P. B., *Arizona Agricultural Pesticide Applicator Training Manual*, Cooperative Extension, University of Arizona, Tucson, 1992.

Bohmont, B. L., *The Standard Pesticide User's Guide,* 4th ed., Prentice Hall, Upper Saddle River, NJ, 1997.

Deutsch, A.E. and A. P. Poole, Eds., *Manual of Pesticide Equipment*, Oregon State University, Corvallis, OR, 1972.

Haskell, P. T., Ed., *Pesticide Application: Principles and Practice,* Oxford University Press, New York, 1985.

Lever, B. G., *Crop Protection Chemicals*, Ellis Horwood, London, 1990.

Matthews, G. A., *Pesticide Application Methods*, 2nd ed., John Wiley & Sons, New York, 1992.

Rice R. P., *Nursery and Landscape Weed Control Manual*, Thomson Publications, Fresno, CA, 1992.

U.S. Environmental Protection Agency, *Applying Pesticides Correctly: A Guide for Private and Commercial Applicators,* revised 1991.

U.S. Environmental Protection Agency, *Applying Pesticides Correctly: A Guide for Private and Commercial Applicators,* 1983.

U.S. Environmental Protection Agency, *Urban Pest Management: A Guide for Commercial Applicators*, EPA 735-B-92-001, 1992.

Ware, G. W., *Complete Guide to Pest Control*, 2nd ed., Thomson Publications, Fresno, CA, 1988.

Ware, G. W., *The Pesticide Book*, 4th ed., Thomson Publications, Fresno, CA, 1994.

Waxman, M. F., *Pesticide Safety Training*, University of Wisconsin Short Course, Madison, 1991.

# CHAPTER 9

# PESTICIDES AND ENVIRONMENTAL PROTECTION

## I. PESTICIDES IN THE ENVIRONMENT

The environment is everything that is around us. It includes not only the natural elements that the word "environment" most often brings to mind, but also people and the man-made components of our world. Nor is the environment limited to the outdoors—it also includes the indoor areas in which we live and work.

The environment, then, is much more than the oceans and the ozone layer. It is air, soil, water, plants, animals, houses, restaurants, office buildings, and factories, and all that they contain. Anyone who uses a pesticide—indoors or outdoors, in a city or in the country—must consider how that pesticide will affect the environment.

The user must ask two questions:

1. How will this pesticide affect the immediate environment at the site where it is being used?
2. What are the dangers that the pesticide will move out of the use site and cause harm to other parts of the environment?

Pesticides can harm all types of environments if they are not used correctly.

Pesticide product labeling statements are intended to alert you to particular environmental concerns that a pesticide product poses. The lack of a particular precautionary statement does not necessarily mean that the product poses no hazard to the environment.

Both the public and the Environmental Protection Agency (EPA) are becoming increasingly concerned about harmful effects on the environment from the use of pesticides. As a result, EPA is looking closely at environmental effects as it considers new applications for registration, and it also is taking another look at existing pesticide registrations. Hazards to humans had been the primary reason for EPA to classify a pesticide as a restricted-use product. Now, more and more pesticide labels list environmental effects, such as contamination of groundwater or toxicity to birds or aquatic invertebrate animals, as a reason for restriction.

## A. SOURCES OF CONTAMINATION

When environmental contamination occurs, it is the result of either point-source or non-point-source pollution. Point-source pollution comes from a specific, identifiable place (point). A pesticide spill that moves into a storm sewer is an example of point-source pollution. Non-point-source pollution comes from a wide area. The movement of pesticides into streams after broadcast applications is an example of non-point-source pollution.

Non-point-source pollution from pesticide applications has most commonly been blamed for pesticide contamination in the outdoor environment. But more and more studies are revealing that, in fact, much environmental contamination does not result from non-point-source pollution. Contamination also results from point sources, such as:

- wash water and spills produced at equipment cleanup sites
- improper disposal of containers, water from rinsing containers, and excess pesticides
- pesticide storage sites where leaks and spills are not correctly cleaned up
- spills that occur while mixing concentrates or loading pesticides into application equipment.

These kinds of tasks are involved with nearly every pesticide use, whether the pesticide is applied outdoors, or in or around an enclosed structure. Figure 9.1 shows a worker cleaning equipment.

As a user of pesticides, especially if you use and supervise the use of restricted-use pesticides, you must become aware of the potential for environmental contamination during every phase of your pesticide operation. Many pesticide uses are restricted because of environmental concerns. Whenever you release a pesticide into the environment—whether intentionally or accidentally—consider:

- whether there are sensitive areas in the environment at the pesticide use site that might be harmed by contact with the pesticide
- whether there are sensitive offsite areas near the use site that might be harmed by contact with the pesticide
- whether there are conditions in the environment at the pesticide use site that might cause the pesticide to move offsite

- whether you need to change any factors in your application or in the pesticide use site to reduce the risk of environmental contamination.

**Figure 9.1** Equipment clean-up site. Contamination can occur if wash water is not collected and disposed of properly (courtesy of University of Wisconsin Extension).

## B. SENSITIVE AREAS

Sensitive areas are sites or living things that are easily injured by a pesticide.

Sensitive areas outdoors include:

- areas where groundwater is near the surface or easily accessed (wells, sinkholes, porous soil, etc.)
- areas in or near surface water
- areas near schools, playgrounds, hospitals, and other institutions
- areas near the habitats of endangered species
- areas near apiaries (honeybee sites), wildlife refuges, or parks
- areas near ornamental gardens, food or feed crops, or other sensitive plantings.

Sensitive areas indoors include:

- areas where people—especially children, pregnant women, the elderly, or the sick—live, work, or are cared for
- areas where food or feed is processed, prepared, stored, or served
- areas where domestic or confined animals live, eat, or are otherwise cared for
- areas where ornamental or other sensitive plantings are grown or maintained.

**Figure 9.2** Potential drift of pesticides onto children near schools or playgrounds.

Sometimes pesticides must be deliberately applied to a sensitive area to control a pest. These applications should be performed by persons who are well trained about how to avoid causing injury in such areas.

At other times, the sensitive area is part of a larger target site. Whenever possible, take special precautions to avoid direct application to the sensitive area. For example, leaving an untreated buffer zone around sensitive areas is often a practical way to avoid contaminating them.

**Figure 9.3** Sensitive area containing assorted wildlife and wetlands.

In still other instances, the sensitive area may be near a site that is used for application, mixing/loading, storage, disposal, or equipment washing. The pesticide users must take precautions to avoid accidental contamination of the sensitive area. For example, a permanent site for mixing/loading or equipment washing could be equipped with a collection pad or tray to catch and contain leaks, spills, or waste water.

Typical pesticide labeling statements that alert you to these concerns include:

- Do not use in hospital patient quarters.
- Remove all animals from building prior to treatment and keep animals out until spray has dried.
- Applications prohibited in areas where food is held, processed, prepared, or served
- Do not use around home gardens, schools, recreational parks, or playgrounds.
- In living areas, make applications in such a manner as to avoid deposits on exposed surfaces or introducing the material into the air.
- Do not use in or around residences.

## C. PESTICIDE MOVEMENT

Pesticides that move away from the release site may cause environmental contamination. Pesticides move away from the release site both indoors and outdoors and may cause harm in both environments. Pesticides move in several ways, including:

- in air, through wind, or through air currents generated by ventilation systems
- in water, through runoff or leaching
- on or in objects, plants, or animals (including humans) that move or are moved offsite.

### 1. Air

Pesticide movement away from the release site in the air is usually called drift. Pesticide particles, dusts, spray droplets, and vapors all may be carried offsite in the air. People who mix, load, and apply pesticides outdoors usually are aware of the ease with which pesticides drift offsite. People who handle pesticides indoors may not realize how easily some pesticides move offsite in the air currents created by ventilation systems and by forced-air heating and cooling systems.

#### a. *Particles and Droplets*

Lightweight particles, such as dusts and wettable powders, are easily carried by moving air. Granules and pellets are much heavier and tend to settle out of air quickly. Small spray droplets also are easily carried in air currents. High-pressure and fine nozzles produce very small spray droplets that are very likely to drift. Lower pressure and coarse nozzles produce larger droplets with less drift potential.

The likelihood that pesticide particles and spray droplets will drift offsite depends partly on the way they are released. Pesticides released close to the ground or floor are not as likely to be caught up in air currents as those released from a greater height. Pesticides applied in an upward direction or from an aircraft are the most likely to be carried on air currents.

#### b. *Vapors*

Pesticide vapors move about easily in air. Fumigant pesticides are intended to form a vapor when they are released. Persons using fumigants must take precautions to make sure the fumigant remains in a sealed container until it is released into the application site, which also must be sealed to prevent the vapor from escaping. Some nonfumigant pesticides also can vaporize and escape into the air. The labeling of volatile pesticides often includes warning statements that the pesticide handler should heed. Any time you release a volatile pesticide in an enclosed area, consider the hazards not only to yourself and to fellow workers, but also to people, animals, and plants that are in or near the release site or that may enter the area soon after the release.

Typical pesticide labeling statements that alert you to avoid drift include:

- Do not apply when weather conditions favor drift from areas treated.
- Do not allow drift onto plants intended for food or feed.
- Drift from treated areas may be hazardous to aquatic organisms in neighboring areas.

c. *Water*

Pesticide particles and liquids may be carried offsite in water. Pesticides can enter water through:

- drift, leaching, and runoff from nearby applications
- spills, leaks, and back-siphoning from nearby mixing, loading, storage, and equipment cleanup sites
- improper disposal of pesticides, rinsates, and containers.

Most pesticide movement in water is across the treated surface (runoff) or downward from the surface (leaching). Runoff and leaching may occur when:

- too much liquid pesticide is applied, leaked, or spilled onto a surface
- too much rainwater, irrigation water, or other water gets onto a surface containing pesticide residue.

Runoff water in the outdoor environment may travel into drainage ditches, streams, ponds, or other surface water where the pesticides can be carried great distances offsite. Pesticides that leach downward through the soil in the outdoor environment sometimes reach the groundwater.

Runoff water in the indoor environment may get into domestic water systems and from there into surface water and groundwater. Runoff can flow into floor drains or other drains and into the water system. Sometimes a careless pesticide handler washes pesticide down a sink drain and into the water system.

Some pesticides can leach downward in indoor environments. In a greenhouse, for example, pesticides may leach through the soil or other planting medium to floors or benches below. Some pesticides used indoors may be absorbed into carpets, wood, and other porous surfaces and remain trapped for a long time.

Typical pesticide labeling statements that alert you to these concerns include:

- Do not contaminate water through runoff, spills, or improper disposal of excess pesticide, spray mixtures, or rinsates.
- Do not allow runoff or spray to contaminate wells, irrigation ditches, or any body of water used for irrigation or domestic purposes.
- Do not apply directly to water and wetlands (swamps, bogs, marshes, and potholes).
- Maintain a buffer zone (lay-off distance) of 100 feet from bodies of water.
- This product is water soluble and can move with surface runoff water. Do not contaminate cropland, water, or irrigation ditches.

### d. *On or in Objects, Plants, or Animals*

Pesticides can move away from the release site when they are on or in objects or organisms that move (or are moved) offsite. Pesticides may stick to shoes or clothing, to animal fur, or to blowing dust and be transferred to other surfaces. When pesticide handlers, applicators, and users bring home or wear home contaminated personal protective equipment, work clothing, or other items, residues can rub off on carpeting, furniture, and laundry items and onto pets and people.

Pesticides may stick to treated surfaces, such as food or feed products that are to be sold. To protect consumers, there are legal limits (tolerances) for how much pesticide residue may safely remain on crops or animal products that are sold for food or feed. Products that exceed these tolerances are illegal and cannot be sold. Crops and animal products will not be over tolerance levels if the pesticides are applied as directed on the product labeling. Illegal pesticide residues levels usually result when:

- too much pesticide is applied to the crop or animal
- the days-to-harvest, days-to-grazing, or days-to-slaughter directions on the pesticide labeling are not obeyed
- pesticides move out of the release site and contaminate plants or animals nearby.

Typical pesticide labeling statements that alert you to these concerns include:

- Do not apply within 5 days of harvest.
- Do not apply under conditions involving possible drift to food, forage, or other plantings that might be damaged or the crops thereof rendered unfit for sale, use, or consumption.
- Remove meat animals from treated areas at least 1 day before slaughter if they were present at application or grazed treated areas within 21 days after application.
- Do not pasture or feed treated hay to lactating dairy cattle within 21 days after application.

## D. HARMFUL EFFECTS ON NONTARGET PLANTS AND ANIMALS

Nontarget organisms may be harmed by pesticides in two ways:

1. The pesticide may cause injury by contacting the nontarget organism directly.
2. The pesticide may leave a residue that causes later injuries.

### 1. Harmful Effects from Direct Contact

Pesticides may harm nontarget organisms that are present during a pesticide application. Poorly timed applications can kill bees and other pollinators that are active in or near the target site. Pesticides may harm other wildlife, too. Even tiny amounts of some pesticides may harm them or destroy their source of food.

Pesticides applied over large areas, such as in mosquito, biting fly, and forest pest control, must be chosen with great care to avoid poisoning nontarget plants and animals in or near the target site. Read the warnings and directions on the pesticide labeling carefully to avoid harming nontarget organisms during a pesticide application.

Drift from the target site may injure wildlife, livestock, pets, sensitive plants, and people. For example, drift of herbicides can damage sensitive nearby plants, including crops, forests, or ornamental plantings. Drift also can kill beneficial parasites and predators that are near the target site.

Pesticide runoff may harm fish and other aquatic animals and plants in ponds, streams, and lakes. Aquatic life also can be harmed by careless tank filling or draining and by rinsing or discarding used containers along or in waterways.

Typical pesticide labeling statements that alert these concerns include:

- Phytotoxic. Do not spray on plants.
- Do not apply this product or allow it to drift to blooming crops or weeds if bees are visiting the treatment area.
- Extremely toxic to aquatic organisms. Do not contaminate water by cleaning of equipment or disposal of wastes.
- This product is toxic to fish, shrimp, crab, birds, and other wildlife. Keep out of lakes, streams, ponds, tidal marshes, and estuaries. Shrimp and crab may be killed at application rates. Do not apply where these are important resources.

## 2. Harmful Effects From Residues

A residue is the part of a pesticide that remains in the environment for a period of time following application or a spill. Pesticides usually break down into harmless components after they are released into an environment. The breakdown time ranges from less than a day to several years. The rate of pesticide breakdown depends primarily on the chemical structure of the pesticide active ingredient. The rate of pesticide breakdown also may be affected by environmental conditions at the release site, such as:

- surface type, chemical composition, and pH
- surface moisture
- presence of microorganisms
- temperature
- exposure to direct sunlight.

**Persistent** pesticides leave residues that stay in the environment without breaking down for long periods of time. These pesticides are sometimes desirable, because they provide longterm pest control and may reduce the need for repeated applications. However, some persistent pesticides that are applied to or spilled on soil, plants, lumber, and other surfaces or into water can later cause harm to sensitive plants or animals, including humans, that contact them. Clues on pesticide labeling that a particular pesticide product is likely to be persistent include:

- Can remain in the soil for 34 months or more and cause injury to certain crops other than those listed as acceptable on the label.

- This product can remain phytotoxic for a year or more.

When using persistent pesticides, consider whether their continued presence in the environment is likely to harm plants and animals.

When pesticides build up in the bodies of animals or in the soil, they are said to **accumulate.** When the same mixing/loading site or equipment cleaning site is used frequently without taking steps to limit and clean up spills, pesticides are likely to accumulate in the soil. When this occurs, plants, animals, and objects that come into contact with the soil may be harmed. When pesticides accumulate in the soil, there is also a higher likelihood that the pesticides will move offsite and contaminate the surrounding environment or move into surface or groundwater.

Sometimes animals can be harmed when they feed on plants or animals that have pesticide residues on or in them. A special concern is for predator birds or mammals that feed on animals that have been killed by pesticides. The predators may be harmed by the pesticide residues remaining on or in the bodies of the dead animals.

Typical pesticide labeling statements that alert you to these concerns include:

- Toxic to fish, birds, and wildlife. This product can pose a secondary hazard to birds of prey and mammals.
- Do not use fish as food or feed within 3 days of application.
- Animals feeding on treated areas may be killed and pose a hazard to hawks and other birds-of-prey. Bury or otherwise dispose of dead animals to prevent poisoning other wildlife.

## E. HARMFUL EFFECTS ON SURFACES

Sometimes surfaces are harmed by pesticides or pesticide residues. Some surfaces may become discolored by contact with certain pesticides. Other surfaces may be pitted or marked by contact with some pesticides. Some pesticides can corrode or obstruct electronic systems or metal. Sometimes a pesticide will leave a visible deposit on the treated surface.

Typical pesticide labeling statements that alert you to these concerns include:

- Do not apply to carpeting, linoleum, or other porous floor coverings, as discoloration may result.
- Do not spray on plastic, painted, or varnished surfaces.
- May cause pitting of automobile and other vehicle paint.

- Do not spray directly into any electronic equipment or into outlets and switches, or any other location where the pesticide may foul or short-circuit contacts and circuits.
- A visible deposit may occur on some dark surfaces.

## II. PROTECTING THE ENVIRONMENT

Concerns about wildlife and the environment are becoming more important in decisions about which pesticides will be registered and what they may be used for. Two environmental concerns are receiving particular attention:

1. Protection of groundwater.
2. Protection of endangered species.

Federal and state efforts to protect groundwater and endangered species are resulting in new instructions and limitations for pesticide handlers and applicators. Whether you apply pesticides indoors or outdoors, in an urban area or in a rural area, you must become aware of the importance of protecting these two vital national resources. Pesticides that are incorrectly or accidentally released into the environment—either during application or during other handling activities, such as mixing, loading, equipment cleaning, storage, transportation, or disposal—pose a threat to groundwater and endangered species.

Whether you must take special action to protect groundwater and endangered species depends mainly on the location of the use site. Groundwater contamination is of greatest concern in release sites where groundwater is close to the surface or where the soil type or the geology allows contaminants to reach groundwater easily. Protection of endangered species usually is required only in locations where they currently live or are being reintroduced. Read the pesticide labeling carefully to determine whether your pesticide use is subject to any special groundwater or endangered species limitations.

The U.S. Environmental Protection Agency may establish specific limitations or instructions for pesticide users in locations where groundwater or endangered species are most at risk. These limitations and instructions are often too long to be included in pesticide labeling. The labeling may state that you must consult another source for the details about the instructions and limitations that apply in your situation. Your legal responsibility for following instructions that are distributed separately is the same as it is for instructions that appear in full on the pesticide labeling.

## A. PROTECTING GROUNDWATER

Groundwater is water located beneath the earth's surface. Many people think that groundwater occurs in vast underground lakes, rivers, or streams. Usually, however, it is located in rock and soil. It moves very slowly through irregular spaces within otherwise solid rock or seeps between particles of sand, clay, and gravel. An exception is in limestone areas, where groundwater may flow through large underground channels or caverns. Surface water may move several feet in a second or a minute. Groundwater may move only a few feet in a month or a year. If the groundwater is capable of providing significant quantities of water to a well or spring, it is called an aquifer. Pesticide contamination of aquifers is very troubling, because these are sources of drinking, washing, and irrigation water.

### 1. Sources of Groundwater

Groundwater is recharged (replaced) primarily from rain or snow that enters the soil. However, some water from lakes and streams and from irrigation also becomes groundwater. Water that is above the ground can move in three ways—it can evaporate into the air; it can move across the surface, as in a stream or river; or it can move downward from the surface. Some of the water that moves downward is absorbed by plants and other organisms. Another portion of the downward-moving water is held in the upper layers of the soil. The rest moves down through the root zone and the relatively dry soil zone until it reaches a zone saturated with water. This saturated zone is the uppermost layer of groundwater and is called the water table. The water table is the "dividing line" between the groundwater and the unsaturated rock or soil above it.

### 2. Pesticide Contamination of Groundwater

When water that is moving downward from the surface contains pesticides—or comes into contact with them as it moves—the pesticides may be carried along with the water until they eventually reach the groundwater. Five major factors determine whether a pesticide will reach groundwater:

- the practices followed by pesticide users
- the presence or absence of water on the surface of the site where the pesticides are released
- the chemical characteristics of the pesticides
- the type of soil in the site where the pesticides are released

- the location of the groundwater—its distance from the surface and the type of geological formations above it.

By being aware of these considerations, you can handle pesticides in ways that will make the potential for groundwater contamination less likely.

**Figure 9.4** Movement of rain water carrying pesticides from treated plants downward through the soil (EPA, Applying Pesticides Correctly, 1991).

### a. *Practices for Pesticide Users*

The best way to keep from contaminating groundwater is to follow labeling directions exactly. Be sure to note whether the labeling requires any special steps to protect groundwater. In addition, remember the following:

- Avoid the temptation to use more pesticide than the labeling directs. Overdosing will increase both the cost of pest control and the odds that the pesticide will reach groundwater. Overdosing is also illegal. Keeping the use of pesticides to a minimum greatly reduces the risk of groundwater contamination.

- Consider whether your application method presents any special risks. For example, soil injection of some pesticides may not be wise when groundwater is close to the surface.
- Take precautions to keep pesticides from back-siphoning into a water source.
- Locate pesticide storage facilities at least 100 feet from wells, springs, sinkholes, and other sites that directly link to groundwater to prevent their contamination from runoff or firefighting water.
- Whenever possible, locate mixload sites and equipment-cleaning sites at least 100 feet from surface water or from direct links to groundwater. This will help prevent back-siphoning, runoff, and spills from contaminating the water sources. If you must locate one of these work sites near a water source, use methods such as dikes, sump pits, and containment pads to keep pesticides from reaching the water.
- Do not contaminate groundwater through improper disposal of unused pesticides, pesticide containers, or equipment and container rinse water. Dispose of all pesticide wastes in accordance with local, state, tribal, and federal laws.

### *b. Water on the Treated Surface*

If there is more water on the soil than the soil can hold, the water (along with any pesticides it contains) is likely to move downward to the groundwater. Prolonged heavy rain or excessive irrigation will produce excess water on the soil surface.

#### Rain

If weather forecasts or your own knowledge of local weather signs cause you to expect heavy rain, delay outdoor handling operations—including mixing and loading, application, and disposal—to prevent wash-off, surface runoff, or leaching.

#### Irrigation

Pesticide movement into groundwater is affected by both the amount of water used in irrigation and how soon before or after a pesticide application the irrigation is done. If irrigation water contains pesticides, be careful to prevent it from flowing into water sources.

*c. Pesticide Factors*

Some pesticide chemicals are more likely than others to move to groundwater. Such movement depends mainly on:

- **solubility**—some pesticides dissolve easily in water and are more likely to move into water systems

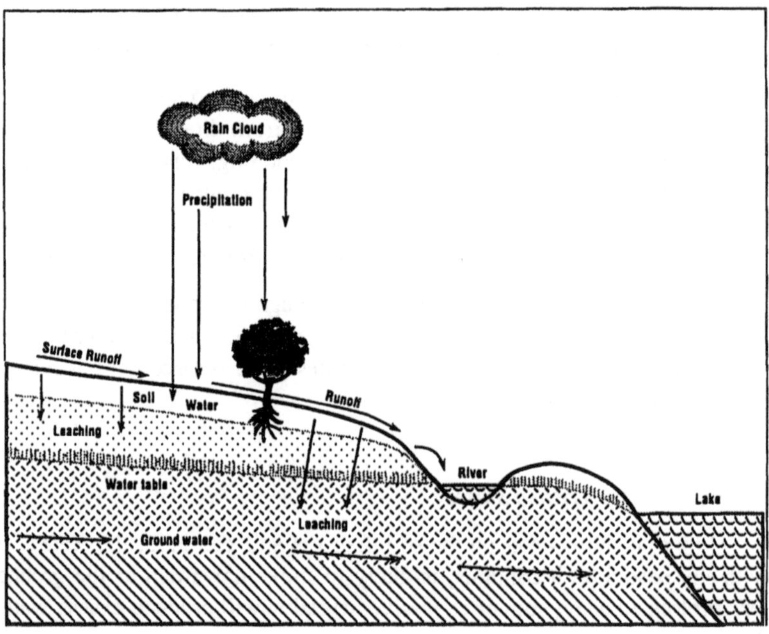


**Figure 9.5** Movement of rainwater and irrigation water on treated surfaces (EPA, Applying Pesticides Correctly, 1991).

- **adsorption**—some pesticides become tightly attached (strongly adsorbed) to soil particles and are not likely to move out of the soil and into water systems
- **persistence**—some pesticides break down slowly and remain in the environment for a long time.

These factors are all related to one another. Pesticides that are most likely to move into groundwater are highly soluble, moderately to highly persistent, and are not strongly adsorbed to soil. A nonpersistent pesticide would be less likely to move to groundwater, even if it is highly soluble or not strongly adsorbed to soil. A pesticide that is strongly adsorbed to soil would be less likely to move to groundwater even if it is persistent.

Pesticide labeling usually does not tell about these properties of the pesticide product. The Soil Conservation Service, Cooperative Extension Service, your trade association, or your pesticide dealer may have specific information about the characteristics of the pesticides you are using.

*d. Soil Factors*

Soil is also an important factor in the breakdown and movement of pesticides. Your local Soil Conservation Service can help determine the types of soil in your area and how they affect breakdown and movement. The three major soil characteristics that affect pesticides are texture, permeability, and organic matter.

**Soil texture** is an indication of the relative proportions of sand, silt, and clay in the soil. Coarse, sandy soils generally allow water to carry the pesticides rapidly downward. Finer textured soils generally allow water to move at much slower rates. They contain more clay, and sometimes organic matter, to which pesticides may cling.

**Soil permeability** is a general measure of how fast water can move downward in a particular soil. The more permeable soils must be managed carefully to keep pesticides from reaching groundwater.

**Soil organic matter** influences how much water the soil can hold before it begins to move downward. Soil containing organic matter has greater ability to stop the movement of pesticides. Soils in which plants are growing are more likely to prevent pesticide movement than bare soils.

*e. Geology*

The distance from the soil surface to the water table is the measure of how deep the groundwater is in a given location. If the groundwater is within a few feet of the soil surface, pesticides are more likely to reach it than if it is farther down. In humid areas, the water table may be only a few feet below the surface of the soil. In arid areas, the water table may lie several hundred feet below the soil surface. The depth to the water table does not stay the same over the course of the year. It varies according to:

- the amount of rain, snow, and irrigation water being added to the soil surface
- the amount of evaporation and plant uptake
- whether the ground is frozen

- how much groundwater is being withdrawn by pumping.

The Soil Conservation Service can provide valuable information on the geology of an area and on the potential for groundwater contamination on your property.

Spring and fall generally are the times when the water table is closest to the soil surface. The water table often moves downward during the summer when evaporation and plant uptake are high and when larger than normal amounts of groundwater are being used for irrigation and other hot weather needs. The water table also moves downward in winter if surface water cannot move down through the frozen soil to recharge the groundwater.

The permeability of geological layers between the soil and groundwater is also important. If surface water can move down quickly, pesticides are more likely to reach groundwater. Gravel deposits are highly permeable. They allow water and any pesticides in it to move rapidly downward to groundwater. Regions with limestone deposits are particularly susceptible to groundwater contamination, because water may move rapidly to the groundwater through caverns or "rivers" with little filtration or chemical breakdown. On the other hand, layers of clay may be totally impermeable and may prevent most water and any pesticides in it from reaching the groundwater.

Sinkholes are especially troublesome. Surface water often flows into sinkholes and disappears quickly into the groundwater. If a pesticide is released into an area that drains into a sinkhole, even a moderate rain or irrigation may carry some of the pesticide directly to the groundwater.

Some pesticides or certain uses of some pesticides may be classified as restricted use because of groundwater concerns. The user and applicator of these pesticides has a special responsibility to handle all pesticides safely in and near use sites where groundwater contamination is particularly likely. Take extra precautions when using techniques that are known to be likely to cause contamination of groundwater, such as chemigation and soil injection.

When a pesticide product has been found in groundwater or has characteristics that may pose a threat of contamination of groundwater, the pesticide product labeling may contain statements to alert you to the concern. Typical pesticide labeling statements include:

- This chemical has been identified in limited groundwater sampling and there is the possibility that it can leach through the soil to groundwater, especially where soils are coarse and groundwater is near the surface.

- This product is readily decomposed into harmless residues under most use conditions. However, a combination of permeable and acidic soil conditions, moderate to heavy irrigation and/or rainfall, use of 20 or more pounds per acre, and soil temperature below 50°F (10°C) at application time tend to reduce degradation and promote movement of residues to groundwater. If the above describes your local use conditions and groundwater in your area is used for drinking, do not use this product without first contacting (registrant's name and telephone number).

## B. PROTECTION OF ENDANGERED SPECIES

An endangered species is a plant or animal that is in danger of becoming extinct. There are two classifications of these plants and animals in danger—"endangered species" and "threatened species." The term "endangered species" is used here to refer to the two classifications collectively. Scientists believe that certain pesticides may threaten the survival of some of America's endangered species if they are used in the places where these plants and animals still exist.

A federal law, the Endangered Species Act, requires the U.S. Environmental Protection Agency (EPA) to ensure that endangered species are protected from pesticides (see Figure 9.6). EPA's goal is to remove or reduce the threat that pesticide use poses to endangered species. Reaching this goal will require some limitations on pesticide use. These limitations usually will apply only in the currently occupied habitat or range of each endangered species at risk. Occasionally the limitations will apply where endangered species are being reintroduced into a habitat they previously occupied.

Habitats, sometimes called "critical habitats," are the areas of land, water, and air space that an endangered species needs for survival. Such areas include breeding sites; sources of food, cover, and shelter; and surrounding territory that gives room for normal population growth and behavior.

### 1. Limitations on Pesticide Use

Read all pesticide labeling carefully to find out whether the use of that product requires any special steps to protect endangered species. The label may direct you to another source for the details about what you must do. When limitations do apply, they usually will be in effect only in some specific geographic locations. Use of a particular pesticide is usually limited in a particular location when:

- the site is designated as the current habitat of an endangered species
- the endangered species that uses the site might be harmed by the use of the pesticide within (or close to) its habitat.

United States
Environmental Protection
Agency

20T-3050
September 1990

Pesticides And Toxic Substances (H 7506C)

# Protecting Endangered Species

The information in this pamphlet is similar to what the U.S Environmental Protection Agency (EPA) expects to distribute once our Endangered Species Protection Program is in effect The limitations on pesticide use are not law at this time, but are being provided now for your use in voluntarily protecting endangered and threatened species from harm due to pesticide use We encourage you to use this information We also welcome your comments

The Endangered Species Act is intended to protect and promote recovery of animals and plants that are in danger of becoming extinct due to the activities of people Under the Act, EPA must ensure that use of pesticides it registers will not result in harm to the species listed as endangered or threatened by the U S Fish and Wildlife Service, or to habitat critical to those species' survival To accomplish this the EPA expects to implement program requirements beginning in 1991 This program will protect endangered and threatened species from harm due to pesticide use

EPA requests your comments regarding the information presented in this publication Please drop us a line to let us know whether the information is clear and correct Also tell us to what extent following the recommended measures would affect your typical pesticide use or productivity This information will be considered by EPA during the final stages of program development

Please submit comments to
Interim Endangered Species
Protection Program (H7506C)
Public Docket and Information Section
U.S. EPA
401 M Street, SW
Washington, DC 20460

## About This Publication

This publication contains a County Map showing the area within the county whe pesticide use should be limited to protec listed species These areas are identified the map by a shaded pattern Each shad pattern corresponds to a species in need protection

The Shading Key shows the name of t species that each shaded pattern represe and describes the shaded area The area may be described in terms of Township Range, and Section or by giving details about the habitat of the species

The first column of the "Table of Pestic Active Ingredients" lists the active ingredients for which there should be limitations on use to protect certain spec The next columns are headed by the shac pattern of the species with Codes listed underneath them

The Code indicates the specific limitatic that is necessary to protect the species T section titled Limitations on Pesticide Use explains the code

## Does This Information Apply To You?

To determine whether this information applies to your use of a pesticide, review the questions below The information applies only if you answer "yes" to both questions

- Do you intend to use pesticides within the shaded area on the county map?
- Are any of the ingredients listed on the front panel of your pesticide product label named in the "Table of Pesticide Active Ingredients"?

If you answer "yes" to both questions you should follow the instructions on "Ho to Use This Information" to determine if you should limit use of the pesticide to he protect listed species

If you answer "no" to either question, you should follow the usage directions on the pesticide product label

Printed on Recycled

**Figure 9.6** The Endangered Act (EPA, Applying Pesticides Correctly, 1991).

## 2. Habitats of Endangered Species

The U.S. Fish and Wildlife Service is responsible for identifying the current habitat or range of each endangered species. For aquatic species, the restricted habitat often will include an additional zone around the body of water to keep any drift, runoff, or leachate in the watershed from reaching the water.

The U.S. Fish and Wildlife Service is attempting to identify the habitats as accurately as possible so that pesticide use will need to be limited only in locations where it is absolutely necessary. For this reason, limitations on pesticide use may apply on one property, while a similar adjoining property may not have these limitations.

### 3. Importance of Protecting Endangered Species

Hundreds of animals (including fish, birds, mammals, reptiles, amphibians, insects, and aquatic invertebrates) and thousands of plants have been named as endangered or threatened species under the provisions of the Endangered Species Act. Some of these animals and plants are ones that everyone knows about, such as the bald eagle. Others are tiny, little-known creatures that may rarely be seen by anyone except trained naturalists.

| Endangered Species Restriction |
|---|
| Under the Endangered Species Act, it is a Federal offense to use any pesticide in a manner that results in the death of a member of an endangered species. Prior to making applications, the user must determine that endangered species are not located immediately adjacent to the site to be treated. If the users are in doubt whether or not endangered species may be affected, they should contact the regional U.S. Fish and Wildlife Service Office (Endangered Specialist) or personnel of the State Fish and Game Office. |

Regardless of the size or apparent significance of these endangered species, it is important that each is allowed to survive—mankind's well-being depends on maintaining **biological diversity.** Biological diversity is the variety and differences among living things, and the complex ways they interact. Diversity is necessary for several reasons.

#### *a. Agriculture*

Nearly all of today's crops started as wild species. Genes from wild species often are used to create new hybrids that have resistance to plant diseases and insects, better climatic tolerance, and higher yields. Having different varieties available is necessary insurance against devastating crop failures caused by climate extremes or major pest outbreaks.

#### *b. Medicine*

Many of today's most important medicines come from obscure plant and animal species. A mold is the source of penicillin, the miracle drug; an herb is the source of quinine, a cure for malaria. Scientists are testing countless plant and animal species around the world for sources of cures for major diseases.

*c. Preserving Choices*

No one can predict which species may be essential to the future of mankind. A species that is allowed to become extinct might have been the key to stopping a global epidemic or to surviving a major climate change.

*d. Interdependence*

The extinction of a single species can set off a chain reaction of harm to other species. The disappearance of a single kind of plant from an area, for example, may lead to the disappearance of certain insects, higher animals, and other plants.

*e. Natural Balance*

Extinction has always been a natural part of an ever-changing process. During most of history, species have formed at a rate greater than the rate of extinctions. Now, however, it appears that human activity is greatly speeding up the rate of extinctions. People, plants, and animals live together in a delicate balance; the disappearance of species could easily upset that balance.

*f. Stability*

The more diversity that exists in an ecosystem, the more stable it is likely to be. There is less likelihood of huge swings in populations of particular organisms. There is also less likelihood of devastation from the introduction of a new species from outside the system.

## REGULATORY COMPLIANCE

Endangered Species Act (ESA) is a federal law administered by the Fish and Wildlife Service (FWS) of the Department of the Interior. The ESA makes it illegal to kill, harm, or collect endangered or threatened wildlife or fish or to remove endangered or threatened plants from areas under federal jurisdiction. It also requires other federal agencies to ensure that any action they carry out or authorize is not likely to jeopardize the continued existence of any endangered or threatened species, or to destroy or adversely modify its critical habitat. As a result, EPA must ensure that no registered pesticide use is likely to jeopardize the survival of any endangered or threatened species

The FWS has the authority to designate land and freshwater species as endangered or threatened and to identify their current habitat or range. The National Marine Fisheries Service has the same authority for marine species.

The FWS has the authority to prosecute persons, including pesticide users, who harm endangered or threatened species. In addition, EPA enforcement personnel have the authority to ensure that pesticide users observe labeling restrictions.

## REFERENCES

Baker, P. B., *Arizona Agricultural Pesticide Applicator Training Manual*, Cooperative Extension, University of Arizona, Tucson, 1992.

Bohmont, B. L., *The Standard Pesticide User's Guide*, 4th ed., Prentice Hall, Upper Saddle River, NJ, 1997.

Cheng, H. H., Ed., *Pesticides in the Soil Environment: Processes, Impacts, and Modeling*, Soil Science Society of America, Madison, WI, 1990.

Huston, D. H. and T. R. Roberts, Eds., *Environmental Fate of Pesticides*, John Wiley & Sons, New York, 1990.

Schnoor, J. L., Ed., *Fate of Pesticides and Chemicals in the Environment*, John Wiley & Sons, New York, 1992.

U.S. Environmental Protection Agency, *Applying Pesticides Correctly: A Guide for Private and Commercial Applicators,* revised 1991.

U.S. Environmental Protection Agency, *Applying Pesticides Correctly: A Guide for Private and Commercial Applicators,* 1983.

Waxman, M. F., *Hazardous Waste Site Operations: A Training Manual for Site Professionals*, John Wiley & Sons, New York, 1996.

# PART TWO

# RESPONSE RESOURCES

# PART TWO

# PESTICIDE AND CHEMICAL TABLES AND GUIDES

Part Two will provide the information required to apply, store, and dispose of specific pesticides and other hazardous chemicals properly. This information can be obtained from the following three tables and the pesticide and chemical guides. Emergency response guidelines can be found in the guides. These guidelines will enable the reader to handle emergency incidents that require a quick, accurate, and safe response.

## I. TABLES USAGE

There are three important tables in Part Two. The first of these is Table 1. This table lists the trade names or brands of pesticides of current importance alphabetically.

The user of a specific product would look up the product's trade or brand name in Table 1. For example, specific information is needed on the product AAtrex. From this table you would learn that the common name of the active ingredient in the product is atrazine. Important information on its health hazards, recommended handling and storage precautions, PPE and emergency guidelines could be found by referring to Pesticide Guide 91. The manufacturer's name is Ciba-Geigy, the type of pest contol product, e.g., herbicide, and the other formulations available containing this active ingredient. Further information could be obtained by using the CAS Number (i.e., the Chemical Abstract Service Number) in selected references. These references include Sax's "Dangerous Properties of Industrial Materials," the "Chris" Manual and RTECs, etc.

Table 2 provides the reader with other important information. Using the common name you can identify the family or group of chemicals that your chemical of concern belongs to or is derived from. If your product is not listed in this table, using the chemical or common name you can find other products with the same active ingredient and use their guide number.

Table 3 lists alphabetically the chemicals that are important to users of pesticides. These chemicals may be hazardous in themselves and, when used with pesticides as adjuvants, carriers, diluents, anti-caking agents, fertilizers, baits, emulsifiers, or detergents, their activity can be potentiated. For example, the product you are using requires petroleum oil or xylene as a carrier. The pesticide by itself may be relatively nontoxic, but the addition of xylene to the product could elevate the product to a Toxicity Category I or II, requiring the use of specific PPE, and may also require specific handling and diposal procedures.

## II. PESTICIDE AND CHEMICAL GUIDES' USAGE

The pesticides listed under trade name in Tables 1 and 2 have each been assigned to a specific pesticide guide. These guides have been developed to assist the user in handling various tasks from mixing and applying pesticides to cleaning up spilled pesticide and disposing of the waste properly. These guides also provide information on how to handle emergencies involving the listed pesticides.

Over 500 formulated pesticides are listed in the Tables 1 and 2. In order to provide usable information, these pesticides are organized into groups of pesticides with similar attributes and hazards, and a guide was developed to provide pertinent information for each group. As a result there are 61 unique pesticide guides.

The hazardous chemicals were treated in a similar fashion and 18 unique chemical guides were developed to provide similar information.

First aid information can be found in Chapter 7, Section E, pages 312-314 and information on the handling, storage, clean up, and disposal of flammable and nonflammable pesticides and chemicals can be found in Chapter 6, Section V, pages 257-282.

## III. HOW TO USE THE PESTICIDE AND CHEMICAL TABLES AND GUIDES

Each of the pesticides listed in Table 1 is referenced to a specific pesticide guide number. In the event of an incident involving a pesticide, the user would use Table 1 to determine which pesticide guide to follow in responding to that specific incident. If the pesticide is suspended in, emulsified with, or dissolved in a chemical that poses a hazard, the user should read the label carefully to identify the carrier, emulsifying agent, or other material(s) and then consult the proper Chemical Guide for that chemical.

The guides are organized as follows:

**Health hazards**, which include routes of exposure, life threat, and environmental concerns.

**Safety guidelines**, which include use classification, solubility properties in water or nonaqueous carriers, EPA toxicity category, handling and storage precautions.

**Protective clothing recommendations**, which suggests the PPE that should be used when mixing, applying, rinsing, or disposing of the pesticide or chemical.

**Emergency guidelines**, which include fire and explosive hazards, spills response and remediation, and first aid.

The following example will demonstrate the use of the guides.

With the advice of the Cooperative Extension Service, you have decided to apply Asana™XL as a treatment for the cutworms on your field corn. Before opening the container, read the label. From the label information you determine that this product is a restricted use pesticide, meaning that you have to be a certified applicator or work under direct supervision of a certified applicator to apply this product. The label also tells what the formulation is. This product is an emulsifiable concentrate that contains 0.66 lbs. of active ingredient per gallon. The chemical name of the active ingredient is also given as (S)-cyano(3-phenoxyphenyl)methyl-(S)-4-chloro-alpha-(1-methyl ethyl) benzeneacetate. Its percentage of the product is 8.4%. The inert or nonactive ingredients are not listed but their percentage of the product is listed as 91.6%. The label also states that the product contains a xylene range aromatic solvent, EPA registration number 352-515, the signal word WARNING, first aid instructions, environmental hazards, storage and disposal recommendations, and directions for use.

To access additional information contained in the Pesticide Guides, you would look up the trade name in Table 1. From this table you learn that Asana is an insecticide, which is manufactured by the Du Pont Agricultural Products and whose common name is Esfenvalerate. The information also includes the fact that the active ingredient is only available as an emulsifiable concentrate. The table also tells you to consult Pesticide Guide 53 for more information.

From Pesticide Guide 53 you find the following facts:

1) Imminent health hazards and environmental concerns

2) Toxicity of the product

3) Handling and storage precautions
4) Protective equipment to use when mixing and applying the product
5) Fire and explosive hazards are present when using this product
6) Spill cleanup procedures
7) Suggested first aid treatments.

In addition to the above information you should also consult the Chemical Guides for the chemical xylene, which is the carrier fluid mentioned on the label. To access the proper guide, first look up xylene in Table 3. From this table you find that Chemical Guide 10 should be consulted.

The following information is given in Guide 10:

1) Imminent health hazards and environmental concerns
2) The NFPA-Hazard Rating
3) Handling and storage precautions
4) Recommendations for protective equipment when using this chemical
5) Fire and explosive hazards when using this product
6) Spill cleanup procedures
7) Suggested first aid treatments.

Read the label very carefully. Most labels contain most of the information needed to use the product safely. However, if the label is vague, missing, or illegible or if an emergency occurs, you will need to consult these tables and guides.

It is not unusual for a pesticide user, even a trained and certified one, to grab a container of pesticide without reading the label. At one time or another, virtually every pesticide user has identified a product by the color or general configuration of the label. It's a relatively common occurrence, but certainly not a safe business practice. Labels of the same color and general makeup may contain very different ingredients. How would you know if the pesticide is too toxic or hazardous to be used safely under your intended conditions? How do you know the formulation is right for your intended use? How do you know what precautions should be taken when mixing the pesticide? How do you know what protective equipment to wear when applying the material? **YOU DON'T** - unless you read the label and, when necessary, consult the following tables and guides.

# PESTICIDE AND CHEMICAL TABLES

# ERRATA

## Agrochemical and Pesticide Safety Handbook

Michael Waxman

Please note that this listing contains revised guide numbers that supercede those on pages 381-398.

| Trade Name | New Guide Number |
| --- | --- |
| 2 Plus 2 | 61 |
| AAtack | 61 |
| Aaterra | 35-DG,L,WP,4-EC |
| AAtrex | 61 |
| Abate | 9-5G, 58-4E,61-ISG,2GD |
| Acarol | 61 |
| Accel | 44 |
| Accelerate | 1 |
| Accent | 61 |
| Access | 61 |
| Access and Triclopyr | 61 |
| Acclaim IEC | 61 |
| Accord | 61 |
| Acrobat | 61 |
| Actellic 5E | 44 |
| Afugan | 30 |
| Agri-Mek | 32 |
| Alanap-L | 61 |
| Aliette | 61 |
| Ally | 61 |
| Ambush | 44 |
| Amdro | 61 |
| Amiben | 61 |
| Amid Thin W | 61 |
| Amitrol T | 61 |
| Amizol Industrial | 61 |
| Ammo 2.5EC | 61 |
| Andalin | 61 |
| Ansar | 61 |
| Anthio | 34 |
| Antor 4ES | 47 |
| Anvil | 61 |
| Apollo SC | 61 |
| Apron 25W | 61 |
| Apron-Terraclor | 61 |
| Aqua Kleen | 61 |
| Aquathol K | 1 |
| Aquazine | 61 |
| Arbotect | 61 |
| Argold | 61 |
| Arrosolo 3-3E | 44 |
| Arsenal | 61 |
| Arsenal Forestry Granule | 61 |
| Asana XL | 53 |
| Aspire | 61 |
| Assert | 4 |
| Assure | 61 |
| Asulox | 61 |
| Atratol 90 | 26 |
| Atrazine | 61 |
| Authority | 61 |
| Avenge | 4 |
| Avirosan | 34 |
| Avitrol | 61-bait,11-PC |
| B-Nine SP | 61 |
| Balan | 61 |
| Banvel | 44 |
| Basagran | 61 |
| Bayleton 50% | 46 |
| Baytan | 61 |
| Bathion | 61 |
| Baythroid 2 | 44 |
| Beacon | 61 |
| Benlate 50 | 61 |
| Betamix | 44 |
| Betanex | 44 |
| Betasan | 44 |
| Bicep | 61 |
| Bidrin 8 | 27 |
| Biobit | 61 |
| Bladex | 52 |
| Blazer | 6 |
| Bolero | 61 |
| Bolstar 6 | 51 |
| Botec | 5 |
| Botran 75 | 61 |
| Bravo | 25 |
| Broadstrike | 61 |
| Bronate | 55 |
| Bronco | 26 |
| Buckle | 61 |
| Buctril | 36 |
| Buctril and Atrazine | 61 |
| Bueno 6 | 61 |
| Bullet | 56 |
| Bumper | 46 |
| Butisan S | 61 |

Table 1. Pesticides Listed Alphabetically by Trade Name

| Trade Name | Common Name | Guide Number[a] | Manufacturer | Category | Formulations[b] | CAS Number[c] |
|---|---|---|---|---|---|---|
| 2 Plus 2 | MCPP and 2,4-D | 87 | Fermenta | Herbicide | Liquid | NA |
| AAtack | Thiram | 91 | Rhone-Poulenc | Fungicide, Seed Protectant | DF, WP | 137-26-8 |
| Aaterra | Etridazole | 49-DG, L, WP, 5-EC | Uniroyal | Soil Fungicide | D, EC, G, WP | 2593-15-9 |
| AAtrex | Atrazine | 91 | Ciba-Geigy | Herbicide | DG, EC, L, WP | 1912-24-9 |
| Abate | Temephos | 15-SG, 83-4E | American Cyanamid Company | Insecticide | EC, G | 3383-96-8 |
| Acarol | Bromopropylate | 100 | Ciba-Geigy | Acaricide | EC | 18181-80-1 |
| Accel | Benzyladenine and Gibberellins | 64 | Abbott | Plant Growth Regulator | LC | NA |
| Accelerate | Endothall | 1 | Altochem | Desiccant | LC | 129-67-9 |
| Accent | Nicosulfuron | 100 | DuPont | Herbicide | WDG | 111991-09-4 |
| Access | Picloram | 100 | DowElanco | Herbicide | EC | NA |
| Access and Triclopyr | Picloram and triclopyr | 99 | DowElanco | Herbicide | EC | NA |
| Acclaim IEC | Fenoxaprop-ethyl | 61 | Hoechst | Herbicide | EC | 66441-23-4 |
| Accord | Glyphosate | 92 | Monsanto | Herbicide | EC | 1071-83-6 |
| Acrobat | Dimethomorph | 91 | American Cyanamid Company | Fungicide | EC, WP | 110488-70-5 |
| Actellic 5E | Pirimiphos-methyl | 60 | ICI | Insecticide | E | 29232-93-7 |
| Afugan | Pyrazophos | 42 | Hoechst | Fungicide | EC | 13457-18-6 |
| Agri-Mek | Abamectin | 44 | Merck | Insecticide, miticide | EC | 71751-41-2 |
| Alanap-L | Naptalam | 106 | Uniroyal | Herbicide | Liquid | 132-66-1 |
| Aliette | Fosetyl-Aluminum | 90 | Rhone-Poulenc | Fungicide | WDG | 39148-24-8 |

Table 1. Pesticides Listed Alphabetically by Trade Name

| Trade Name | Common Name | Guide Number[a] | Manufacturer | Category | Formulations[b] | CAS Number[c] |
|---|---|---|---|---|---|---|
| Ally | Metsulfuron methyl | 89 | DuPont | Herbicide | DF | 74223-64-6 |
| Ambush | Permethrin | 60 | ICI | Insecticide | EC | 52645-53-1 |
| Amdro | Hydramethylnon | 100 | American Cyanamid Company | Insecticide | G | 67485-29-4 |
| Amiben | Chloramben | 101 | Rhone-Poulenc | Herbicide | DS, G | 133-90-4 |
| Amid Thin W | Naphthaleneacet-amide (NAAm and NAD) | 89 | Rhone-Poulenc | Plant Growth Regulator | W | 86-86-2 |
| Amitrol T | Amitrole | 92 | Rhone-Poulenc | Herbicide | SP, DG, LS, WP, F | 61-82-5 |
| Amizol Industrial | Amitrole | 92 | Rhone-Poulenc | Herbicide | SP, DG, LS. WP. F | 61-82-5 |
| Ammo 2.5 EC | Cypermethrin | 89 | FMC | Insecticide | EC | 52315-07-8 |
| Andalin | Flucycloxuron | 89 | Uniroyal | Insecticide, ascaricide | DC | 113036-88-7 |
| Ansar | MSMA | 92 | Fermenta | Herbicide | L. SG | 2163-80-6 |
| Anthio | Formothion | 48 | Sandoz | Ascaricide, Insecticide | EC, ULV | 2540-82-1 |
| Antor 4ES | Diethatyl Ethyl | 69 | NOR-AM | Herbicide | ES | 38727-55-8 |
| Anvil | Hexaconazole | 103 | Zeneca | Fungicide | EC | 79983-71-4 |
| Apollo SC | Clofentezine | 89 | NOR-AM | Miticide | SC | 74115-24-5 |
| Apron 25W | Metalaxyl | 91 | Ciba-Geigy, Uniroyal | Fungicide | D. FL,W | 57837-19-1 |
| Apron-Terraclor | Metalaxyl and PCNB | 103 | Uniroyal | Fungicide | D | NA |
| Aqua Kleen | 2,4-D Butoxyethyl ester | 94 | Phone-Poulenc | Herbicide | G | 1929-73-3 |
| Aquathol K | Endothall. dipotassium salt | 1 | Atochem | Herbicide | G | 129-67-9 |
| Aquazine | Simazine | 105 | Ciba-Geigy | Herbicide | WP. WDG. liquid. G | 122-34-9 |
| Arbotect | Thiabendazole | 92 | Merck | Fungicide | Liquid | 148-79-8 |
| Argold | Cinmethylin | 91 | American Cyanamid Company | Herbicide | EC | 87818-31-3 |
| Arrosolo 3-3E | Molinate and Propanil | 60 | ICI | Herbicide | EC | NA |

Table 1. Pesticides Listed Alphabetically by Trade Name

| Trade Name | Common Name | Guide Number [a] | Manufacturer | Category | Formulations [b] | CAS Number [c] |
|---|---|---|---|---|---|---|
| Arsenal | Imazapyr | 100 | American Cyanamid Company | Herbicide | Aqueous liquid | 81334-34-1 |
| Arsenal Forestry Granule | Imazapyr | 100 | American Cyanamid Company | Herbicide | Aqueous liquid | 81334-34-1 |
| Asana XL | Esfenvalerate | 77 | DuPont | Insecticide | EC | 66230--04-4 |
| Aspire | Candida oleophila | 89 | Ecogen | Biological Fungicide | Extruded granule | NA |
| Assert | Imazamethabenz-methyl | 6 | American Cyanamid Company | Herbicide | LC | 81405-85-8 |
| Assure | Quizalofop ethyl | 104 | DuPont | Herbicide | EC, S C | 76578-14-8 |
| Asulox | Asulam | 101 | Rhone-Poulenc | Herbicide | Liquid aqueous solution, FWP | 3337-71-1 |
| Atratol 90 | Atrazine | 36 | Ciba-Geigy | Herbicide | WDG | 1912-24-9 |
| Atrazine | Atrazine | 91 | DuPont | Herbicide | DFP, FL, Liquid, WDG, WP | 1912-24-9 |
| Authority | Sulfentrazone | 96 | FMC | Herbicide | AF, G, WP | 122836-35-5 |
| Avenge | Difenzoquat Methyl Sulfate | 6 | American Cyanamid Company | Herbicide | WML, WP | 43222-48-6 |
| Avirosan | Piperophos and Dimethametryn | 47 | Ciba-Geigy | Herbicide | EC, Granules | NA |
| Avitrol | 4-aminopyridine | 106-bait, 17-PC | Avitrol Corp. | Avi-repellent, avicide | Grain baits, PC | 504-24-5 |
| B-Nine SP | Daminozide | 101 | Uniroyal | Plant Growth Regulator | WSP | 1596-84-5 |
| Balan | Benefin | 89 | DowElanco | Herbicide | DF, EC | 1861-40-1 |
| Banvel | Dicamba | 64 | Sandoz | Herbicide | Liquid, granular | 1918-00-9 |
| Basagran | Bentazon | 92 | BASF | Herbicide | Soluble concentrate | 25057-89-0 |
| Bayleton 50% | Triadimefon | 66 | Mobay | Fungicide | DF, EC. G, WP | 43121-43-3 |
| Baytan | Triadimenol | 89 | Bayer | Fungicide | EC,F, G, WDG, water-oil emulsion, WP | 55219-65-3 |

Table 1. Pesticides Listed Alphabetically by Trade Name

| Trade Name | Common Name | Guide Number [a] | Manufacturer | Category | Formulations [b] | CAS Number [c] |
|---|---|---|---|---|---|---|
| Baythion | Phoxim | 89 | Bayer | Insecticide | Aerosol, bait, DP, EC, G, ULV liquid | 14816-18-3 |
| Baythroid 2 | Cyfluthrin | 60 | Mobay | Insecticide | Aerosol, EC, liquid, oil-in-water emulsion | 68359-37-5 |
| Beacon | Primisulfuron-methyl | 100 | Ciba-Geigy | Herbicide | Wettable granules in WSP | 86209-51-0 |
| Benlate 50 | Benomyl | 108 | DuPont | Fungicide | Oil-Dispersible-WP | 17804-35-2 |
| Betamix | Desmedipham and Phenmedipham | 62 | NOR-AM | Herbicide | EC, Oil-SC | NA |
| Betanex | Desmedipham | 62 | NOR-AM | Herbicide | EC | 13684-56-5 |
| Betasan | Bensulide | 60 | ICI | Herbicide | EC, G | 741-58-2 |
| Bicep | Atrazine and Metolachlor | 91 | Ciba-Geigy | Herbicide | Liquid | NA |
| Bidrin 8 | Dicrotophos | 37 | DuPont | Insecticide, acaricide | WSC, ULV spray | 141-66-2 |
| Biobit | Bacillus thuringiensis, var. kurstaki | 89 | DuPont | Insecticide | FC, WP, | 68038-71-1 |
| Bladex | Cyanazine | 76 | DuPont | Herbicide | DF, liquid | 21725-46-2 |
| Blazer | Acifluorfen-sodium | 12 | BASF | Herbicide | LC | 62476-59-9 |
| Bolero | Thiobencarb or benthiocarb | 89 | Valent | Herbicide | EC, G | 28249-77-6 |
| Bolstar 6 | Sulprofos | 75 | Mobay | Insecticide | EC | 35400-43-2 |
| Botec | Captan and Dichloran | 8 | NOR-AM | Fungicide, Seed Protectant | Dust | NA |
| Botran 75 | Dichloran (DCNA) | 108 | NOR-AM | Fungicide | WDG | NA |
| Bravo | Chlorothalonil | 35 | Fermenta | Fungicide | Dust, F, liquid, WDG, WP, | 1897-45-6 |
| Broadstrike | Flumetsulam | 91 | DowElanco | Herbicide | EC | 98967-40-9 |
| Bronate | Bromoxynil and MCPA | 80 | Rhone-Poulenc | Herbicide | EC | NA |
| Bronco | Alachor and Glyphosate | 36 | Monsanto | Herbicide | EC | NA |

Table 1. Pesticides Listed Alphabetically by Trade Name

| Trade Name | Common Name | Guide Number [a] | Manufacturer | Category | Formulations [b] | CAS Number [c] |
|---|---|---|---|---|---|---|
| Buckle | Triallate and Trifluralin | 89 | Monsanto | Herbicide | Granular | NA |
| Buctril and Atrazine | Atrazine and Bromoxynil | 89 | Rhone-Poulenc | Herbicide | EC | NA |
| Buctril | Bromoxynil | 50 | Rhone-Poulenc | Herbicide | EC | 1689-84-5 |
| Bueno 6 | MSMA | 92 | Fermenta | Herbicide | Liquid, WSG | 2163-80-6 |
| Bullet | Alachor and Atrazine | 81 | Monsanto | Herbicide | Microencapsulated liquid | NA |
| Bumper | Propiconazole | 66 | Makhteshim-Agan | Fungicide | EC, SCW, WP | 60207-90-1 |
| Butisan S | Metazachlor | 89 | BASF | Herbicide | SC | 67129-08-2 |
| Butyrac | 2,4-DB, dimethyl-amine salt | 57 | Rhone-Poulenc | Herbicide | EC | 94-82-6 |
| Cannon | Alachor and Trifluralin | 85 | Monsanto | Herbicide | EC | NA |
| Canopy | Chlorimuron ethyl and Metribuzin | 89 | DuPont | Herbicide | DG | NA |
| Caparol 4L | Prometryn | 105 | Ciba-Geigy | Herbicide | Liquid, WP | 7287-19-6 |
| Captan 50 | Captan | 8 | ICI | Fungicide, Seed Protectant | Dust, F, WP | 133-06-2 |
| Captan-DCNA 60-20 | Captan and DCNA | 8 | Uniroyal | Fungicide | D | NA |
| Capture 2EC | Bifenthrin | 76 | FMC | Insecticide, Miticide | EC, WP | 82657-04-3 |
| Carbamult | Promecarb | 45 | Hoechst | Insecticide | EC, WP | 2631-37-0 |
| Carzol SP | Formetanate Hydrochloride | 16 | NOR-AM | Insecticide, Miticide | Soluble powder | 23422-53-9 |
| Casoron 2G, 4G | Dichlobenil | 91 | Uniroyal | Herbicide | G, WP | 1194-65-6 |
| Cerone | Ethephon | 6 | Rhone-Poulenc | Plant Growth Regulator | AS, LC | 16672-87-0 |
| Chiptox MCPA Sodium | MCPA | 37 | Rhone-Poulenc | Herbicide | EC, WSC | 94-74-6 |
| Chopper | Imazapyr | 100 | American Cyanamid Company | Herbicide | Aqueous liquid, RTU | 81334-34-1 |
| Classic | Chlorimuron ethyl | 89 | DuPont | Herbicide | DG | 90982-32-4 |

Table 1. Pesticides Listed Alphabetically by Trade Name

| Trade Name | Common Name | Guide Number [a] | Manufacturer | Category | Formulations [b] | CAS Number [c] |
|---|---|---|---|---|---|---|
| Cobra Herbicide | Lactofen | 4 | Valent | Herbicide | EC | 77501-63-4 |
| Comite | Propargite | 11 | Uniroyal | Miticide, acaricide | WP, EC, EW | 2312-35-8 |
| Command 4EC | Clomazone | 60 | FMC | Herbicide | EC | 81777-89-1 |
| Commence EC | Clomazone and Trifluralin | 60 | FMC | Herbicide | EC | NA |
| Compete | Fluoroglycofen-ethyl | 89 | Rohm and Haas | Herbicide | WP | 77501-60-1 |
| Concep II | Oxabetrinil | 103 | Ciba-Geigy | Seed Protectant | WP | 74782-23-3 |
| Concert | Chlorimuron-ethyl and thifensulfuron-methyl | 89 | DuPont | Herbicide | DF | NA |
| Confirm | Tebufenozide | 100 | Rohm and Haas | Insecticide | EC | 11240-23-8 |
| Cotoran | Fluometuron | 91 | Ciba-Geigy | Herbicide | D G(DF), L, WP | 2164-17-2 |
| Counter | Terbufos | 19 | American Cyanamid Company | Insecticide, Nematicide | G | 13071-79-9 |
| Croneton | Ethiofencarb | 44 | Bayer | Insecticide | EC, G | 29973-13-5 |
| Cropstar GB | Alachor | 84 | Monsanto | Herbicide | EC, G, Microencapsulated | 15972-60-8 |
| Crossbow | 2,4-D and Triclopyr | 99 | DowElanco | Herbicide | EC | NA |
| Crusade | Fonofos | 45 | ICI | Insecticide | G | NA |
| Curacron 6E | Profenofos | 78 | Ciba-Geigy | Insecticide, Miticide | EC | 41198-08-7 |
| Curtail | Clopyralid and 2,4-D | 92 | DowElanco | Herbicide | EC | NA |
| Curtail M (petro distill) | Clopyralid and MCPA | 6 | DowElanco | Herbicide | EC | NA |
| Cycle | Cyanazine and Metochlor | 89 | Ciba-Geigy | Herbicide | Liquid | NA |
| Cyclone | Paraquat dichloride | 28 | ICI | Herbicide | SC | 4685-14-7 |
| Cycocel | Chlormequat | 92 | American Cyanamid Company | Plant Growth Regulator | Aqueous solution | 999-81-5 |
| Cygon 400 | Dimethoate | 45 | Cyanimid | Insecticide, Miticide | EC | 60-51-5 |
| Cymbush 3E | Cypermethrin | 31 | ICI | Insecticide | EC | 52315-07-8 |

Table 1. Pesticides Listed Alphabetically by Trade Name

| Trade Name | Common Name | Guide Number [a] | Manufacturer | Category | Formulations [b] | CAS Number [c] |
|---|---|---|---|---|---|---|
| Cymbush 12.5 WP | Cypermethrin | 89 | ICI | Insecticide | WP | 52315-07-8 |
| Cyprex 65-W | Dodine | 7 | American Cyanamid Company | Fungicide | WP | 2439-10-3 |
| Cythion | Malathion | 45-EC, 91-D, RTU, ULV | American Cyanamid Company | Insecticide | Dust, ULV, EC, RTU | 121-75-5 |
| D.Z.N | Diazinon | 63 | Ciba-Geigy | Insecticide, nematicide | D, EC, G, ULV,WP, Microencapsulated | 333-41-5 |
| Dacamine 4D | 2,4-D | 94 | Fermenta | Herbicide | EC | 94-75-7 |
| Daconate 6 | MSMA | 92 | Fermenta | Herbicide | Liquid | 2163-80-6 |
| Daconil 2787 | Chlorothalonil | 35 | Fermenta | Fungicide | F, WP, WDP | 1897-45-6 |
| Dacthal W-75 | DCPA | 100 | Fermenta | Herbicide | WP | 1861-32-1 |
| Def 6 | Butifos | 74 | Mobay | Desiccant | EC | 78-48-8 |
| Demon | Cypermethrin | 47 | ICI | Insecticide | WP, WSP | 52315-07-8 |
| Des-I-Cate | Endothall | 1 | Atochem | Desiccant and/or Herbicide | G | 129-67-9 |
| Desiccant L-10 | Arsenic Acid | 40 | Atochem | Desiccant and Herbicide | Liquid | 7778-39-4 |
| DeVine | Phytophthora | 106 | Abbott | Herbicide | Dry E | NA |
| Devrinol | Napropamide | 5-EC, 100-G, DF, WP | ICI | Herbicide | EC, G, DF, WP | 15299-99-7 |
| Di-Syston | Disulfoton | 23 | Mobay | Insecticide | EC, G | 298-04-4 |
| Dibrom 8 | Naled | 72 | Valent | Insecticide | EC, liquid, LVC | 300-76-5 |
| Dimilin | Diflubenzuron | 103 | Uniroyal | Insect growth regulator | F, WP | 35367-38-5 |
| DiPel | Bacillus thuringiensis, var. kurstaki | 89 | Abbott | Insecticide | ES, WP, F, G | 68038-71-1 |
| Diquat Herbicide H/A | Diquat dibromide | 46 | Valent | Herbicide | Liquid | 2764-72-9 |
| Dithane | Mancozeb | 100 | Rhom and Haas | Fungicide | DF | 8018-01-7 |

Table 1. Pesticides Listed Alphabetically by Trade Name

| Trade Name | Common Name | Guide Number [a] | Manufacturer | Category | Formulations [b] | CAS Number [c] |
|---|---|---|---|---|---|---|
| Dropp 50 | Thidiazuron | 94 | NOR-AM | Plant growth regulator, defoliant | F, WP | 51707-55-2 |
| DSMA | Disodium methanearsonate | 92 | Fermenta | Herbicide | Liquid,WSG, | 144-21-8 |
| Dual | Metolachlor | 63-EC, 91-G | Ciba-Geigy | Herbicide | EC,G | 51218-45-2 |
| Duplosan KV | Mecoprop-P | 63 | BASF | Herbicide | AS,EC | NA |
| Dyfonate EC/G | Fonofos | 18 | ICI | Insecticide | EC,G | 944-22-9 |
| Dylox 80% SP | Trichlorfon | 33 | Mobay | Insecticide, penetrant | DP, EC, G, SP, SC, WP | 52-68-6 |
| Dyrene | Anilazine | 60 | Mobay | Fungicide | Suspension C, WP | 101-05-3 |
| Endocide | Endosulfan | 2 | Platte Chemical | Insecticide | D, EC, G, ULV, WP | 115-29-7 |
| Endothal Turf Herbicide | Endothall | 1 | Atochem | Desiccant and Herbicide | G | 129-67-9 |
| Envert 171 | 2,4-D and Dichlorprop (2,4-DP) | 50 | Rhone-Poulenc | Herbicide | EC | 120-36-5 |
| Eptam | EPTC | 89 | ICI | Herbicide | EC, G | 759-94-4 |
| Eradicane E | EPTC | 89 | ICI | Herbicide | EC | 759-94-4 |
| Escort | Metsulfuron methyl | 89 | DuPont | Herbicide | DF | 74223-64-6 |
| Esteron 99 | 2,4-D | 94 | Rhone-Poulenc | Herbicide | EC | 94-75-7 |
| Etan 3G | Lindane | 71 | Diachem | Insecticide | EC, F, WP, G, D | 58-89-9 |
| Ethrel Plant Regulator | Ethephon | 6 | Rhone-Poulenc | Plant Growth Regulator | Aqueous Solution, LC | 16672-87-0 |
| Evik | Ametryn | 89 | Ciba-Geigy | Herbicide | EC, F, WP | 834-12-8 |
| Evital | Norflurazon | 103 | Sandoz | Herbicide | DF, G | 27314-13-2 |
| Express | Tribenuron methyl | 104 | DuPont | Herbicide | DF | 101200-48-0 |
| Extrazine 11 | Atrazine and Cyanazine | 76 | DuPont | Herbicide | DF (dispersible granule), L | NA |

Table 1. Pesticides Listed Alphabetically by Trade Name

| Trade Name | Common Name | Guide Number [a] | Manufacturer | Category | Formulations [b] | CAS Number [c] |
|---|---|---|---|---|---|---|
| Fallow Master | Dicamba and Glyphosate | 6 | Monsanto | Herbicide | Liquid | NA |
| Far-Go | Triallate | 89 | Monsanto | Herbicide | EC, G | 2303-17-5 |
| Finesse | Chlorsulfuron and Metasulfuron methyl | 89 | DuPont | Herbicide | DF | NA |
| Folex 6-EC | Tribufos | 51 | Rhone-Poulenc | Cotton defoliant | EC | 78-48-8 |
| Force | Tefluthrin | 82 | ICI | Insecticide | EC,G | 79538-32-2 |
| Formula 40 | 2,4-D | 12 | Rhone-Poulenc | Herbicide | EC | 94-75-7 |
| Freedom | Alachor and Trifluralin | 85 | Monsanto | Herbicide | EC | NA |
| Frigate | Ethoxylated Fatty Amines | 5 | Fermenta | Adjuvant, activator, penetrant | Miscible liquid | NA |
| Frontier | Dimethenamid | 54 | Sandoz | Herbicide | EC | 8767468-8 |
| Fruitone CPA | Cloprop (3-CPA) | 92 | Rhone-Poulenc | Plant Growth Regulator | EC | 101-10-0 |
| Fruitone N | NAA | 92 | Rhone-Poulenc | Plant Growth Regulator | L, P | 86-87-3 |
| Furadan | Carbofuran | 25 | FMC, Mobay | Insecticide, Nematicide | F, G | 1563-66-2 |
| Fusilade 2000 | Fluazifop-P-butyl | 106 | ICI | Herbicide | EC | 79241-46-6 |
| Galaxy | Acifluorfen and Bentazon | 12 | BASF | Herbicide | LC, SC | NA |
| Galben | Benalaxyl | 89 | ISAGRO | Fungicide | EC, G, WP | 71626-11-4 |
| Gallery | Isoxaben | 107 | DowElanco | Herbicide | DF, SC | 82558-50-7 |
| Germate Plus | Carboxin, Diazinon, and Lindane | 95 | Uniroyal | Insecticide, Fungicide | Dust | NA |
| Glean | Chlorsulfuron | 89 | DuPont | Herbicide | DF | 64902-72-3 |
| Gnatrol | Bacillus thuringiensis, var. israelensis | 89 | Abbott | Insecticide | DF | NA |
| Goal | Oxyfluorfen | 60 | Rohm and Haas | Herbicide | EC, G | 42874-03-3 |
| Gramoxone | Paraquat dichloride | 28 | ICI | Herbicide | Liquid | 1910-42-5 |

Table 1. Pesticides Listed Alphabetically by Trade Name

| Trade Name | Common Name | Guide Number[a] | Manufacturer | Category | Formulations[b] | CAS Number[c] |
|---|---|---|---|---|---|---|
| Guthion | Azinphos-Methyl | 20 | Mobay | Insecticide | EC, WP | 86-50-0 |
| Harmony Extra | Tribenuron methyl and Thifensulfuron methyl | 100 | DuPont | Herbicide | DF | NA |
| Harvade 5F | Dimethipin | 89 | Uniroyal | Plant Growth Regulator | F | 55290-64-7 |
| Herbicide 273 | Endothall | 1 | Atochem | Desiccant/Herbicide | Liquid | 129-67-9 |
| Hinder | Ammonium soaps of higher fatty acids | 64 | Uniroyal | Repellent | EC | NA |
| Hivol-44 | 2,4-D | 94 | Uniroyal | Herbicide | EC | 94-75-7 |
| Hoelon 3EC | Diclofop-methyl | 38 | Hoechst | Herbicide | EC | 51338-27-3 |
| Horizon IEC | Fenoxaprop-ethyl | 61 | Hoechst | Herbicide | EC | 6641-23-4 |
| Hydrothol | Endothall | 1 | Atochem | Herbicide | L, G | 129-67-9 |
| Hyvar | Bromacil | 89 | DuPont | Herbicide | DF, WP, WSL | 314-40-9 |
| Illoxan | Diclofop-methyl | 38 | Hoechst | Herbicide | EC | 51338-27-3 |
| Image | Imazaquin | 101 | American Cyanamid Company | Herbicide | LC | NA |
| Imidan 50 | Phosmet | 47 | ICI | Insecticide | WP | 732-11-6 |
| Javelin WG | Bacillus thuringiensis, var. kurstaki | 89 | Sandoz | Insecticide | Aqueous F | 68038-71-1 |
| Karate | Lambdacyhalothrin | 30 | ICI | Insecticide | EC | 91465-08-6 |
| Karmex DF | Diuron | 60 | DuPont | Herbicide | DF | 330-54-1 |
| Kelthane | Dicofol | 59 | Rohm and Haas | Insecticide, Miticide | EC, WP | 115-32-2 |
| Kerb 50W | Pronamide | 104 | Rohm and Haas | Herbicide | WP, G | 23950-58-5 |
| Knox Out 2FM | Diazinon | 100 | Atochem | Insecticide | Flowable, microencapsulated | 333-41-5 |
| Krenite S | Fosamine | 70 | DuPont | Herbicide | Water soluble liquid | 25954-13-6 |
| Krovar DF | Bromacil and Diuron | 89 | DuPont | Herbicide | DF (dispersible granule) | NA |

Table 1. Pesticides Listed Alphabetically by Trade Name

| Trade Name | Common Name | Guide Number [a] | Manufacturer | Category | Formulations [b] | CAS Number [c] |
|---|---|---|---|---|---|---|
| Kryocide | Cryolite | 89 | Atochem | Insecticide | Dust, WP | 15096-52-3 |
| Laddok | Atrazine and Bentazone | 36 | BASF | Herbicide | F | NA |
| Landmaster | 2,4-D, isopropylamine salt and Glyphosate | 6 | Monsanto | Herbicide | Liguid | NA |
| Lannate | Methomyl | 26 | DuPont | Insecticide | WSP, L, WSL | 16752-77-5 |
| Lariat | Alachor and Atrazine | 79 | Monsanto | Herbicide | EC | NA |
| Larvin | Thiodicarb | 49 | Rhone-Poulenc | Insecticide | DF, F, WP | 59669-26-0 |
| Lasso | Alachor | 84 | Monsanto | Herbicide | EC, G, Microencapsu-lated | 15972-60-8 |
| Lexone | Metribuzin | 97 | DuPont | Herbicide | DF | 21087-64-9 |
| Londax | Bensulfuron methyl | 69 | DuPont | Herbicide | DF | 83055-99-6 |
| Lorox | Linuron | 89 | DuPont | Herbicide | DF, F, WP | 330-55-2 |
| Lorsban | Chlorpyrifos | 56 | DowElanco | Insecticide | EC, DF, WP | 2921-88-2 |
| Maneb | Maneb | 100 | Atochem | Fungicide | F, FS, WP | 12427-38-2 |
| Manzate | Benomyl and Mancozeb | 108 | DuPont | Fungicide | DF | 12427-38-2 |
| Marksman | Atrazine and Dicamba | 92 | Sandoz | Herbicide | F | NA |
| Metasystox-R | Oxydemeton-methyl | 73 | Mobay | Insecticide | EC | 301-12-2 |
| Microthiol Special | Sulfur | 104 | Atochem | Fungicide, Miticide, Ascricide | Dust, DF,SC,WP | 7704-34-9 |
| Mitac | Amitraz | 52 | NOR-AM | Insecticide, Miticide | EC, WP | 33089-61-1 |
| Mocap | Ethopop | 21 | Rhone-Poulenc | Insecticide, Nemati-cide | EC, G | 13194-48-4 |
| Mocap Plus EC | Disulfoton and Ethopop | 19 | Rhone-Poulenc | Insecticide | EC, G | NA |
| Monitor | Methamidophos | 22 | Mobay, Valent | Insecticide | SC | 10265-92-6 |
| Morestan | Chinomethionat (Oxythioquinox) | 65 | Bayer, Mobay | Fungicide, Miticide | WP(WSP) | 2439-01-2 |
| N-Serve E | Nitrapyrin | 4 | DowElanco | Nitrification Inhibitor | EC | 1929-82-4 |

Table 1. Pesticides Listed Alphabetically by Trade Name

| Trade Name | Common Name | Guide Number [a] | Manufacturer | Category | Formulations [b] | CAS Number [c] |
|---|---|---|---|---|---|---|
| NAA-800 | NAA(Naphthalene-acetic acid) | 64 | Rhone-Poulenc | Plant Growth Regulator | EC, P | 86-87-3 |
| Nemacur EC/G | Fenamiphos | 19 | Mobay | Insecticide, Nematicide | EC, G | 22224-92-6 |
| New Lorox Plus | Chlorimuron ethyl and Linuron | 89 | DuPont | Herbicide | WDG | NA |
| Nimrod | Bupirimate | 88 | Zeneca | Fungicide | EC, WP | 41483-43-6 |
| Nortron | Ethofumesate | 13 | NOR-AM | Herbicide | EC | 26225-79-6 |
| Nova 40W | Myclobutanil | 60 | Rohm and Haas | Fungicide | WP | 88671-89-0 |
| Nutra-Spray Copophos | Basic Copper Sulfate | 92 | Uniroyal | Fungicide | Spray | NA |
| Omite | Propargite | 11 | Uniroyal | Miticide | EC, WP, WSB | 2312-35-8 |
| Option IEC | Fenoxaprop-p-ethyl | 61 | Hoechst | Herbicide | EC | 113158-40-0 |
| Orbit | Propiconazole | 66 | Ciba-Geigy | Fungicide | EC, WP, SWC | 60207-90-1 |
| Ordram EC/G | Molinate | 60-EC, 100-G | ICI | Herbicide | EC, G | 2212-67-1 |
| Ornamite | Propargite | 11 | Uniroyal | Miticide | EC, WP, WSB | 2312-35-8 |
| Orthene | Acephate | 92 | Valent | Insecticide | SP, G | 30560-19-1 |
| Oust | Sulfometuron Methyl | 104 | DuPont | Herbicide | WDG | 74222-97-2 |
| Paarlan E.C. | Isopropalin | 89 | DowElanco | Herbicide | EC | 33820-53-0 |
| Passport | Imazethapyr | 4 | American Cyanamid Company | Herbicide | Aqueous concentrate | 101917-66-2 |
| Penncap-M | Methyl Parathion | 76 | Atochem | Insecticide | Microencapsulated suspension | NA |
| Penncozeb | Mancozeb | 100 | Atochem | Fungicide | DF, WP | NA |
| Pinnacle | Trifensulfuron methyl | 104 | DuPont | Herbicide | DF | 79277-27-3 |
| Pix | Mepiquat Chloride | 98 | BASF | Plant Growth Regulator | SC | 24307-26-4 |
| Plantvax 75W | Oxycarboxin | 91 | Uniroyal | Fungicide | WP | 5259-88-1 |

Table 1. Pesticides Listed Alphabetically by Trade Name

| Trade Name | Common Name | Guide Number [a] | Manufacturer | Category | Formulations [b] | CAS Number [c] |
|---|---|---|---|---|---|---|
| Poast | Sethoxydim | 66 | BASF | Herbicide | EC | 74051-80-2 |
| Pondmaster | Glyphosate | 64 | Monsanto | Herbicide | EC | 1071-83-6 |
| Pounce G/EC | Permethrin | 89 | FMC | Insecticide | EC, G, WP | 52645-53-1 |
| Pramitol 25E | Prometon | 11 | Ciba-Geigy | Herbicide | EC | 1610-18-0 |
| Pramitol 5PS | Prometon and Simazine | 10 | Ciba-Geigy | Herbicide | Pellets | 1610-18-0 |
| Prefar 4-E | Bensulide | 60 | ICI | Herbicide | EC, G | 741-58-2 |
| Prep | Ethephon | 6 | Rhone-Poulenc | Plant Growth Regulator | AS, LC | 16672-87-0 |
| Preview | Chlorimuron ethyl and Metribuzin | 89 | DuPont | Herbicide | WDG | NA |
| Prime + | Flumetralin | 9 | Ciba-Geigy | Plant Growth Regulator | EC | 62924-70-3 |
| Princep | Simazine | 105 | Ciba-Geigy | Herbicide | G, L. WDG, WP | 122-34-9 |
| Probe | Methazole | 63 | Sandoz | Herbicide | G | 20354-26-1 |
| ProGibb Plus | Gibberellic Acid | 93 | Abbott | Plant Growth Regulator | SP, L | 77-06-5 |
| Promalin | Benzyladenine and Gibberellins | 64 | Abbott | Plant Growth Regulator | LC | NA |
| Prostar 4E | Flutolanil | 60 | AgrEvo | Herbicide | WP, WSB | 66332-96-5 |
| ProVide | Gibberellins | 92 | Abbott | Plant Growth Regulator | LC | 77-06-5 |
| Prowl | Pendimethalin | 60-EC, 89-DG | American Cyanamid Company | Herbicide | EC, DG | 40487-42-1 |
| Pursuit | Imazethapyr | 89 | American Cyanamid Company | Herbicide | Aqueous concentrate | 101917-66-2 |
| Pursuit Plus | Imazethapyr and Pendimethalin | 89 | American Cyanamid Company | Herbicide | Aqueous concentrate | NA |
| Pyramin | Chloridazon or pyrazon | 89 | BASF | Herbicide | DF, FL | 1698-60-8 |
| Rally 40W | Myclobutanil | 60 | Rohm and Haas | Fungicide | EC, WP, WSP | 88671-89-0 |

Table 1. Pesticides Listed Alphabetically by Trade Name

| Trade Name | Common Name | Guide Number [a] | Manufacturer | Category | Formulations [b] | CAS Number [c] |
|---|---|---|---|---|---|---|
| Ramrod | Propachlor | 58 | Monsanto | Herbicide | F, G | 1918-16-7 |
| Ramrod and Atrazine | Propachlor and Atrazine | 86 | Monsanto | Herbicide | F | NA |
| Ranger | Glyphosate | 6 | Monsanto | Herbicide | EC | 1071-83-6 |
| Reflex 2LC | Fomesafen | 64 | ICI | Herbicide | LC(aqueous) | 72178-02-0 |
| Reldan D | Chlorpyrifos-methyl | 89 | Uniroyal | Insecticide | EC | 5598-13-0 |
| Release | Gibberellic Acid | 93 | Abbott | Plant Growth Regulator | SP | 77-06-5 |
| Rely | Triclopyr | 3 | Hoechst | Herbicide | EC | 77182-82-2 |
| Rescue | 2,4-DB, sodium salt and Naptalam | 55 | Uniroyal | Herbicide | EC | 132-66-1 |
| Rhomene | MCPA, dimethylamine salt | 37 | Rhone-Poulenc | Herbicide | EC | 94-74-6 |
| Rhonox | MCPA, isooctyl ester | 53 | Rhone-Poulenc | Herbicide | EC | 94-74-6 |
| Ridomil plus Bravo W | Chlorothalonil and Metalaxyl | 3 | Ciba-Geigy | Fungicide | WP | NA |
| Ridomil | Metalaxyl | 63 | Ciba-Geigy | Fungicide | EC, G | 57837-19-1 |
| Ridomil MZ | Mancozeb and Metalaxyl | 67 | Ciba-Geigy | Fungicide | EC, WP, WSP | NA |
| Ridomil PC | Metalaxyl and PCNB (Quintozene) | 103 | Ciba-Geigy | Fungicide | G | NA |
| Ro-Neet 6-E | Cycloate | 89 | ICI | Herbicide | EC | 1134-23-2 |
| Ronilan | Vinclozolin | 104 | BASF | Fungicide | DF, F, WP | 50471-44-8 |
| Rotate | Bendiocarb | 89 | NOR-AM | Insecticide | G | 22781-23-3 |
| Roundup | Glyphosate | 64-EC, 92-D-Pak | Monsanto | Herbicide | EC | 1071-83-6 |
| Rovral | Ipodione | 91 | Rhone-Poulenc | Fungicide | F, WP | 36734-19-7 |
| Roxion | Dimethoate | 45 | American Cyanamid Company | Insecticide, Acaricide | dust, EC, ULVC | 60-51-5 |
| Royal MH-30 | Maleic Hydrazide | 101 | Uniroyal | Plant Growth Regulator | EC, WSG | 51542-52-0 |

Table 1. Pesticides Listed Alphabetically by Trade Name

| Trade Name | Common Name | Guide Number [a] | Manufacturer | Category | Formulations [b] | CAS Number [c] |
|---|---|---|---|---|---|---|
| RTU-PCNB | PCNB (Quintozene) | 103 | Uniroyal | Fungicide | EC | 82-68-8 |
| RTU-Vitavax-Thiram | Carboxin and Thiram | 54 | Uniroyal | Fungicide, Seed Protectant | EC, LF | NA |
| Rubigan | Fenarimol | 60 | DowElanco | Fungicide | AS, EC, WP | 60168-88-9 |
| RyzUp | Gibberellic Acid | 93 | Abbott | Plant Growth Regulator | LC | 77-06-5 |
| Salute 4EC | Metribuzin and Trifluralin | 91 | Mobay | Herbicide | EC | NA |
| Scepter | Imazaquin | 101 | American Cyanamid Company | Herbicide | EC,DG | NA |
| Scepter OT | Acifluorfen and Imazaquin | 12 | American Cyanamid Company | Herbicide | EC | NA |
| Scout X-TRA | Tralomethrin | 34 | Hoechst | Insecticide | EC | 66841-25-6 |
| Sencor | Metribuzin | 97 | Mobay | Herbicide | DF, F | 21087-64-9 |
| Sevin | Carbaryl | 43 | Rhone-Poulenc | Insecticide | AD, EC, F, WP, oil | 63-25-2 |
| Showdown | Triallate | 89 | Monsanto | Herbicide | EC, G | 2303-17-5 |
| Sinbar | Terbacil | 104 | DuPont | Herbicide | WP | 5902-51-2 |
| Solicam | Norflurazon | 103 | Sandoz | Herbicide | DF | 27314-13-2 |
| Sonalan | Ethalfluralin | 61 | DowElanco | Herbicide | EC, G | 55283-68-6 |
| Spike 20P | Tebuthiuron | 89 | DowElanco | Herbicide | P, WP | 34014-18-1 |
| Squadron | Imazaquin and Pendimethalin | 13 | American Cyanamid Company | Herbicide | EC | NA |
| Stam 80EDF | Propanil | 60 | Rohm and Haas | Herbicide | DF | 709-98-8 |
| Stampede CM | MCPA and Paraquat dichloride | 89 | Rohm and Haas | Herbicide | EC | 709-98-8 |
| Starfire | Paraquat dichloride | 28 | ICI | Herbicide | SC | 4685-14-7 |
| Stinger | Clopyralid | 101 | DowElanco | Herbicide | EC | 1702-17-6 |
| Stomp | Pendimethalin | 60 | American Cyanamid Company | Herbicide | EC, WDG | 40487-42-1 |

Table 1. Pesticides Listed Alphabetically by Trade Name

| Trade Name | Common Name | Guide Number [a] | Manufacturer | Category | Formulations [b] | CAS Number [c] |
|---|---|---|---|---|---|---|
| Storm Herbicide | Acifluorfen and Bentazon | 12 | BASF | Herbicide | SL | NA |
| Strel 4E | Propanil | 60 | Rohm and Haas | Herbicide | EC | 709-98-8 |
| Sulfur Alfa | Sulfur | 104 | ICI | Fungicide | WP | 7704-34-9 |
| Sulv | 2,4-D, dimethylamine salt | 12 | Uniroyal | Herbicide | EC | 94-75-7 |
| Supracide 2 | Methidathion | 24 | Ciba-Geigy | Insecticide | EC, WP | 950-37-8 |
| Surflan | Oryzalin | 101 | DowElanco | Herbicide | AS, DF, WP | 19044-88-3 |
| Sutan + | Butylate | 89 | ICI | Herbicide | EC, G | 2008-41-5 |
| Sutazine + | Atrazine and Butylate | 36 | ICI | Herbicide | EC, G | NA |
| Talon-G | Brodifacoum | 44 | ICI | Rodenticides | Pellets, minipellets | 56073-10-0 |
| Tandem | Tridiphane | 56 | DowElanco | Herbicide | EC | 58138-08-2 |
| Telar | Chlorsulfuron | 89 | DuPont | Herbicide | DF | 64902-72-3 |
| Telone C-17 | Dichloropropene and Chloropicrin | 32 | DowElanco | Fungicide, Nematicide | EC | NA |
| Telone II | Dichloropropene | 29 | DowElanco | Nematicide, Soil Fumigants | EC | 542-75-6 |
| Temik 15G | Aldicarb | 41 | Rhone-Poulenc | Insecticide, Miticide, Nematicide | G | 116-06-3 |
| Terraclor | PCNB | 102 | Uniroyal | Fungicide | EC | 82-68-8 |
| Terrazole 25% Ornamental | Etridiazole | 5 | Uniroyal | Fungicide | EC | 2593-15-9 |
| Terrazole 35% Turf | Etridiazole | 49 | Uniroyal | Fungicide | WP | 2593-15-9 |
| Tersan 1991 | Benomyl | 108 | DuPont | Fungicide | DF | 17804-35-2 |
| Thimet 15-G | Phorate | 20 | American Cyanamid Company | Insecticide | G | 298-02-2 |
| Thiodan EC/WP | Endosulfan | 2 | FMC | Insecticide | EC, WP | 115-29-7 |
| Thiolux DF | Sulfur | 104 | Sandoz | Fungicide, Miticide | DF | 7704-34-9 |
| Thistrol | MCPB | 92 | Rhone-Poulenc | Herbicide | AS | 94-81-5 |

Table 1. Pesticides Listed Alphabetically by Trade Name

| Trade Name | Common Name | Guide Number[a] | Manufacturer | Category | Formulations[b] | CAS Number[c] |
|---|---|---|---|---|---|---|
| Thistrol and 2,4-D | MCPB and 2,4-D | 87 | Rhone-Poulenc | Herbicide | AS | 94-81-5 |
| Tillam 6-E | Pebulate | 89 | ICI | Herbicide | EC, G | 1114-71-2 |
| Tiller | Fenoxaprop-ethyl, 2,4-D, and MCPA | 60 | Hoechst | Herbicide | EC | NA |
| Tilt | Propiconazole | 66 | Ciba-Geigy | Fungicide | EC, WP | 60207-90-1 |
| Topsin M 70W | Thiophanate-methyl | 104 | Atochem | Fungicide | DF, EC, F | 23564-06-9 |
| Tordon 22K | Picloram, potassium salt | 100 | DowElanco | Herbicide | WSL | 191802-1 |
| Tordon RTU | 2,4-D and Picloram | 68 | DowElanco | Herbicide | WSL | 191802-1 |
| Torpedo | Permethrin | 60 | ICI | Insecticide | EC | 70451-80-2 |
| Tre-Hold Sprout Inhibitor | NAA(Naphthaleneacetic acid) | 92 | Rhone-Poulenc | Plant Growth Regu-lator | EC | 86-87-3 |
| Treflan EC | Trifluralin | 47 | DowElanco | Herbicide | EC | 1582-09-8 |
| Treflan | Trifluralin | 94 | DowElanco | Herbicide | G, MFT, DC | 1582-09-8 |
| TRI-4 | Trifluralin | 94 | American Cyanamid Company | Herbicide | EC | 1582-09-8 |
| Tri-Scept | Imazaquin and Trifluralin | 60 | American Cyanamid Company | Herbicide | EC | NA |
| Trident II | Bacillus thuringiensis | 89 | Sandoz | Insecticide | WDL | NA |
| Trigard | Cyromazine | 91 | Ciba-Geigy | Insecticide | WP | 66215-27-8 |
| Triple-Noctin L | Thiram + Nitrogen fixing bacteria | 91 | Uniroyal | Fungicide, Seed Protectant | LS | 137-26-8 |
| Tupersan | Siduron | 100 | DuPont | Herbicide | WP | 1982-49-6 |
| Turbo 8EC | Metolachlor and Metribuzin | 91 | Mobay | Herbicide | EC | NA |
| Turfcide | PCNB | 102 | Uniroyal | Fungicide | EC | 82-68-8 |
| Vapam | Metam-Sodium | 92 | ICI | Soil Fumigants | AS | 137-42-8 |
| VectoBac | Bacillus thuringiensis | 89 | Abbott | Insecticide | AS, G | NA |

Table 1. Pesticides Listed Alphabetically by Trade Name

| Trade Name | Common Name | Guide Number [a] | Manufacturer | Category | Formulations [b] | CAS Number [c] |
|---|---|---|---|---|---|---|
| Velpar | Hexazinone | 5 | DuPont | Herbicide | SP, WDL, SG | 51235-04-2 |
| Vendex | Fenbutalin-oxide | 14 | DuPont | Mitocide | LC, WP, WSL | 13356-08-6 |
| Vitavax-200 Vitavax-75W | Carboxin and Thiram | 54-200, 91-75W | Uniroyal | Fungicide, Seed Protectant | F | NA |
| Vitavax-34 | Carboxin | 91 | Uniroyal | Fungicide | F | 5234-68-4 |
| Vitavax-PCNB | Carboxin and PCNB | 54 | Uniroyal | Fungicide | F | NA |
| Volck Supreme Spray | Petroleum Oil | 100 | Valent | Insecticide. Miticide | Oil spray | NA |
| Vorlex | Dichloropropene and Methylisothiocyanate | 29 | NOR-AM | Soil Fumigants | EC | NA |
| Vydate L | Oxamyl | 27 | DuPont | Insecticide, Miticide, Nematicide | WSL | 23135-22-0 |
| WeatherBlok | Brodifacoum | 44 | ICI | Rodenticides | Pellets. minipellets | 56073-10-0 |
| Weedar 64 | 2,4-D, dimethylamine salt | 12 | Rhone-Poulenc | Herbicide | EC | 94-75-7 |
| Weedmaster | 2,4-D and Dicamba | 106 | Sandoz | Herbicide | WSL | NA |
| Weedone 170 | 2,4-D and dichlorprop (2,4-DP) | 50 | Rhone-Poulenc | Herbicide | EC | NA |
| Weedone 2,4-DP | 2,4-DP | 50 | Rhone-Poulenc | Herbicide | EC | 120-36-5 |
| Weedone LV6 | 2,4-D | 94 | Rhone-Poulenc | Herbicide | EC | 94-75-7 |
| Whip 1EC | Fenoxaprop-ethyl | 61 | Hoechst | Herbicide | EC | 66441-23-4 |
| Whip 360 | Fenoxaprop-p-ethyl | 61 | Hoechst | Herbicide | EC | 113158-40-0 |
| Ziram | Ziram | 39-FL. 65-WP | Atochem | Fungicide | FL, WP | 137-30-4 |
| Zorial Rapid 80 | Norflurazon | 103 | Sandoz | Herbicide | DF | 27314-13-2 |

[a] The Pesticide Guide Number is assigned to a particular group of pesticides with a specific set of properties. This guide contains all the information required to handle those pesticides properly and safely.

[b] The formulations commercially available for this brand. See Chapter 4 for the meaning of the abbreviation used for the formulations.

[c] A unique numbering system established by the Chemical Abstract Service. Each chemical formulation is given a specific number. Therefore. different formulations of the same chemical will have different CAS numbers.

[d] NA used to designate that the CAS number is not available.

Table 2. Pesticides Listed Alphabetically by Common Name

| Common Name [a] | Trade Name [b] | Family Name [c] | Chemical Name [d] |
|---|---|---|---|
| Abamectin | Agri-Mek | Avermectin | Avermectin B |
| Acephate | Orthene SP/S/Insect Spray | Organophosphate | O,S-dimethyl acetylphosphoramidothioate |
| Acifluorfen and Bentazon | Galaxy | Diphenyl ethers and benzothiadiazole | Sodium 5-[2-chloro-4-(trifluoro methyl)phenoxy]-2-nitrobenzoate and sodium 3-isopropyl-1H-2,1,3-benzothiadiazin-4(3H)-one-2,2-dioxide |
| Acifluorfen and Bentazon | Storm Herbicide | Diphenyl ethers and benzothiadiazole | Sodium 5[2-chloro-4(trifluoromethyl) phenoxy]-2-nitrobenzoate and 3-isopropyl-1H-2,1,3-benzothiadiazin -4(3H)-one 2,2-dioxide |
| Acifluorfen and Imazaquin | Scepter OT | Diphenyl ethers and imidazole | Sodium 5-[2-chloro-4-(trifluoro methyl)phenoxy]-2-nitrobenzoate and (±)2-[4,5-dihydro-4-methyl-4-(1-methylethyl)-5-oxo-1H-imidazol-2-yl] -3-quinolinecarboxylic acid |
| Acifluorfen-sodium | Blazer | Diphenyl ethers | Sodium 5[2-chloro-4(trifluoromethyl) phenoxy]-2-nitrobenzoate |
| Alachor | Cropstar GB | Acetanilide | 2-chloro-2',6'-diethyl-N-(methoxy-methyl)acetanilide |
| Alachor | Lasso | Acetanilide | 2-chloro-2',6'-diethyl-N-(methoxy-methyl)acetanilide |
| Alachor and Atrazine | Bullet | Acetanilide and triazine | 2-chloro-2',6'-diethyl-N-(methoxy-methyl)acetanilide and 6-chloro-N2-ethyl-N4-isopropyl-1,3,5-triazine-2,4-diamine |
| Alachor and Atrazine | Lariat | Acetanilide and triazine | 2-chloro-2',6'-diethyl-N-(methoxy-methyl)acetanilide and 6-chloro-N2-ethyl-N4-isopropyl-1,3,5-triazine-2,4-diamine |
| Alachor and Glyphosate | Bronco | Acetanilide and phosphonate | 2-chloro-2',6'-diethyl-N-(meth-oxymethyl)acetanilide and isopropyl-amine salt of N-(phosphonomethyl)-glycine |
| Alachor and Trifluralin | Cannon | Acetanilide and dinitroaniline | 2-chloro-2',6'-diethyl-N-(methoxy-methyl)acetanilide and a,a,a trifluoro-2,6-dinitro-N,N-dipropyl-p-toluidine |
| Alachor and Trifluralin | Freedom | Acetanilide and dinitroaniline | 2-chloro-2',6'-diethyl-N-(methoxy-methyl)acetanilide and a,a,a trifluoro-2,6-dinitro-N,N-dipropyl-p-toluidine |
| Aldicarb | Temik 15G | Carbamate | 2-methyl-2-(methylthio)propionaldehyde O-(methylcarbamoyl)oxime |
| Ametryn | Evik | Triazine | 2-ethylamino-4-isopropylamino-6-methylthio-s-triazine |
| Aminopyridine | Avitrol | Pyridine | 4-aminopyridine |
| Amitraz | Mitac WP | Triazapentadiene | N'-(2,4-dimethylphenyl)-N-[[(2,4-dimethylphenyl)imino]methyl]-N-methylmethanimidamide |
| Amitrole | Amitrol T | Triazole | 1,2,4-triazol-3-ylamine |

Table 2. Pesticides Listed Alphabetically by Common Name

| Common Name [a] | Trade Name [b] | Family Name [c] | Chemical Name [d] |
|---|---|---|---|
| Amitrole | Amizol Industrial | Triazole | 1,2,4-triazol-3-ylamine |
| Ammonium soaps of higher fatty acids | Hinder | Fatty acid salt | Ammonium soaps of higher fatty acids |
| Anilazine | Dyrene | Triazine | 4,6-dichloro-N-(2-chlorophenyl)-1,3,5-triazin-2-amine |
| Arsenic Acid | Desiccant L-10 | Inorganic | Arsenic acid |
| Asulam | Asulox | Carbamate | Methyl sulfanilylcarbamate |
| Atrazine | AAtrex | Triazines | 6-chloro-$N^2$-ethyl $N^4$-isopropyl-1,3,5-triazine-2.4-diamine |
| Atrazine | Atratol 90 | Triazine | 6-chloro-N2-ethyl-N4-isopropyl-1.3,5-triazine-2,4-diamine () |
| Atrazine | Atrazine | Triazine | 6-chloro-N2-ethyl-N4-isopropyl-1,3,5-triazine-2,4-diamine () |
| Atrazine and Bentazone | Laddok | Triazine and benzothiadia-zole | 6-chloro-N2-ethyl-N4-isopropyl-1,3,5-triazine-2,4-diamine and sodium 3-isopropyl-1H-2,1,3-benzothiadiazin-4(3H)-one-2,2-dioxide |
| Atrazine and Bro-moxynil | Buctril + Atrazine | Triazine and benzonitrile | 6-chloro-N2-ethyl-N4-isopropyl-1,3,5-triazine-2,4-diamine and 3,5-dibromo-4-hydroxybenzonitrile |
| Atrazine and Butylate | Sutazine + | Triazine and Thiocar-bamate | 6-chloro-N2-ethyl N4-isopropyl-1,3,5-triazine-2,4-diamine and S-ethyl diisobutylthio-carbamate |
| Atrazine and Cyanazine | Extrazine 11 DF/L | Triazine | 6-chloro-N2-ethyl N4-isopropyl-1,3,5-triazine-2,4-diamine and 2-[[4-chloro-6-(ethylamino)-1,3,5-triazin-2-yl]amino]-2-methyl-propionitrile |
| Atrazine and Dicamba | Marksman | Triazine and ethylene bisdithiocarbamate | 6-chloro-N2-ethyl-N4-isopropyl-1,3,5-triazine-2,4-diamine and 3,6-dichloro-2-methoxybenzoic acid |
| Atrazine and Meto-lachlor | Bicep | Triazine. acetanilide | 6-chloro-N2-ethyl-N4-isopropyl-1,3,5-triazine-2,4-diamine and 2-chloro-N-(2-ethyl-6-methylphenyl)-N-(2-methoxy-1-methylethyl) acetamide |
| Azinphos-Methyl | Guthion S/E/WP | Organophosphate | O,O-dimethyl S[(4-oxo-1,2,3-benzotriazin-3(4H)-yl)methyl]phosphorodithioate |
| Bacillus thuringiensis | Trident II | Biological | Bacillus thuringiensis, var. tenebrionis |
| Bacillus thuringiensis | VectoBac AS/G | Biological | Bacillus thuringiensis, serotype H-14 |
| Bacillus thuringiensis, var. israelensis | Gnatrol | Biological (bacteria) | Bacillus thuringiensis, var. israelensis |
| Bacillus thuringiensis, var. kurstaki | Biobit FC/WP | Biological | Biological (bacteria) |
| Bacillus thuringiensis. var. kurstaki | DiPel | Biological | Bacillus thuringiensis, var. kurstaki |

Table 2. Pesticides Listed Alphabetically by Common Name

| Common Name [a] | Trade Name [b] | Family Name [c] | Chemical Name [d] |
|---|---|---|---|
| Bacillus thuringiensis, var. kurstaki | Javelin WG | Biological | Bacillus thuringiensis, var. kurstaki |
| Basic Copper Sulfate | Nutra-Spray Copo-phos | Inorganic | Copper Sulfate, phosphoric acid, Zn |
| Benalaxyl | Galben | Triazole | Methyl N-phenylacetyl-N-2,6-xylyl-DL-alaninate |
| Bendiocarb | Rotate | Carbamate | 2,2-dimethyl-1,3-benzodioxol-4-yl methylcarbamate |
| Benefin | Balan DF/EC | Dinitroaniline | N-butyl-N-ethyl-a,a,a-trifluoro-2,6-dinitro-p-toluidine |
| Benomyl | Benlate 50 DF | Benzimidazole | Methyl 1-(butylcarbamoyl)benzimidazol-2-ylcarbamate |
| Benomyl | Tersan 1991 DF | Benzimidazole | Methyl 1-(butylcarbamoyl)benzimidazol -2-ylcarbamate |
| Benomyl and Mancozeb | Manzate | Benzimidazole and dithio-carbamate | Methyl 1-(butylcarbamoyl) benzimidazol-2-ylcarbamate and Zn and Mn ethylene bisdithiocarbamate |
| Bensulfuron methyl | Londax | Sulfonylurea | Methyl 2-[[[[(4,6-dimethoxypyrimidin-2-yl)amino]carbonyl]amino]-sulfonyl]methyl]benzoate |
| Bensulide | Betasan 2 9-EF/G | Organophospate | O,O-diisopropyl phosphorodithioate S-ester with N-(2-mercaptoethyl) benzenesul-fonamide |
| Bensulide | Prefar 4-E | Amide | O,O-diisopropyl phosphorodithioate S-ester with N-(2-mercaptoethyl) benzenesul-fonamide |
| Bentazon | Basagran | Benzothiadiazoles | 3-isopropyl-1H-2,1,3-benzothiadiazin-4(3H)-one 2,2-dioxide |
| Benzyladenine and Gibberellins | Promalin | Gibberellins | N-(phenylmethyl)-1H-purine-6-amine and Giberellins A4A7 |
| Bifenthrin | Capture 2EC | Pyrethroid | (2 methyl[1,1'-biphenyl]-3-yl)methyl 3-(2-chloro-3,3,3-trifuoro-1-propenyl)-2,2-dimethyl-cyclopropane-carboxylate |
| Brodifacoum | Talon-G | Coumarin | 3-[3-(4'-bromo[1,1'-biphenyl]-4-yl)-1,2,3,4-tetrahydro-1-napthalenyl]-4-hydroxy-2H-1-benzopyran-2-one |
| Brodifacoum | WeatherBlok | Coumarin | 3-[3-(4'-bromo[1,1'-biphenyl]-4-yl)-1,2,3,4-tetrahydro-1-napthalenyl]-4-hydroxy-2H-1-benzopyran-2-one |
| Bromacil | Hyvar | Uracil | 5-bromo-3-sec-butyl-6-methyluracil |
| Bromacil and Diuron | Krovar DF | Uracil and substituted urea | 5-bromo-3-sec-butyl-6-methyluracil and N'-(3,4-dichlorophenyl)-N,N-dimethylurea |
| Bromopropylate | Acarol | Miscellaneous organic | Isopropyl 4,4'-dibromobenzilate |
| Bromoxynil | Buctril | Benzonitrile | 3,5-dibromo-4-hydroxybenzonitrile |
| Bromoxynil and MCPA | Bronate | Benzonitrile | 3,5-dibromo-4-hydroxybenzonitrile and 2-methyl-chlorophenoxyacetic acid |

Table 2. Pesticides Listed Alphabetically by Common Name

| Common Name [a] | Trade Name [b] | Family Name [c] | Chemical Name [d] |
|---|---|---|---|
| Bupirimate | Nimrod | Pyrimidine | 5-butyl-2-ethylamino-6 methylpyrimidin-4-yl dimethyl-sulfamate |
| Butylate | Sutan + 6.7-E | Thiocarbamate | S-ethyl diisobutylthiocarbamate |
| Candida oleophila | Aspire | Biological (yeast) | yeast |
| Captan | Captan 50 | Thiophthalimide | (N-trichloromethylthio-4-cyclohexene-1,2-dicarboximide) |
| Captan and DCNA | Captan-DCNA 60-20 | Substituted aromatic and thiophthalimide | (N-trichloromethylthio-4-cyclohexene-1,2-dicarboximide) and 2,6-dichloro-4-nitroaniline |
| Captan and Dichloran | Botec | Sulfenimide and substituted aromatic | (N-trichloromethylthio-4-cyclohexene-1,2-dicarboximide) and 2,6-dichloro-4-nitroaniline |
| Carbaryl | Sevin | Carbamate | 1-naphthyl methylcarbamate |
| Carbofuran | Furadan | Carbamate | 2,3-dihydro-2,2-dimethyl-7-benzofuranyl methylcarbamate |
| Carboxin | Vitavax-34 | Oxathiin | 5,6-dihydro-2-methyl-N-phenyl-1,4-oxathiin-3-carboxamide |
| Carboxin and PCNB | Vitavax-PCNB | Oxathiin and substituted aromatic | 5,6-dihydro-2-methyl-N-phenyl-1,4-oxathiin-3-carboxamide and pentachloronitrobenzene |
| Carboxin and Thiram | RTU-Vitavax-Thiram | Oxathiin and thiocarbamate | 5,6-dihydro-2-methyl-N-phenyl-1,4-oxathiin-3-carboxamide and bis(dimethylthiocarbamoyl) disulfide or tetramethylthiuram disulfide |
| Carboxin and Thiram | Vitavax-200 | Oxathiin and thiocarbamate | 5,6-dihydro-2-methyl-N-phenyl-1,4-oxathiin-3-carboxamide and bis(dimethylthiocarbamoyl) disulfide |
| Carboxin, Diazinon, and Lindane | Germate Plus | Oxathiin, organophosphate, and organochlorine | 5,6-dihydro-2-methyl-N-phenyl-1,4-oxathiin-3-carboxamide, O,O-diethyl O-[6-methyl-2-(1-methylethyl)-4-pyrimidinyl] phosphorothioate and 1,2,3,4,5,6-hexachloro-cyclohexane, |
| Chinomethionat (Oxythioquinox) | Morestan | Dithiocarbonate | 6-methyl-1,3-dithiolo[4,5-b]quinoxalin-2-one |
| Chloramben | Amiben | Benzoic acid | 3-Amino-2,5-dichlorobenzoic acid |
| Chloridazon or pyrazon | Pyramin | Pyridazinone | 5-amino-4-chloro-2-phenyl-3-(2H)-pyridazinone |
| Chlorimuron ethyl | Classic | Sulfonylurea | Ethyl 2-[[[[(4-chloro-6-methoxy-2-pyrimidindinyl)amino]carbonyl]-amino]sulfonyl]benzoate |
| Chlorimuron ethyl and Linuron | New Lorox Plus | Substituted urea and sulfonyl urea | Ethyl 2-[[[[(4-chloro-6-methoxy-2-pyrimidindinyl)amino]carbonyl]-amino]sulfonyl]benzoate and 3-(3,4-dichlorophenyl)-1-methoxy-1-methylurea |
| Chlorimuron ethyl and Metribuzin | Canopy | Sulfonylurea and triazinone | Ethyl 2-[[[[(4-chloro-6-methoxy-2-pyrimidindinyl)amino]carbonyl]-amino]sulfonyl]benzoate and 4-amino-6-(1,1-dimethylethyl)-3-(methylthio)-1,2,4-triazin-5(4H)-one |

Table 2. Pesticides Listed Alphabetically by Common Name

| Common Name [a] | Trade Name [b] | Family Name [c] | Chemical Name [d] |
|---|---|---|---|
| Chlorimuron ethyl and Metribuzin | Preview | Sulfonylurea and triazinone | Ethyl 2-[[[[(4-chloro-6-methoxy-2-pyrimidindinyl)amino]carbonyl]-amino]sulfonyl]benzoate and 4-amino-6-(1,1-dimethylethyl)-3-(methylthio)-1,2,4-triazin-5(4H)-one |
| Chlorimuron-ethyl and thifensulfuron-methyl | Concert | Sulfonylurea | Ethyl 2-[[[[(4-chloro-6-methoxy-2-pyrimidindinyl)amino]carbonyl]-amino]sulfonyl]benzoate and Methyl 3-[[[[(4-methoxy-6-methyl-1,3,5-triazin-2-yl)amino]carbonyl]amino]sulfonyl]-2-thiophencarboxylate |
| Chlormequat | Cycocel | Methyl ammonium chloride | 2-chloroethyltrimethylammonium chloride |
| Chlorothal | Dacthal W-75 | Phthalate | Dimethyl tetrachloroterephthalate |
| Chlorothal | Dacthal W-75 Turf | Phthalate | Dimethyl tetrachloroterephthalate |
| Chlorothalonil | Bravo | Substituted aromatic | Tetrachloroisophthalonitrile |
| Chlorothalonil | Daconil 2787† | Substituted aromatic | Tetrachloroisophthalonitrile |
| Chlorothalonil and Metalaxyl | Ridomil Bravo W | Substituted aromatic and phenylamide | Tetrachloroisophthalonitrile and N-(2,6-dimethylphenyl)-N-(methoxyacetyl(-DL-alanine methyl ester |
| Chlorpyrifos | Lorsban E/G/WP | Organophosphate | O,O-diethyl O-(3,5,6-trichloro-2-pyridinyl)phosphorothioate |
| Chlorpyrifos-methyl | Reldan D | Organophosphate | O,O-dimethyl O-(3,5,6-trichloro-2-pyridinyl) phosphorothioate |
| Chlorsulfuron | Glean | Sulfonyl urea | 2-chloro-N-[[(4-methoxy-6-methyl-1,3,5-triazin-2-yl)amino]carbonyl] benzenesulfonamide |
| Chlorsulfuron | Telar | Sulfonyl urea | 2-chloro-N-[[(4-methoxy-6-methyl-1,3,5-triazin-2-yl)amino]carbonyl] benzenesulfonamide |
| Chlorsulfuron and Metasulfuron methyl | Finesse | Sulfonyl urea | 2-chloro-N-[[(4-methoxy-6-methyl-1,3,5-triazin-2-yl)amino]carbonyl] benzenesulfonamide and methyl 2-[[[[(4-methoxy-6-methyl-1,3,5-triazin-2 yl)-amino]carbonyl] amino]sulfonyl] benzoate |
| Cinmethylin | Argold | Cineole | exo-1-methyl-4-(1-methylethyl)-2-[(2-methylphenyl)methoxy]-7-oxabicyclo[2.2.1] heptane |
| Clofentezine | Apollo SC | Tetrazine | 3,6-bis(2-chlorophenyl)-1,2,4,5-tetrazine |
| Clomazone | Command 4EC | Pyridazinone | 2-[(2-chlorophenyl)methyl]-4,4-dimethyl-3-isoxazolidinone |
| Clomazone and Trifluralin | Commence EC | Pyridazinone and dinitroaniline | 2-[(2-chlorophenyl)methyl]-4,4-dimethyl-3-isoxazolidinone and a,a,a trifluoro-2,6-dinitro-N,N-dipropyl-p-toluidine |
| Cloprop (3-CPA) | Fruitone CPA | Chlorophenoxy acid | 2-(3-chlorophenoxy)propionic acid |
| Clopyralid | Stinger | Picolinic acid | 3,6-dichloropicolinic acid |

Table 2. Pesticides Listed Alphabetically by Common Name

| Common Name [a] | Trade Name [b] | Family Name [c] | Chemical Name [d] |
|---|---|---|---|
| Clopyralid and 2,4-D | Curtail | Picolinic acid and chlorinated phenoxy | 3,6-dichloropicolinic acid and 2,4-dichlorophenoxyacetic acid |
| Clopyralid and MCPA | Curtail M (petro distill) | Picolinic acid and chlorinated phenoxy | 3,6-dichloropicolinic acid and (4-chloro-2-methyl)phenoxyacetic acid |
| Cryolite | Kryocide | Fluoride | Sodium aluminofluoride |
| Cyanazine | Bladex L/DF | Triazine | 2-[[4-chloro-6-(ethylamino)-1,3,5-triazin-2-yl]amino]-2-methyl-propionitrile |
| Cyanazine and Metochlor | Cycle | Triazine and acetanilide | 2-[[4-chloro-6-(ethylamino)-1,3,5-triazin-2-yl]amino]-2-methyl-propionitrile and 2-chloro-N-(2-ethyl-6-methylphenyl)-N-(2-methoxy-1-methylethyl) Acetamide |
| Cycloate | Ro-Neet 6-E | Thiocarbamate | S-ethyl cyclohexyl(ethyl)thiocarbamate |
| Cyfluthrin | Baythroid 2 | Pyrethroid | Cyano(4-fluoro-3-phenoxyphenyl) methyl 3-(2,2-dichloroethenyl)-2,2-dimethylcyclopropane-carboxylate |
| Cypermethrin | Ammo 2.5 EC | Pyrethroid | (±)-a-cyano-3-phenoxybenzyl (±)-cis,trans-3-(2,2-dichlorovinyl)-2,2-dimethylcyclopropane- carboxylate |
| Cypermethrin | Cymbush 3E | Pyrethroid | (±)-a-cyano-3-phenoxybenzyl (±)-cis,trans-3-(2,2-dichlorovinyl)-2,2-dimethylcyclopropanecarboxylate |
| Cypermethrin | Demon | Pyrethroid | (±)-a-cyano-3-phenoxybenzyl (±)-cis,trans-3-(2,2-dichlorovinyl)-2,2-dimethylcyclopropanecarboxylate |
| Cyromazine | Trigard | Triazine | N-cyclopropyl-1,3,5-triazine-2,4,6-triamine |
| 2,4-D | Dacamine 4D | Chlorophenoxy acid | 2,4-dichlorophenoxyacetic acid |
| 2,4-D | Esteron 99 | Chlorophenoxy acid | 2,4-dichlorophenoxyacetic acid |
| 2,4-D | Formula 40 | Chlorophenoxy acid | 2,4-dichlorophenoxyacetic acid |
| 2,4-D | Hivol-44 | Chlorophenoxy acid | 2,4-dichlorophenoxyacetic acid |
| 2,4-D | Weedone LV6 | Chlorophenoxy acid | 2,4-dichlorophenoxyacetic acid |
| 2,4-D and Dicamba | Weedmaster | Chlorophenoxy and benzoic acid | 2,4-dichlorophenoxyacetic acid and 3,6-dichloro-2-methoxybenzoic acid |
| 2,4-D and Picloram | Tordon RTU | Pyridinoxy and Chlorophenoxy acids | 2,4-dichlorophenoxyacetic acid and 4-amino-3,5,6-trichloropicolinic acid |
| 2,4-D and Triclopyr | Crossbow | Chlorophenoxy and chloropyridinyl | 2,4-dichlorophenoxyacetic acid and (3,5,6-trichloro-2-pyridinyloxy) acetic acid |
| 2,4-D Butoxy-ethyl ester | Aqua Kleen | Chlorophenoxy acid | 2,4-dichlorophenoxyacetic acid butoxyethyl ester |

Table 2. Pesticides Listed Alphabetically by Common Name

| Common Name [a] | Trade Name [b] | Family Name [c] | Chemical Name [d] |
|---|---|---|---|
| 2,4-D, and dichlor-prop(2,4-DP) | Weedone 170 | Chlorophenoxy acid | 2,4-dichlorophenoxyacetic acid and (RS)-2-(2,4-chlorophenoxy)propionic acid |
| 2,4-D, and dichlor-prop(2,4-DP) | Envert 171 | Chlorophenoxy acid | 2,4-dichlorophenoxyacetic acid and (RS)-2-(2,4-dichlorophenoxy) propionic acid |
| 2,4-D, dimethyl-amine salt | Sulv | Chlorophenoxy Herbicide | 2.4-dichlorophenoxyacetic acid |
| 2,4-D, dimethyl-amine salt | Weedar 64 | Chlorophenoxy acid | 2,4-dichlorophenoxyacetic acid |
| 2,4-D, isopropylamine salt and Glyphosate | Landmaster | Chlorophenoxy acid and phosphonate | 2,4-dichlorophenoxyacetic acid and isopropyl-amine salt of N-(phosphonomethyl)-glycine |
| 2,4-DB (dimethyl-amine salt) | Butyrac | Chlorophenoxy acid | 4-(2,4-dichlorophenoxy)butyric acid |
| 2,4-DB and Naptalam | Rescue | Chlorophenoxy acid amide | 4-(2,4-dichlorophenoxy) butyric acid and N-1-naphthylphthalamic acid |
| 2,4-DP | Weedone 2,4-DP | Chlorophenoxy acid | (RS)-2-(2,4-chlorophenoxy)propionic acid |
| Daminozide | B-Nine SP | Miscellaneous organic | Butanedioic acid mono (2,2-dimethyl hydrazine) |
| Desmedipham | Betanex | Carbamate | Ethyl 3'-phenylcarbamoyloxycarbanilate |
| Desmedipham and Phenmedipham | Betamix | Carbamate | Ethyl 3'-phenylcarbamoyloxycarbanilate and methyl m-hydroxy-carbaniliate m-methylcarbanilate ester |
| Diazinon | D.Z.N | Organophosphate | O,O-diethyl O-[6-methyl-2-(1-methylethyl)-4-pyrimidinyl] phosphorothioate |
| Diazinon | Knox Out 2FM | Organophosphate | O,O-diethyl O-[6-methyl-2-(1-methylethyl)-4-pyrimidinyl] phosphorothioate |
| Dicamba | Banvel | Benzoic acid | 3,6-dichloro-2-methoxybenzoic acid |
| Dicamba and Glyphosate | Fallow Master | Benzoic acid and phos-phonate | 3,6-dichloro-2-methoxybenzoic acid and isopropylamine salt of N-(phosphonomethyl)-glycine |
| Dichlobenil | Casoron 2G | Benzonitrile | 2,6-dichlorobenzonitrile |
| Dichlobenil | Casoron 4G | Benzonitrile | 2,6-dichlorobenzonitrile |
| Dichloran (DCNA) | Botran 75WDG | Substituted aromatic | 2,6-dichloro-4-nitroaniline |
| Dichloropropene | Telone II | Chlorinated hydrocarbon fumigant | 1.3-dichloropropene |
| Dichloropropene and Chloropicrin | Telone C-17 | Chlorinated hydrocarbon fumigant | 1.3-dichloropropene and trichloronitromethane |

Table 2. Pesticides Listed Alphabetically by Common Name

| Common Name [a] | Trade Name [b] | Family Name [c] | Chemical Name [d] |
|---|---|---|---|
| Dichloropropene and Methylisothiocyanate | Vorlex | Chlorinated hydrocarbon and isothiocyanate | 1,3-dichloropropene, methylisothio cyanate and other chlorinated C3 hydrocarbons |
| Diclofop-methyl | Hoelon 3EC | Oxyphenoxy acid ester | (RS)-2-[4-(2,4-dichlorophenoxy) phenoxy]propionate |
| Diclofop-methyl | Illoxan | Oxyphenoxy acid ester | (RS)-2-[4-(2,4-dichlorophenoxy) phenoxy]propionate |
| Dicofol | Kelthane | Organochlorine | 1,1-bis(chlorophenyl)-2,2,2-trichloroethanol |
| Dicrotophos | Bidrin 8 | Organophospate | (E)-2-dimethylcarbamoyl-1-methylvinyl dimethyl phosphate |
| Diethatyl Ethyl | Antor 4ES | Chloracetanilide | Ethyl,N-(chloroacetyl)-N-(2,6-diethylphenyl) glycinate |
| Difenzoquat Methyl Sulfate | Avenge | Pyrazolium salt | 1,2-dimethyl-3,5-diphenyl-1H-pyrazolium methyl sulfate |
| Diflubenzuron | Dimilin | Substituted urea | N{[(4-chlorophenyl)amino]carbonyl]-2,6-difluorobenzamide |
| Dimethenamid | Frontier | Amide | 2-chloro-N-(2,4-dimethyl-3-thienyl)-N-(2- methoxy-1-methylethyl) acetamide |
| Dimethipin | Harvade 5F | Substituted dithiin | 2,3-dihydro-5,6-dimethyl-1,4-dithiin 1,14,4-tetraoxide |
| Dimethoate | Cygon 400 | Organophosphate | O,O-dimethyl S-methylcarbamoylmethyl phosphorodithioate |
| Dimethoate | Roxion | Organophosphate | O,O-dimethyl S-methylcarbamoylmethyl phosphorodithioate |
| Dimethomorph | Acrobat | Morpholine | (E,Z)-4-[3-(4-chlorophenyl)-3-(3,4-dimethoxyphenyl)acryloyl]morpholine |
| Diquat dibromide | Diquat Herbicide H/A | Dipyridylium | [6,7-dihydrodipyrido(1,2-a:2',1'-c)pyrazinediium dibromide |
| Disodium methanear-sonate | DSMA | Organic arsenical | Disodium methanearsonate |
| Disulfoton | Di-Syston | Organophosphate | O,O-diethyl S-[2-(ethylthio)ethyl]phosphorodithioate |
| Disulfoton and Ethopop | Mocap Plus EC | Organophospate | O,O-diethyl S-[2-(ethylthio)ethyl] phosphorodithioate and O-ethyl S,S-dipropyl phos-phorodithioate |
| Diuron | Karmex DF | Substituted urea | N'-(3,4-dichlorophenyl)-N,N-dimethylurea |
| Dodine | Cyprex 65-W | Aliphatic nitrogen | 1-dodecylguanidine acetate |
| Endosulfan | Endocide | Organochlorine | 6,7,8,9,10,10-hexachloro-1,5,5a,6,9,9a-hexahydro6,9-methano-2,4,3-benzodioxathiepin 3-oxide |
| Endosulfan | Thiodan EC/WP | Organochlorine | 6,7,8,9,10,10-hexachloro-1,5,5a,6,9,9a-hexahydro6,9-methano-2,4,3-benzodioxathiepin 3-oxide |

Table 2. Pesticides Listed Alphabetically by Common Name

| Common Name [a] | Trade Name [b] | Family Name [c] | Chemical Name [d] |
|---|---|---|---|
| Endothall | Accelerate | Phthalate | 7-oxabicyclo[2,2,1]heptane-2,3-dicarboxylic acid () used as sodium, potassium or amine salts |
| Endothall | Aquathol K | Phthalate | 7-oxabicyclo[2,2,1]heptane-2,3-dicarboxylic acid |
| Endothall | Des-I-Cate | Phthalate | 7-oxabicyclo[2,2,1]heptane-2,3-dicarboxylic acid () used as sodium, potassium or amine salts |
| Endothall | Endothal Turf Herbicidet | Phthalate | 7-oxabicyclo[2,2,1]heptane-2,3-dicarboxylic acid () used as sodium salts |
| Endothall | Herbicide 273 | Phthalate | 7-oxabicyclo[2,2,1]heptane-2,3-dicarboxylic acid () used as sodium salts |
| Endothall | Hydrothol | Phthalate | 7-oxabicyclo[2,2,1]heptane-2,3-dicarboxylic acid () used as sodium salts |
| EPTC | Eptam t | Carbamate | S-ethyl dipropylthiocarbamate |
| EPTC | Eradicane E | Carbamate | S-ethyl dipropylthiocarbamate |
| Esfenvalerate | Asana XL | Pyrethroid | (S)-a-cyano-3-phenoxybenzyl(S)-2-(4-chlorophenyl)-3-methylbutyrate |
| Ethalfluralin | Sonalan G/EC | Dinitroaniline | N-ethyl-N-(2-methyl-2-propenyl)-2,6-dinitro-4-(trifluoromethyl)-benzenamine |
| Ethephon | Cerone | Miscellaneous organic | (2-chloroethyl)phosphonic acid |
| Ethephon | Ethrel Plant Regulator | Phosphonic acid | (2-chloroethyl)phosphonic acid |
| Ethephon | Prep | Phosphonic acid | (2-chloroethyl)phosphonic acid |
| Ethiofencarb | Croneton | Carbamate | 2-[(ethylthio)methyl]phenyl methylcarbamate |
| Ethofumesate | Nortron | Benzofuran (heterocyclic) | (±)2-ethoxy-2,3-dihydro-3,3-dimethyl-5-benzofuranylmethane sulfonate |
| Ethopop | Mocap EC/F | Organophospate | O-ethyl S,S-dipropyl phosphorosithioate |
| Ethoxylated fatty amines | Frigate | Penetrant | Ethoxylated fatty amines |
| Etridazole | Aaterra | Thiazole | 5-ethoxy-3-trichloromethyl-1,2,4-thiadiazole |
| Etridiazole | Terrazole 25% Ornamental | Thiazole | 5-ethoxy-3-trichloromethyl-1,2,4-thiadiazole |
| Etridiazole | Terrazole 35% Turf | Thiazole | 5-ethoxy-3-trichloromethyl-1,2,4-thiadiazole |
| Fenamiphos | Nemacur EC/G | Organophosphate | Ethyl 3-methyl-4-(methylthio)phenyl (1-methylethyl)phosphor-amidate |
| Fenarimol | Rubigan | Pyrimidine | a-(2-chlorophenyl)-a-(chlorophenyl)-5-pyrimidinemethanol |

Table 2. Pesticides Listed Alphabetically by Common Name

| Common Name [a] | Trade Name [b] | Family Name [c] | Chemical Name [d] |
|---|---|---|---|
| Fenbutalin-oxide | Vendex L/WP | Organotin | bis{tris(2-methyl-2-phenylpropyl)tin] oxide |
| Fenoxaprop-ethyl | Acclaim IEC | Chlorophenoxy acid | (±)-ethyl 2-[4-[(6-chloro-2-benzoxazolyl)oxy]-phenoxy]propanoate |
| Fenoxaprop-ethyl | Horizon IEC | Oxyphenoxy acid ester | (±)-ethyl-2-[4-[(6-chloro-2-benzoxazolyl)oxy]phenoxy]propanoate |
| Fenoxaprop-ethyl | Whip 1EC | Oxyphenoxy acid ester | (±)-ethyl 2-[4-[(6-chloro-2-benzoxazolyl)oxy]-phenoxy]propanoate |
| Fenoxaprop-ethyl, 2,4-D, and MCPA | Tiller | Chlorophenoxy acid | (±)-ethyl 2-[4-[(6-chloro-2-benzoxazolyl)oxy]-phenoxy]propanoate and 2,4-dichlorophenoxyacetic acid and (4-chloro-2-methyl)phenoxyacetic acid |
| Fenoxaprop-p-ethyl | Option IEC | Oxyphenoxy acid ester | (R)-2-[4-(6-chloro-1,3-benzoxazol-2-yloxy)-phenoxy]propionic acid |
| Fenoxaprop-p-ethyl | Whip 360 | Oxyphenoxy acid ester | (R)-2-[4-[(6-chloro-2-benzoxazolyl)oxy]phenoxy]propanoic acid |
| Fluazifop-P-butyl | Fusilade 2000 | Oxyphenoxy acid ester | butyl (R)-2-[4-[[5-(trifluoromethyl) -2-pyridinyl]oxy]phenoxy]propanoate |
| Flucycloxuron | Andalin | Substituted benzoylurea | N-[[[4-[[[[(4-chlorophenyl)cyclopropyl-methylene]amino]oxy]methyl]phenyl]carbonyl]-2,6-difluorobenzamide |
| Flumetralin | Prime + | Dinitroaniline | 2-chloro-N-[2,6-dinitro-4-(trifluoromethyl)phenyl]-N-ethyl-6-fluoro-benzenemethanamine |
| Flumetsulam | Broadstrike | Triazolopyrimidine | 2-(2,6-difluorophenylsulphamoyl)-5-methyl[1,2,4]-triazolo[1,5-a] pyrimidine |
| Fluometuron | Cotoran WP/L/DG | Substituted urea | 1,1-dimethyl-3(a,a,a-trifluoro-m-tolyl)urea |
| Fluoroglycofen-ethyl | Compete | Diphenyl ether | Ethyl O-[5-(2-chloro-a,a,a-trifluoro-p-tolyloxy)-2-nitrobenzoyl]-glycolate |
| Flutolanil | Prostar 4E | Anilide | 3'isopropoxy-2-(trifluoromethyl)benzanilide |
| Fomesafen | Reflex 2LC | Diphenyl ether | 5-[2-chloro-4-(trifluoromethyl)phenoxy]-N-(methylsulfonyl)-2-nitrobenzamide |
| Fonofos | Crusade | Organophosphate | O-ethyl s-phenyl ethylphosphono dithioate |
| Fonofos | Dyfonate EC/G | Organophosphate | O-ethyl S-phenyl ethylphosphon-odithioate |
| Formetanate hydrochloride | Carzol SP | Carbamate | N,N-dimethyl-N'[3-[[(methylamino) carbonyl]oxy]phenyl]methanimidamide monohydrochloride |
| Formothion | Anthio | Organophosphonate | S[2-(formylmethylamino)-2-oxoethyl] O,O-dimethyl phosphorodithioate |
| Fosamine | Krenite S | Organophosphate | Ammonium ethyl carbamoylphosphonate |
| Fosetyl-Al | Aliette | Organophosphate | Aluminum tris (O-ethyl phosphonate) |
| Gibberellic acid | ProGibb Plus | Gibberellin | 1a,2b,4aa,4bb,10b)-2,4a,7-trihydroxy-1-methyl-8-methylenegibb-3-ene-1,10-carboxylic acid 1,4a-lactone |

Table 2. Pesticides Listed Alphabetically by Common Name

| Common Name [a] | Trade Name [b] | Family Name [c] | Chemical Name [d] |
|---|---|---|---|
| Gibberellic acid | Release | Gibberellin | 1a,2b,4aa,4bb,10b)-2,4a,7-trihydroxy-1-methyl-8-methylenegibb-3-ene-1,10-carboxylic acid 1,4a-lactone |
| Gibberellic acid | RyzUp | Gibberellin | 1a,2b,4aa,4bb,10b)-2,4a,7-trihydroxy-1-methyl-8-methylenegibb-3-ene-1,10-carboxylic acid 1,4a-lactone |
| Gibberellins | ProVide | Gibberellin | Gibberellins A4A7 |
| Gibberellins and benzyladenine | Accel | Gibberellin and cytokinin | Gibberellins A4A7 (gibberellic acid) and 6-benzyladenine (BAP) |
| Glyphosate | Accord | Phosphonate | Isopropylamine salt of N-(phosphonomethyl)-glycine |
| Glyphosate | Pondmaster | Phosphonate | Isopropylamine salt of N-(phosphonomethyl)-glycine |
| Glyphosate | Ranger | Phosphonate | Isopropylamine salt of N-(phosphonomethyl)-glycine |
| Glyphosate | Roundup | Phosphonate | Isopropylamine salt of N-(phosphonomethyl)-glycine |
| Hexaconazole | Anvil | Triazole | (RS)-2-(2,4-dichlorophenyl)-1-(IH-1,2,4-triazol-1-yl)-hexan-2-ol |
| Hexazinone | Velpar SP/G/WDL | Triazine | 3-cyclohexyl-6-(dimethylamino)-1-methyl-1,3,5-triazine-2,4-(1H,3H) dione |
| Hydramethylnon | Amdro | Miscellaneous organic | Tetrahydro-5,5-dimethyl-2(1H)-pyrimidinone [3-[4-(trifluoromethyl)phenyl]-1-[2-[4-(trifluoro-methyl)phenyl]rthenyl]-2-propenylidene] hydrazone |
| Imazamethabenz-methyl | Assert | Imidazole | (±)-6-(4-isopropyl-4-methyl-5-oxo-2-imidazolin-2-yl) methyl ester, and (±)-2-(4-isopropyl-4-methyl-5-oxo-2-imidazolin-2-yl) methyl ester |
| Imazapyr | Arsenal | Imidazole | 2-(4-isopropyl-4-methyl-5-oxo-2-imidazolin-2-yl)nicotinic acid |
| Imazapyr | Arsenal Forestry Granule | Imidazole | 2-(4-isopropyl-4-methyl-5-oxo-2-imidazolin-2-yl)nicotinic acid |
| Imazapyr | Chopper | Imidazole | 2-(4-isopropyl-4-methyl-5-oxo-2-imidazolin-2-yl)nicotinic acid |
| Imazaquin | Image | Imidazole | 2-[4,5-dihydro-4-methyl-(1-methylethyl)-5-oxo-1H-imidazol-2-yl]-3-quinolinecarboxylic acid |
| Imazaquin | Scepter | Imidazole | (±)2-[4,5-dihydro-4-methyl-4-(1-methylethyl)-5-oxo-1H-imidazol-2-yl] -3-quinolinecarboxylic acid |
| Imazaquin and pendimethalin | Squadron | Imidazolinone and dinitroanaline | (±)2-[4,5-dihydro-4-methyl-4-(1-methylethyl)-5-oxo-1H-imidazol-2-yl] -3-quinolinecarboxylic acid and N-(1-ethylpropyl)-3,4-dimethyl-2,6-dinitrobenzenamine |
| Imazaquin and trifluralin | Tri-Scept | Imidazole and dinitroaniline | 2-[4,5-dihydro-4-methyl-(1-methylethyl)-5-oxo-1H-imidazol-2-yl]-3-quinolinecarboxylic acid and a,a,a trifluoro-2,6-dinitro-N,N-dipropyl-p-toluidine |
| Imazethapyr | Pursuit | Imidazole | (±)-2-[4,5-dihydro-4-methyl-4-(1-methylethyl)-5-oxo-1H-imidazol2-yl]-5-ethyl-3-pyridine carboxylic acid |

Table 2. Pesticides Listed Alphabetically by Common Name

| Common Name [a] | Trade Name [b] | Family Name [c] | Chemical Name [d] |
|---|---|---|---|
| Imazethapyr | Passport | Imidazolinone | (±)-2-[4,5-dihydro-4-methyl-4-(1-methylethyl)-5-oxo-1H-imidazol2-yl]-5-ethyl-3-pyridine carboxylic acid |
| Imazethapyr and pendimethalin | Pursuit Plus | Imidazole and dinitroaniline | (±)-2-[4,5-dihydro-4-methyl-4-(1-methylethyl)-5-oxo-1H-imidazol2-yl]-5-ethyl-3-pyridine carboxylic acid and N-(1-ethylpropyl)-3,4-dimethyl-2,6-dinitrobenzenamine |
| Ipodione | Rovral | Dicarboximide | 3-(3,5-dichlorophenyl)-N-(1-methylethyl)-2,4-dioxo-1-imidazolidine-carboxamide |
| Isopropalin | Paarlan E.C. | Dinitroaniline | 2,6-dinitro-N,N-dipropylcumidine |
| Isoxaben | Gallery | Benzamide | N-[3-(1-ethyl-1-methylpropyl)-5-isoxazolyl]-2,6-dimethoxybenzamide |
| Lactofen | Cobra herbicide | Diphenyl ethers | 1-(carboethoxy)ethyl 5-[2-chloro-4(trifluoromethyl)phenoxy]-2-nitrobenzoate |
| Lambda cyhalothrin | Karate | Pyrethroid | a-cyano-3-phenoxybenzyl 3-(2-chloro-3,3,3-trifluoroprop-1-enyl)-2,2-dimethylcyclopropanecarboxylate |
| Lamda cyhalothrin | Force | Pyrethroid | 2,3,5,6-tetrafluoro-4-methylphenyl)-methyl-1a,3a,)-(Z)-(±)-3-(2-chloro-3,3,3-trifluoro-1-propenyl)-2,2-dimethylcyclopropanecarboxylate |
| Lindane | Etan 3G | Organochlorine | 1,2,3,4,5,6-hexachloro-cyclohexane |
| Linuron | Lorox DF/L | Substituted urea | 3-(3,4-dichlorophenyl)-1-methoxy-1-methylurea |
| Malathion | Cythion | Organophosphate | O,O-dimethyl phosphorodithioate of diethyl mercaptosuccinate |
| Maleic Hydrazide | Royal MH-30 | Hydrazide | 1,2-dihydro-3,6-pyridazinedione (potassium salt) |
| Mancozeb | Dithane | Dithiocarbamate | Zn ion and manganese ethylene bisdithiocarbamate |
| Mancozeb | Penncozeb WP/DF | Dithiocarbamate | Zn ion and manganese ethylene bisdithiocarbamate |
| Mancozeb and metalaxyl | Ridomil MZ | Dithiocarbamate and phenylamide | Zn ion and manganese ethylene bisdithiocarbamate and N-(2,6-dimethylphenyl)-N-(methoxyacetyl(-DL-alanine methyl ester |
| Maneb | Maneb | Dithiocarbamate | Manganese ethylenebisdithiocarbamate |
| MCPA | Chiptox MCPA sodium | Chlorophenoxy acid | (4-chloro-2-methyl)phenoxyacetic acid |
| MCPA and propanil | Stampede CM | Chlorophenoxy and amide | (4-chloro-2-methyl)phenoxyacetic acid and N-(3,4-dichlorophenyl)propanamide |
| MCPA, dimethylamine salt | Rhomene | Chlorophenoxy acid | (4-chloro-2-methyl)phenoxyacetic acid |
| MCPA, isooctyl ester | Rhonox | Chlorophenoxy acid | (4-chloro-2-methyl)phenoxyacetic acid |
| MCPB | Thistrol | Chlorophenoxy acid | 4-(2-methyl-4-chlorophenoxy)butyric acid |

Table 2. Pesticides Listed Alphabetically by Common Name

| Common Name [a] | Trade Name [b] | Family Name [c] | Chemical Name [d] |
|---|---|---|---|
| MCPP and 2,4-D | 2 Plus 2 (MCPP+2,4-D) | Chlorophenoxy acid | (RS) -2-(2-methyl-4-chlorophenoxy) propionate and 2,4-dichlorophenoxy-acetic acid |
| Mecoprop-P | Duplosan KV | Propionic acid | (+)-(R)-2-(4-chloro-2-methylphenoxy) propionic acid |
| Mepiquat chloride | Pix | Quaternary ammonium salt | 1,1-dimethylpiperidinium chloride |
| Metalaxyl | Apron 25W | Phenylamide | N-(2,6-dimethylphenyl)-N-(methoxyacetyl(-DL-alanine methyl ester |
| Metalaxyl | Ridomil E/G | Phenylamide | N-(2.6-dimethylphenyl)-N-(methoxyacetyl(-DL-alanine methyl ester |
| Metalaxyl and PCNB | Apron-Terraclor | Phenylamide and substituted aromatic fungicide | N-(2,6-dimethylphenyl)-N-(methoxyacetyl(-DL-alanine methyl ester and pentachloronitro-benzene |
| Metalaxyl and PCNB (Quintozene) | Ridomil PC | Phenylamide and substituted aromatic | N-(2,6-Dimethylphenyl)-N-(methoxyacetyl(-DL-alanine methyl ester and pentachloronitrobenzene |
| Metam-Sodium | Vapam | Isothiocyanate | Sodium N-methyldithiocarbamate |
| Methamidophos | Monitor L/Spray | Organophospate | O,S-dimethyl phosphoramidothioate |
| Methazole | Probe | Oxadiazolidinedione | 2-(3,4-dichlorophenyl)-4-methyl-1,2,4-oxadiazolidine-3,5-dione |
| Methidathion | Supracide 2E | Organophospate | O,O-dimethyl phosphorodithioate, S-ester with 4-(mercaptomethyl)-2-methoxy D2-1,3,4-thiadiazolin-5-one |
| Methomyl | Lannate L/LV | Carbamate | S-methyl N-[(methylcarbamoyl)oxy]thioacetimidate |
| Methyl parathion | Penncap-M | Organophosphate | O,O-dimethyl O-(4-nitrophenyl) phosphorothioate |
| Metolachlor | Butisan S | Acetanilide | 2-Chloro-N-(2,6-dimethylphenyl)-N-(1H-pyrazol-1-ylmethyl)acetamide |
| Metolachlor | Dual (Dual II) | Acetanilide | 2-chloro-N-(2-ethyl-6-methylphenyl)-N-(2-methoxy-1-methylethyl) acetamide |
| Metolachlor and metribuzin | Turbo 8EC | Acetanilide and triazinone | 2-chloro-N-(2-ethyl-6-methylphenyl)-N-(2-methoxy-1-methylethyl) acetamide and 4-amino-6-(1,1-dimethylethyl)-3-(methylthio)-1,2,4-triazin-5(4H)-one |
| Metribuzin | Lexone DF | Triazinone | 4-amino-6-(1,1-dimethylethyl)-3-(methylthio)-1,2.4-triazin-5(4H)-one |
| Metribuzin | Sencor DF | Triazinone | 4-amino-6-(1,1-dimethylethyl)-3-(methylthio)-1,2,4-triazin-5(4H)-one |
| Metribuzin and trifluralin | Salute 4EC | Triazinone and dinitroanaline | 4-amino-6-(1,1-dimethylethyl)-3-(methylthio)-1,2,4-triazin-5(4H)-one and a,a,a trifluoro-2,6-dinitro-N,N-dipropyl-p-toluidine |
| Metsulfuron methyl | Ally | Sulfonylureas herbicides | Methyl 2-[[[[(4-methoxy-6-methyl-1,3,5-triazin-2-yl)-amino]carbonyl] amino]sulfonyl]benzoate |
| Metsulfuron methyl | Escort | Sulfonylurea | Methyl 2-[[[[(4-methoxy-6-methyl-1,3,5-triazin-2-yl)-amino]carbonyl] amino]sulfonyl]benzoate |

Table 2. Pesticides Listed Alphabetically by Common Name

| Common Name [a] | Trade Name [b] | Family Name [c] | Chemical Name [d] |
|---|---|---|---|
| Molinate | Ordram EC/G | Thiocarbamate | S-Ethyl hexahydro-1H-azepine-1-carbothioate |
| Molinate and propanil | Arrosolo 3-3E | Thiocarbamate and amide | S-Ethyl hexahydro-1H-azepine-1-carbothioate and N-(3,4-dichlorophenyl)propanamide |
| MSMA | Ansar | Organic arsenical | Monosodium methane arsonate |
| MSMA | Daconate 6 | Organic arsenical | Monosodium methane arsonate |
| MSMA | Bueno 6 | Organic arsenical | Monosodium methane arsonate |
| Myclobutanil | Nova 40W | Triazole | a-butyl-a-(4-chlorophenyl)-1H-1,2,4-triazole-1-propanenitrile |
| Myclobutanil | Rally 40W | Triazole | a-butyl-a-(4-chlorophenyl)-1H-1,2,4-triazole-1-propanenitrile |
| Naled | Dibrom 8 | Organophosphate | 1,2-dibromo-2,2-dichloroethyl dimethyl phosphate |
| Naphthalene acetamide naphthalene acetic acid | Amid Thin W | Auxins | 2-(1-naphthyl)acetamide and b-naphtaleneacetic acid (NAD, NAA) |
| Naphthaleneacetic acid | Fruitone N | Auxin | 2-(1-naphthyl)acetic acid |
| Naphthaleneacetic acid | Tre-Hold Sprout Inhibitor | Auxin | 2-(1-naphthaleneacetic acid) ethyl ester |
| Naphthaleneacetic acid (NAA) potassium salt | NAA-800 | Auxin | 2-(1-naphthaleneacetic acid) |
| Napropamide | Devrinol | Amide | (RS)-N,N-diethyl-2-(1-naphthyloxy) propionamide |
| Naptalam | Alanap-L | Amide herbicides | Sodium 2-[(1-naphthalenylamino)carbonyl] benzoate |
| Nicosulfuron | Accent | Sulfonylureas | 2-(4,6-dimethoxypyrimidin-2-yl carbamoylsulfamoyl)-N,N-dimethyl-nicotinamide () |
| Nitrapyrin | N-Serve E | Miscellaneous organic | 2-chloro-6-(trichloromethyl)pyridine |
| Norflurazon | Evital | Pyridazinone | 4-chloro-5-methylamino-2-(a,a,a-trifluoro-m-tolyl)pyridazin-3(2H)-one |
| Norflurazon | Solicam DF | Pyridazinone | 4-chloro-5-methylamino-2-(a,a,a-trifluoro-m-tolyl)pyridazin-3(2H)-one |
| Norflurazon | Zorial Rapid 80 | Pyridazinone | 4-chloro-5-methylamino-2-(a,a,a-trifluoro-m-tolyl)pyridazin-3(2H)-one |
| Oryzalin | Surflan A S | Dinitroaniline | 3,5-dinitro-N4,N4-dipropyl sulfanilamide |
| Oxabetrinil | Concep II | Oxime ether | a-[1.3-dioxolan-2-yl)methoxyimino] benzeneacetonitrile |
| Oxamyl | Vydate L | Carbamate | S-methyl N,N'-dimethyl-N-(methylcarbamoyloxy)-1-thio-oxamimidate |
| Oxycarboxin | Plantvax 75W | Oxathiin | 5,6-dihydro-2-methyl-N-phenyl-1,4-oxathiin-3-carboxamide4,4-dioxide |

Table 2. Pesticides Listed Alphabetically by Common Name

| Common Name [a] | Trade Name [b] | Family Name [c] | Chemical Name [d] |
|---|---|---|---|
| Oxydemeton-methyl | Metasystox-R | Organophospate | S-[2-(ethylsulfinyl)-1-methylethyl] O,O-dimethyl phosphorodithioate |
| Oxyfluorfen | Goal | Diphenyl ether | 2-chloro-1-(3-ethoxy-4-nitro phenoxy)-4-trifluoromethylbenzene |
| Paraquat dichloride | Cyclone | Bipyridylium | 1,1-dimethyl-4,4'-bipyridinium dichloride |
| Paraquat dichloride | Gramoxone | Dipyridylium | 1,1-dimethyl-4,4'-bipyridinium dichloride |
| Paraquat dichloride | Starfire | Bipyridylium | 1,1-dimethyl-4,4'-bipyridinium dichloride |
| PCNB | Turfcide | Substituted aromatic | Pentachloronitrobenzene |
| PCNB | Terraclor | Substituted aromatic | Pentachloronitrobenzene |
| PCNB (Quintozene) | RTU-PCNB | Substituted aromatic | pentachloronitrobenzene |
| Pebulate | Tillam 6-E | Carbamate | S-propyl butyl(ethyl)thiocarbamate |
| Pendimethalin | Prowl | Dinitroaniline | N-(1-ethylpropyl)-3,4-dimethyl-2,6-dinitrobenzenamine |
| Pendimethalin | Stomp | Dinitroaniline | N-(1-ethylpropyl)-3,4-dimethyl-2,6-dinitro benzenamine |
| Permethrin | Ambush | Pyrethroid | (3-phenoxyphenyl methyl) (±)-cis,trans-3-(2,2-dichloroethenyl) -2,2-dimethylcyclopropane- carboxylate |
| Permethrin | Pounce G/EC | Pyrethroid | (3-phenoxyphenyl methyl) (±)-cis,trans-3-(2,2-dichloroethenyl) -2,2-dimethylcyclopropane- carboxylate |
| Permethrin | Torpedo | Pyrethroid Insecticides | (3-phenoxyphenyl) methyl (±)-cis,trans-3-(2,2-dichloroethenyl) -2,2-dimethylcyclopropanecarboxylate |
| Petroleum oil | Volck (Valent) Supreme Spray | Miscellaneous organic | Petroleum oil |
| Phorate | Thimet 15-G | Organophosphate | O,O-diethyl S-[(ethylthio)methyl]phosphorodithioate |
| Phosmet | Imidan 50-WP | Organophosphate | O,O-dimethyl phosphorodithioatemS-ester with N-(mercaptomethyl) phthalimide |
| Phoxim | Baythion | Organophospate | a-[[(diethoxyphosphinothioyl)oxy] imino] benxeneacetonitrile |
| Phytophthora | DeVine | Biological agent | Phytophthora spores |
| Picloram and triclopyr | Access | Pyridinoxy acid | 4-amino-3,5,6-trichloropicolinic acid () and (3,5,6-trichloro-2-pyridinyloxy)acetic acid (CAS 9CI) |
| Picloram, potassium salt | Tordon 22K | Pyridinoxy acid | 4-amino-3,5,6-trichloropicolinic acid |
| Piperophos and di-methametryn | Avirosan | Miscellaneous organic | O,O-dipropyl S-2-methylpiperidino-carbonyl-methyl phosphorodithioate and 4-(1,2-dimethyl-n-propylamino)-2-ethylamino-6-methylthio-s-triazine |

Table 2. Pesticides Listed Alphabetically by Common Name

| Common Name [a] | Trade Name [b] | Family Name [c] | Chemical Name [d] |
|---|---|---|---|
| Pirimiphos-methyl | Actellic 5E | Organophosphate | O-(2-diethylamino-6-methylpyrimidin-4-yl)-O.O-dimethyl phosphorothioate |
| Primisulfuron-methyl | Beacon | Sulfonylurea | 3-[4,6-bis(difluoromethoxy)pyrimidi-2-yl]-1-(2-methoxycarbonylphenyl-sulfonyl)urea |
| Profenofos | Curacron 6E | Organophosphate | O-4-bromo-2-chlorophenyl O-ethyl S-propyl phosphorothioate |
| Promecarb | Carbamult | Carbamate | 3-isopropyl-5-methylphenyl methylcarbamate |
| Prometon | Pramitol 25E | Triazine | 2,4-bis(isopropylamino)-6-methoxy-s-triazine |
| Prometon and simazine | Pramitol 5PS | Triazine | 2,4-bis(isopropylamino)-6-methoxy-s-triazine and 2-chloro-4,6-bis(ethylamino)-s-triazine |
| Prometryn | Caparol 4L | Substituted triazine | 2,4-bis(isopropylamino)-6-(methylthio)-s-triazine |
| Pronamide | Kerb 50W | Benzamide | 3.5-dichloro-N-(1.1-dimethyl-2-propynyl)benzamide |
| Propachlor | Ramrod | Anilides | 2-chloro-N-isopropylacetanalide |
| Propachlor and atrazine | Ramrod and Atrazine | Anilide and triazine | 2-chloro-N-isopropylacetanalide and 6-chloro-N2-ethyl-N4-isopropyl-1,3.5-triazine-2,4-diamine |
| Propanil | Stam 80EDF | Amide | N-(3,4-dichlorophenyl)propanamide |
| Propanil | Strel 4E | Amide | N-(3,4-dichlorophenyl)propanamide |
| Propargite | Comite | Organosulfur | 2-[4-(1,1-dimethylethyl)phenoxy] cyclohexyl 2-propynyl sulfite |
| Propargite | Omite | Organosulfur | 2-[4-(1,1-dimethylethyl)phenoxy]cyclohexyl2-propynyl sulfite |
| Propargite | Ornamite | Organosulfur | 2-[4-(1.1-dimethylethyl)phenoxy]cyclohexyl2-propynyl sulfite |
| Propiconazole | Bumper | Triazole | 1-[[2-(2,4-dichlorophenyl)-4-propyl-1.3-dioxolan-2-yl]methyl]-1H-1.2,4-triazole |
| Propiconazole | Orbit | Triazole | 1-[[2-(2,4-dichlorophenyl)-4-propyl-1,3-dioxolan-2-yl]methyl]-1H-1.2,4-triazole |
| Propiconazole | Tilt | Triazole | 1-[[2-(2.4-dichlorophenyl)-4-propyl-1,3-dioxolan-2-yl]methyl]-1H-1,2,4-triazole |
| Pyrazophos | Afugan | Organophosphate | Ethyl 2-diethoxythiophosphoryloxy-5-methyl-pyrazolo(1,5-a)pyrimidine-6-carboxylate () |
| Quizalofop ethyl | Assure | Oxyphenoxy acid ester | Ethyl-2-[4-(6-chloroquinoxalin-2-yloxy)phenoxyl]propionate |
| Sethoxydim | Poast | Cyclohexenone | 2-[1-(ethoxyimino)butyl]-5-[2-(ethylthio)propyl]-3-hydroxy-2-cyclohexen-1-one |
| Siduron | Tupersan | Substituted urea | 1-(2-methylcyclohexyl)-3-phenylurea |
| Simazine | Aquazine | Triazine | 2-chloro-4.6-bis(ethylamino)-s-triazine |

Table 2. Pesticides Listed Alphabetically by Common Name

| Common Name [a] | Trade Name [b] | Family Name [c] | Chemical Name [d] |
|---|---|---|---|
| Simazine | Princep | Triazine | 2-chloro-4,6-bis(ethylamino)-s-triazine |
| Sulfentrazone | Authority | Aryl triazolinone | N-[2,4-dichloro-5-[4-(difluoro-methyl)-4,5-dihydro-3-methyl-5-oxo-iH-1,2,4-triazol-1-yl]phenyl] methanesulfonamide |
| Sulfometuron methyl | Oust | Sulfonylurea | Methyl 2[[[[(4,6-dimethyl-2-pyrimidinyl)amino] carbonyl]amino]-sulfonyl]benzoate |
| Sulfur | Microthiol Special | Inorganic | Sulfur |
| Sulfur | Sulfur Alfa | Inorganic | Sulfur |
| Sulfur | Thiolux DF | Inorganic | Sulfur |
| Sulprofos | Bolstar 6 | Organophosphospate | O-ethyl O-[4-(methylthio)phenyl]S-propyl phosphorodithioate |
| Tebufenozide | Confirm | Miscellaneous organic | N-tert-butyl-N'-(4-ethylbenzoyl)-3,5-dimethylbenzohydrazine |
| Tebuthiuron | Spike 20P | Substituted urea | N-[5-(1,1-dimethylethyl)-1,3,4-thiadiazol-2-yl]-N,N'-dimethylurea |
| Temephos | Abate | Organophosphate | O,O'-thiodi-4,1-phenylene O,O,O'O'-tetramethyl phosphorothioate |
| Terbacil | Sinbar | Uracil | 3-tert-butyl-5-chloro-6-methyluracil |
| Terbufos | Counter | Organophospate | S-[(1,1-dimethylethyl)thio] O,O-di-ethyl phosphorodithioate |
| Thiabendazole | Arbotect | Benzimidazole | 2-(4-thiazolyl)-1H-benzimidazole |
| Thidiazuron | Dropp 50WP | Substituted urea | N-phenyl-N'-1,2,3-thiadiazol-5-ylurea |
| Thiobencarb | Bolero EC/G | Thiocarbamate | S-4-chlorobenzyl diethylthiocarbamate () |
| Thiodicarb | Larvin | Carbamate | Dimethyl N,N-(thiobis(methylimino)carbonyloxy)bis(ethanimido-thioate) |
| Thiophanate-methyl | Topsin M 70W | Carbamate | Dimethyl[(1,2-phenylene)bis(imido-carbonothioyl)]bis carbamate] |
| Thiram | AAtack | Thiocarbamate | Bis(dimethylthiocarbamoyl) disulfide or tetramethylthiuram disulfide |
| Thiram + nitrogen fixing bacteria | Triple-Noctin L | Thiocarbamate | Bis(dimethylthiocarbamoyl) disulfide |
| Tralomethrin | Scout X-TRA | Pyrethroid | (1R,3S)3[(1'RS)(1',2',2',2',-tetrabromomethyl)]-2,2-dimethylcyclo-propanecarboxylic acid (S)-a-cyano-3-phenoxybenzyl ester |
| Triadimefon | Bayleton 50% | Triazole | 1-(4-chlorophenoxy)-3,3-dimethyl-1-(1H-1.2.4-triazoyl-1-yl)-2-butanone |
| Triadimenol | Baytan | Triazole | b-(4-chlorophenoxy)-a-(1,1-dimethyl -ethyl)-1H-1,2,4-triazole-1-ethanol |
| Triallate | Far-Go | Carbamate | S(2.3,3-trichloroallyl) diisopropyl-thiocarbamate |

Table 2. Pesticides Listed Alphabetically by Common Name

| Common Name [a] | Trade Name [b] | Family Name [c] | Chemical Name [d] |
|---|---|---|---|
| Triallate | Showdown | Thiocarbamate | S(2,3,3-trichloroallyl) diisopropyl-thiocarbamate |
| Triallate and trifluralin | Buckle | Carbamate and dinitroaniline | S(2,3,3-trichloroallyl) diisopropyl-thiocarbamate and a,a,a trifluoro-2,6-dinitro-N,N-dipropyl-p-toluidine |
| Tribenuron methyl | Express | Sulfonyl urea | Methyl 2-[[[[4-methoxy-6-methyl- 1,3,5-triazine-2-yl)methylamino] carbonyl]amino]sulfonyl]benzoate |
| Tribenuron methyl and thifensulfuron methyl | Harmony Extra | Sulfonyl urea | Methyl 2-[[[[4-methoxy-6-methyl- 1,3,5-triazine-2-yl)methylamino] carbonyl]amino]sulfonyl]benzoate and methyl 3-[[[[(4-methoxy-6-methyl-1,3,5-triazin-2-yl)amino]carbonyl]amino]sulfonyl]-2-thiophencarboxylate |
| Tribufos | Def 6 | Organophosphate | S,S,S-tributyl phosphorotrithioate |
| Tribufos, merphos | Folex 6-EC | Organophosphate | S,S,S-tributyl phosphorotrithioate |
| Trichlorfon | Dylox 80% SP | Organophosphate | Dimethyl (2,2,2-trichloro-1-hydroxyethyl)phosphonate |
| Triclopyr | Rely | Pyridinoxy acid | (3,5,6-trichloro-2-pyridinyloxy)acetic acid |
| Tridiphane | Tandem | Miscellaneous organic | (±)-2-3,5-dichlorophenyl)-2-(2,2,2-trichloroethyl)oxirane |
| Trifensulfuron methyl | Pinnacle DF | Sulfonylurea | Methyl 3-[[[[(4-methoxy-6-methyl-1,3,5-triazin-2-yl)amino]carbonyl]-amino]sulfonyl]-2-thiophencarboxylate |
| Trifluralin | Treflan E C | Dinitroaniline | a,a,a trifluoro-2,6-dinitro-N,N-dipropyl-p-toluidine |
| Trifluralin | TRI-4 | Dinitroaniline | a,a,a trifluoro-2,6-dinitro-N,N-dipropyl-p-toluidine |
| Vinclozolin | Ronilan | Dicarboximide | 3-(3,5-dichlorophenyl)-5-methyl-5-vinyl-1,3-oxazolidine-2,4-dione |
| Ziram | Ziram | Dithiocarbamate | Zinc bis(dimethyldithiocarbamate |

[a] The common name is the name coined for the active ingredient(s) of the pest control chemical. In the U.S., the American National Standards Institute (ANSI) provides a procedure whereby manufacturers, formulators, or other interested parties may obtain acceptance for a common name.
[b] The trade name is the brand name under which the material is sold. Different formulations may have the same trade name.
[c] The family name is the name of the chemical group the pesticide is derived from or belongs to. Examples of families are amides, carbamates, chlorophenoxy acids, etc.
[d] Chemical name is the name given to a specific chemical. The name is derived from the chemical structure of the material.

**Table 3.** Agrochemicals

| Chemical Name | Chemical ID Number [a] | CAS Number [b] | Guide Number [c] |
|---|---|---|---|
| Acetic acid, glacial | 2789 | 64-19-7 | 1 |
| Acetic acid (solution) | 2790 | | 2 |
| Acetone | 1090 | 67-64-1 | 2 |
| Acetylene | 1001 | 74-86-2 | 5 |
| Acid, liquid | 1760 | | 2 |
| Adhesives, flammable | 1133 | | 3 |
| Alcohol, denatured | 1987 | | 3 |
| Ammonia, Anhydrous | 1005 | 7664-41-7 | 6 |
| Ammonia solution | 1005, 2073, 2672, 3318 | | 6 |
| Ammonium chloride | 9085 | 12125-02-9 | 7 |
| Ammonium laureth sulfate | | | 7 |
| Ammonium nitrate fertilizer | 2072 | 6484-52-2 | 8 |
| Amorphous silica (fumed silica) | | 112945-52-5 | 7 |
| Anti-freeze | 1142 | | 3 |
| Aromatic oils | | | 4 |
| Battery fluid, acid | 2796 | | 9 |
| Battery fluid, alkali | 2797 | | 2 |
| Benzene | 1114 | 71-43-2 | 10 |
| Borax | | 1303-96-4 | 7 |
| Brake fluid, hydraulic | 1118 | | 10 |
| 2-Butanone | 1193 | 78-93-3 | 3 |
| Butyl acetate | 1123 | 123-86-4 | 3 |
| Calcium carbonate (limestone, calcite) | | 471-34-1 | 7 |
| Calcium hydroxide (hydrated lime) | | 1305-62-0 | 2 |
| Calcium hypochlorite | 2880 | 7778-54-3 | 11 |
| Carbon tetrachloride | 1846 | 56-23-5 | 15 |
| Clay, kitty litter, absorbent clay | | | 7 |
| Cleaners (combustible or flammable) | 1142 | | 3 |
| Cleaning compound, acid (phosphoric acid) | 1805 | 7664-38-2 | 2 |
| Cleaning compound, alkali (sodium hydroxide) | 1824 | 1310-73-2 | 2 |
| Coal tar oil and distillate | 1136, 1137 | | 4 |
| Coal tar naphtha | 2553 | 65996-79-4 | 4 |
| Cresol (o, m, p) | 2076 | 1319-77-3 | 12 |
| Creosote, coal tar | 1993 | 8001-58-9 | 4 |
| Diatomaceous earth | | | 7 |
| 1,2-dichloroethylene | 1150 | 25323-30-2 | 1 |

**Table 3.** Agrochemicals

| Chemical Name | Chemical ID Number [a] | CAS Number [b] | Guide Number [c] |
|---|---|---|---|
| 1,2-dichloropropane (see propylene dichloride) | 1279 | 78-87-5 | 15 |
| Diesel fuel | 1993, 1202 | | 4 |
| Engine starting fluid | 1960 | | 13 |
| Ethanol/ethyl alcohol | 1170 | 64-17-5 | 3 |
| Ether/ethyl ether | 1155 | 60-29-7 | 3 |
| Ethyl acetate | 1173 | 141-78-6 | 3 |
| Ethylene dibromide | 1605 | 106-93-4 | 2 |
| Ethylene dichloride (1,2-dichloroethane) | 1184 | 107-06-2 | 3 |
| Ethylene glycol | | 107-21-1 | 3 |
| Ethylene oxide | 1040 | 75-21-8 | 18 |
| Expanded minerals (silica, perlite, vermiculite) | | | 7 |
| Ferrous sulfate | 9125 | 7782-63-0 | 7 |
| Fertilizer, ammoniating solution | 1043 | | 6 |
| Formaldehyde solution (formalin) | 1198, 2209 | 50-00-0 | 1 |
| Fuel oil | 1202, 1993 | | 4 |
| Fumed silica (see amphorphous silica) | | 112945-52-5 | 7 |
| Gasoline/motor fuel/petrol | 1203 | 8006-61-9 | 4 |
| Hydrated lime | | 1305-62-0 | 7 |
| Hydrogen peroxide | 2015, 2984 | 7722-84-1 | 8 |
| Hydrochloric acid (solution) | 1789 | 7647-01-0 | 2 |
| Hypochlorite solution (sodium hypochlorite) | 1791 | 7681-52-9 | 11 |
| Isopropyl alcohol (isopropanol) | 1219 | 67-63-0 | 3 |
| Isopropylamine | 1221 | 75-31-0 | 1 |
| Kerosene | 1223 | 8008-20-6 | 4 |
| Lacquer | 1263 | | 4 |
| Lighter fluid | 1226 | | 3 |
| Liquefied natural gas | 1972 | | 13 |
| Liquefied petroleum gas (LPG) | 1075 | 74-98-6 | 13 |
| Lye (potassium hydroxide or sodium hydroxide | 1813, 1814 | 1310-73-2 | 2 |
| Mercury oxide | 1641 | 21908-53-2 | 12 |
| Methanol/methyl alcohol | 1230 | 67-56-1 | 14 |
| Methyl ethyl ketone | 1193 | 78-93-3 | 3 |
| Methylene chloride (dichloromethane) | 1593 | 75-09-2 | 15 |
| Mineral oils (petroleum oils, refined petroleum distillate) | 1268, 1270 | 64741-89-5<br>64741-88-4 | 4 |
| Mineral spirits (petroleum spirits or Stoddard solvent) | 1271 | 8052-41-3 | 4 |

**Table 3.** Agrochemicals

| Chemical Name | Chemical ID Number [a] | CAS Number [b] | Guide Number [c] |
|---|---|---|---|
| Muriatic acid | 1789 | | 2 |
| Naphtha/petroleum naphtha | 1255, 1256, 2553 | 8030-30-6 | 4 |
| Naphthalene | 1334 | 91-20-3 | 16 |
| Oil, petroleum, see petroleum oil | 1270 | | 4 |
| Paint and related materials | 1263 | | 4 |
| PCBs (polychlorinated biphenyls) | 2315 | 1336-36-3 | 7 |
| Pentachlorophenol (PCP) | 2020 | 87-86-5 | 12 |
| Perchloroethylene (tetrachloroethylene) | 1897 | 127-18-4 | 15 |
| Petroleum spirits (mineral spirits) | 1271 | 8030-30-6 | 4 |
| Phenol | 1671, 2821 | 108-95-2 | 12 |
| Phosgene | 1076 | 75-44-5 | 6 |
| Phosphoric acid | 1805 | 7664-38-2 | 2 |
| Potassium hydroxide (caustic potash) | 1813, 1814 | 1310-58-3 | 2 |
| Potassium permanganate | 1490 | 7722-64-7 | 11 |
| Propane | 1978 | 74-98-6 | 13 |
| Propylene dichloride (1,2-dichloropropane) | 1279 | 78-87-5 | 15 |
| Refined petroleum distillate (mineral oils, petroleum oils) | 1268, 1270 | 64741-89-5/ 64741-88-4 | 4 |
| Shellac | 1263 | | 4 |
| Silicates (synthetic dry) used as carrier | | | 7 |
| Sodium bisulfate | 1821 | 7681-38-1 | 2 |
| Sodium dichloro-s-triazinetrione (sodium dichlorocyanurate) | 2465 | 2893-78-9 | 17 |
| Sodium dichromate | 1479 | 10588-01-9 | 11 |
| Sodium hydroxide/lye/caustic soda | 1823, 1824 | 1310-73-2 | 2 |
| Sodium hypochlorite | 1791 | 7681-52-9 | 11 |
| Sodium laureth sulfate | | | 7 |
| Stoddard solvent | 1271 | 8052-41-3 | 4 |
| Sulfur dioxide | 1079 | 7446-09-5 | 6 |
| Sulfuric acid | 1830, 1832 | 7664-93-9 | 9 |
| Tank and sprayer cleaners (see Tank Kleen) | | | 7 |
| Tetrachloroethylene (perchlorethylene) | 1897 | 127-18-4 | 15 |
| Thinner (paint, varnish) | 1263 | | 4 |
| Toluene | 1294 | 108-88-3 | 10 |
| Trichloroacetic acid-sodium salt (TCA) | 1839, 2564 | 650-51-1 | 12 |
| 1,1,1-trichloroethane | 2831 | 71-55-6 | 15 |
| Trichloroethylene | 1710 | 79-01-6 | 15 |

**Table 3.** Agrochemicals

| Chemical Name | Chemical ID Number[a] | CAS Number[b] | Guide Number[c] |
|---|---|---|---|
| Turpentine | 1299 | 8006-64-7 | 4 |
| Vermiculite | | 1318-00-9 | 7 |
| Xylene/xylol | 1307 | 1330-20-7 | 10 |

[a] The Chemical ID Number is the UN/UA number assigned to specific hazardous chemicals by the U.S. Department of transportation (DOT). Not all hazardous chemicals are assigned a number.

[b] A unique numbering system established by the Chemical Abstract Service. Each chemical formulation is given a specific number. Therefore, different formulations of the same chemical will have different CAS numbers.

[c] The Guide Number that is assigned to a particular group of pesticides with a specific set of properties. This guide contains all the information required to handle those pesticides properly and safely.

# PESTICIDE GUIDES

# PESTICIDE GUIDE 1

## HEALTH HAZARDS

***Hazard Code: Danger***

Danger: poisonous; ingestion may be fatal.

Danger: absorption through skin may be fatal.

Danger: inhalation of the dry material or mists may be fatal.

Caution: avoid eye contact; may cause severe eye irritation and injury.

Caution: contact may cause severe skin irritation.

Fire may produce irritating or poisonous gases.

Runoff from fire or spills may cause pollution.

## SAFETY GUIDELINES

***Restricted Use Pesticide:*** No

***Solubility***: Soluble or miscible in water

***Toxicity Category:*** Class I Oral Hazard, Class I Dermal Hazard, Class I Inhalation Hazard

Formulations may contain xylene, ethylbenzene, and trimethylbenzene.

***Restricted-entry Interval (REI):*** 48 hours

## HANDLING AND STORAGE PRECAUTIONS

***Use Precautions***

Avoid contact with susceptible crops and other desirable broadleaf plants.

Avoid spray drift.

Select and wear the proper PPE.

Mix ingredients in areas of sufficient ventilation or outdoors to minimize exposure. Avoid spilling material while mixing.

Wash with soap and water immediately after use or exposure. Follow good personal hygiene practices.

Do not reuse spray equipment for any purpose unless thoroughly cleaned with a suitable cleaner.

Reentry—do not enter treated areas without protective clothing until sprays have dried or after 48 hrs.

Notification of a pesticide application must be given to all workers who will be within 1/4 mile of a field during the application or before the REI expires. This notification may be either oral or written, e.g., a posted sign at the

entrances to treated sites. For some products, both oral and written notifications are necessary.

### *Storage and Disposal Precautions*

The reader should consult Chapter 6, Part V, Sections A and B, pages 251-265 for recommendations on the proper storage and disposal precautions.

## PROTECTIVE CLOTHING

Protective eyewear such as goggles or face shield is strongly recommended; chemical-resistant gloves and boots, long-sleeved shirt, long pants, and shoes and socks are strongly recommended. When mixing, use a splash apron, and if dust or vapor is a problem, use approved air purifying respirators for pestcides.

## EMERGENCY GUIDELINES

### *Fire or Explosive Hazard*

Evacuate nonessential personnel from fire area.

The products are not considered flammable. They will support combustion and may decompose under fire conditions and give off toxic products.

Some of these materials may react violently with water.

Segregate from oxidizers and incompatible materials.

The use of self-contained breathing apparatus and chemical protective clothing (chemical-resistant gloves, boots, hood, and suit) is recommended.

The use of high-pressure water may cause spread of hazardous materials.

*Small fires:* Dry chemical, $CO_2$, Halon, water spray, or foam.

*Large fires:* Water spray, fog, or standard or alcohol foam is recommended.

Move containers of chemicals from fire area if you can do so without risk.

Cool containers that are exposed to flames with water until fire is out and heat of container is reduced.

Stay upwind of fire.

### *Spills*

The reader should consult Chapter 6, Part V, Section C, pages 265-276 for the proper methods, adsorbent recommendations, and directions to control, contain and clean-up spills, and for disposal of pesticide wastes.

### *First Aid*

The reader should consult Chapter 7, Section E, pages 312-314 for emergency first aid information.

# PESTICIDE GUIDE 2

## HEALTH HAZARDS

***Hazard Indicator: Danger***

Danger: poisonous; ingestion may be fatal.

Warning: absorption through skin may be fatal.

Warning: inhalation of the dry material or mists may cause respiratory irritation and may be fatal.

Caution: may cause moderate eye irritation and eye injury.

Caution: contact may cause moderate skin irritation.

Fire may produce irritating or poisonous gases.

Runoff from fire or spills may cause pollution.

## SAFETY GUIDELINES

***Restricted Use Pesticide:*** No

***Solubility***: Very slightly soluble in water

***Toxicity Category:*** Class I Oral Hazard, Class II Dermal Hazard, Class II Inhalation Hazard

Formulations may contain xylene, ethylbenzene, and trimethylbenzene.

***Restricted-entry Interval (REI):*** 48 hours

## HANDLING AND STORAGE PRECAUTIONS

***Use Precautions***

Avoid contact with susceptible crops and beneficial insects.

Avoid spray drift.

Select and wear the proper PPE.

Mix ingredients in areas of sufficient ventilation or outdoors to minimize exposure. Avoid spilling material while mixing.

Wash with soap and water immediately after use or exposure. Follow good personal hygiene practices.

Do not reuse spray equipment for any purpose unless thoroughly cleaned with a suitable cleaner.

Reentry—do not enter treated areas without protective clothing until sprays have dried or after 48 hrs.

Notification of a pesticide application must be given to all workers who will be within 1/4 mile of a field during the application or before the REI expires. This notification may be either oral or written, e.g., a posted sign at the en-

trances to treated sites. For some products, both oral and written notifications are necessary.

### *Storage and Disposal Precautions*

The reader should consult Chapter 6, Part V, Sections A and B, pages 251-265 for recommendations on the proper storage and disposal precautions.

## PROTECTIVE CLOTHING

Protective eyewear, chemical-resistant gloves and boots, long-sleeved shirt, long pants, and shoes and socks are strongly recommended. When mixing, use a splash apron and, if dust or vapor is a problem, use approved air purifying respirators for pesticides.

## EMERGENCY GUIDELINES

### *Fire or Explosive Hazard*

Evacuate nonessential personnel from fire area.

The products are not considered flammable. They will support combustion and may decompose under fire conditions and give off toxic products.

Some of these materials may react violently with water.

Segregate from oxidizers and incompatible materials.

The use of self-contained breathing apparatus and chemical protective clothing (chemical-resistant gloves, boots, hood, and suit) is recommended.

The use of high-pressure water may cause spread of hazardous materials.

*Small fires*: Dry chemical, $CO_2$, Halon, water spray, or foam.

*Large fires:* Water spray, fog, or standard or alcohol foam is recommended.

Move containers of chemicals from fire area if you can do so without risk.

Cool containers that are exposed to flames with water until fire is out and heat of container is reduced.

Stay upwind of fire.

### *Spills*

The reader should consult Chapter 6, Part V, Section C, pages 265-276 for the proper methods, adsorbent recommendations and directions to control, contain and clean-up spills, and for disposal of pesticide wastes.

### *First Aid*

The reader should consult Chapter 7, Section E, pages 312-314 for emergency first aid information.

# PESTICIDE GUIDE 3

## HEALTH HAZARDS

***Hazard Indicator: Danger***

Danger: corrosive, contact may cause irreversible eye damage.

Warning: ingestion may be fatal.

Caution: inhalation of the dry material or mists may cause respiratory irritation.

Caution: absorption through skin may be harmful.

Caution: contact may cause moderate skin irritation.

May cause eye irritation.

Fire may produce irritating or poisonous gases.

Runoff from fire or spills may cause pollution.

## SAFETY GUIDELINES

***Restricted Use Pesticide:*** No

***Solubility:*** Practically insoluble in water

***Toxicity Category:*** Class I Eye Hazard, Class II Oral Hazard, Class III Dermal Hazard

***Restricted-entry Interval (REI):*** 48 hours

## HANDLING AND STORAGE PRECAUTIONS

***Use Precautions***

Avoid contact with susceptible crops, other desirable broadleaf plants, and beneficial insects.

Avoid spray drift.

Select and wear the proper PPE.

Mix ingredients in areas of sufficient ventilation or outdoors to minimize exposure. Avoid spilling material while mixing.

Wash with soap and water immediately after use or exposure. Follow good personal hygiene practices.

Do not reuse spray equipment for any purpose unless thoroughly cleaned with a suitable cleaner.

Reentry—do not enter treated areas without protective clothing until sprays have dried or after 48 hrs.

Notification of a pesticide application must be given to all workers who will be within 1/4 mile of a field during the application or before the REI expires.

This notification may be either oral or written, e.g., a posted sign at the entrances to treated sites. For some products, both oral and written notifications are necessary.

### *Storage and Disposal Precautions*

The reader should consult Chapter 6, Part V, Sections A and B, pages 251-265 for recommendations on the proper storage and disposal precautions.

## PROTECTIVE CLOTHING

Protective eyewear such as goggles or face shield is strongly recommended. Chemical-resistant gloves and boots, long-sleeved shirt, long pants, and shoes and socks are also strongly recommended. When handling undiluted product use face shield or goggles, impermeable gloves, splash apron, protective clothing and boots and, if dust or vapor is a problem, use approved air purifying respirators for pesticides.

## EMERGENCY GUIDELINES

### *Fire or Explosive Hazard*

Evacuate nonessential personnel from fire area.

The product is not considered flammable, but will both support combustion and may decompose under fire conditions and give off toxic products.

Some of these materials may react violently with water.

Segregate from oxidizers and incompatible materials.

The use of self-contained breathing apparatus and chemical protective clothing (chemical-resistant gloves, boots, hood, and suit) is recommended.

The use of high-pressure water may cause spread of hazardous materials.

*Small fires:* Dry chemical, $CO_2$, Halon, water spray, or standard foam.

*Large fires:* Water spray or fog is recommended.

Move containers of chemicals from fire area if you can do so without risk.

Cool containers that are exposed to flames with water until fire is out and heat of container is reduced.

Stay upwind of fire.

### *Spills*

The reader should consult Chapter 6, Part V, Section C, pages 265-276 for the proper methods, adsorbent recommendations and directions to control, contain and clean-up spills, and for disposal of pesticide wastes.

### *First Aid*

The reader should consult Chapter 7, Section E, pages 312-314 for emergency first aid information.

# PESTICIDE GUIDE 4

## HEALTH HAZARDS

***Hazard Indicator: Danger***

Danger: corrosive; contact may causes irrevesible eye damage, severe skin irritation and tissue damage.

Caution: ingestion may be harmful.

Caution: inhalation of dry material or mists may cause respiratory irritation.

Caution: absorption through skin may be harmful.

Fire may produce irritating or poisonous gases and runoff from fire or spills may cause pollution.

## SAFETY GUIDELINES

***Restricted Use Pesticide:*** No

***Solubility:*** Soluble to insoluble in water

***Toxicity Category:*** Class I Eye Hazard, Class III Oral Hazard, Class III Dermal Hazard

Some formulations are flammable and may contain aromatic and other chlorinated and non-chlorinated organic hydrocarbons. Some of these organics may be possible carcinogens; therefore, avoid exposure.

***Restricted-entry Interval (REI):*** 48 hours

## HANDLING AND STORAGE PRECAUTIONS

***Use Precautions***

Avoid contact with susceptible crops and other desirable broadleaf plants.

Avoid spray drift.

Select and wear the proper PPE.

Mix ingredients in areas of sufficient ventilation or outdoors to minimize exposure. Avoid spilling material while mixing.

Wash with soap and water immediately after use or exposure. Follow good personal hygiene practices.

Do not reuse spray equipment for any purpose unless thoroughly cleaned with a suitable cleaner.

Reentry—do not enter treated areas without protective clothing until sprays have dried or after 48 hrs.

Notification of a pesticide application must be given to all workers who will be within 1/4 mile of a field during the application or before the REI expires. This notification may be either oral or written, e.g., a posted sign at the en-

trances to treated sites. For some products, both oral and written notifications are necessary.

### *Storage and Disposal Precautions*

The reader should consult Chapter 6, Part V, Sections A and B, pages 251-265 for recommendations on the proper storage and disposal precautions.

## PROTECTIVE CLOTHING

Protective eyewear such as goggles or face shield is strongly recommended. Chemical-resistant gloves and boots, long-sleeved shirt, long pants, and shoes and socks are strongly recommended. When mixing, use a splash apron and, if dust or vapor is a problem, use approved air purifying respirators for pesticides.

## EMERGENCY GUIDELINES

### *Fire or Explosive Hazard*

Evacuate nonessential personnel from fire area.

Most products are not considered flammable. They will support combustion and may decompose under fire conditions and give off toxic products. However, several products are flammable. For information on product flammability the reader should consult Table 1 Pesticides Listed Alphabetically by Trade Name under formulations (see footnotes). The reader should review Appendix A for information on storage, disposal, and handling of flammable products.

Some of these materials may react violently with water.

Segregate from oxidizers and incompatible materials.

The use of self-contained breathing apparatus and chemical protective clothing (chemical-resistant gloves, boots, hood, and suit) is recommended.

The use of high-pressure water may cause spread of hazardous materials.

*Small fires:* Dry chemical, $CO_2$, Halon, water spray, or standard foam.

*Large fires:* Water spray, fog, or alcohol foam is recommended.

Move containers of chemicals from fire area if you can do so without risk.

Cool containers that are exposed to flames with water until fire is out and heat of container is reduced.

Stay upwind of fire.

### *Spills*

The reader should consult Chapter 6, Part V, Section C, pages 265-276 for the proper methods, adsorbent recommendations and directions to control, contain and clean-up spills, and for disposal of pesticide wastes.

### *First Aid*

The reader should consult Chapter 7, Section E, pages 312-314 for emergency first aid information.

# PESTICIDE GUIDE 5

## HEALTH HAZARDS

***Hazard Indicator: Danger***

Danger: avoid eye contact; can cause irreversible damage to eyes.

Danger: inhalation of the dry material or mists may be toxic.

Caution: could be harmful if absorbed through skin.

Caution: ingestion may be harmful.

Caution: contact may irritate skin.

Fire may produce irritating or poisonous gases.

Runoff from fire or spills may cause pollution.

## SAFETY GUIDELINES

***Restricted Use Pesticide:*** No

***Solubility:*** Insoluble in water

***Toxicity Category***: Class I Eye Hazard, Class I Inhalation Hazard, Class III Oral Hazard, Class III Dermal Hazard

***Restricted-entry Interval (REI):*** 48 hours

## HANDLING AND STORAGE PRECAUTIONS

***Use Precautions***

Avoid contact with 2,4-D susceptible crops and beneficial insects.

Avoid spray drift.

Select and wear the proper PPE.

Mix ingredients in areas of sufficient ventilation or outdoors to minimize exposure.

Wash with soap and water immediately after use or exposure. Follow good personal hygiene practices.

Do not reuse spray equipment for any purpose unless thoroughly cleaned with a suitable cleaner.

Reentry—do not enter treated areas without protective clothing until sprays have dried or after 48 hrs.

Notification of a pesticide application must be given to all workers who will be within 1/4 mile of a field during the application or before the REI expires. This notification may be either oral or written, e.g., a posted sign at the entrances to treated sites. For some products, both oral and written notifications are necessary.

### *Storage and Disposal Precautions*

The reader should consult Chapter 6, Part V, Sections A and B, pages 251-265 for recommendations on the proper storage and disposal precautions.

## PROTECTIVE CLOTHING

Protective eyewear such as goggles and face shield, chemical-resistant gloves and boots, long-sleeved shirt, long pants, and shoes and socks are strongly recommended. When mixing, use a splash apron and use only MSHA/NIOSH approved air purifying respirators for pesticides.

## EMERGENCY GUIDELINES

### *Fire or Explosive Hazard*

Evacuate nonessential personnel from fire area.

The products are not considered flammable. They may support combustion and may decompose under fire conditions and give off toxic products.

Some of these materials may react violently with water.

Segregate from oxidizers and incompatible materials.

The use of self-contained breathing apparatus and chemical protective clothing (chemical-resistant gloves, boots, hood, and suit) is recommended.

The use of high-pressure water may cause spread of hazardous materials.

*Small fires:* Dry chemical, $CO_2$, Halon, water spray, or standard foam.

Large fires: Water spray, fog, or standard foam is recommended.

Move containers of chemicals from fire area if you can do so without risk.

Cool containers that are exposed to flames with water until fire is out and heat of container is reduced.

Stay upwind of fire.

### *Spills*

The reader should consult Chapter 6, Part V, Section C, pages 265-276 for the proper methods, adsorbent recommendations and directions to control, contain and clean-up spills, and for disposal of pesticide wastes.

### *First Aid*

The reader should consult Chapter 7, Section E, pages 312-314 for emergency first aid information.

# PESTICIDE GUIDE 6

## HEALTH HAZARDS

***Hazard Indicator: Danger***

Danger: corrosive; may cause severe eye irritation and corneal injury.

Caution: ingestion may be harmful.

Caution: inhalation of dry material or mists may cause respiratory irritation.

Caution: absorption through skin may be harmful.

Caution: contact may cause moderate skin irritation.

Fire may produce irritating or poisonous gases.

Runoff from fire or spills may cause pollution.

## SAFETY GUIDELINES

***Restricted Use Pesticide:*** No

***Solubility:*** Soluble to insoluble in water

***Toxicity Category:*** Class I Eye Hazard, Class III Oral Hazard, Class III Dermal Hazard, Class III Inhalation Hazard

Several products are considered flammable; the formulations may contain N-butanol, ethylbenzene, isopropanol, petroleum naphtha, phenyl glycol ether, trimethylbenzene, or xylene.

***Restricted-entry Interval (REI):*** 48 hours

## HANDLING AND STORAGE PRECAUTIONS

***Use Precautions***

Avoid contact with susceptible crops and other desirable broadleaf plants.

Avoid spray drift and select and wear the proper PPE.

Mix ingredients in areas of sufficient ventilation or outdoors to minimize exposure. Avoid spilling material while mixing.

Wash with soap and water immediately after use or exposure. Follow good personal hygiene practices.

Do not reuse spray equipment for any purpose unless thoroughly cleaned with a suitable cleaner.

Reentry—do not enter treated areas without protective clothing until sprays have dried or after 48 hrs.

Notification of a pesticide application must be given to all workers who will be within 1/4 mile of a field during the application or before the REI expires. This notification may be either oral or written, e.g., a posted sign at the en-

trances to treated sites. For some products, both oral and written notifications are necessary.

### *Storage and Disposal Precautions*

The reader should consult Chapter 6, Part V, Sections A and B, pages 251-265 for recommendations on the proper storage and disposal precautions.

## PROTECTIVE CLOTHING

Protective eyewear such as goggles or face shield is strongly recommended. Chemical-resistant gloves and boots, long-sleeved shirt, long pants, and shoes and socks are strongly recommended. When mixing, use a splash apron and, if dust or vapor is a problem, use approved air purifying respirators for pesticides.

## EMERGENCY GUIDELINES

### *Fire or Explosive Hazard*

Evacuate nonessential personnel from fire area.

Most products are not considered flammable. They will support combustion and may decompose under fire conditions and give off toxic products. However, several products are flammable. For information on product flammability, the reader should consult Table 1 under formulations (see footnotes). The reader should review Chapter 6, Section V, for information on storage, disposal, and handling of flammable products.

Some of these materials may react violently with water.

Segregate from oxidizers and incompatible materials.

The use of self-contained breathing apparatus and chemical protective clothing (chemical-resistant gloves, boots, hood, and suit) is recommended.

The use of high-pressure water may cause spread of hazardous materials.

Small fires: Dry chemical, $CO_2$, Halon, water spray, or standard foam.

Large fires: Water spray, fog, or standard foam is recommended.

Move containers of chemicals from fire area if you can do so without risk.

Cool containers that are exposed to flames with water until fire is out and heat of container is reduced.

Stay upwind of fire.

### *Spills*

The reader should consult Chapter 6, Part V, Section C, pages 265-276 for the proper methods, adsorbent recommendations and directions to control, contain and clean-up spills, and for disposal of pesticide wastes.

### *First Aid*

The reader should consult Chapter 7, Section E, pages 312-314 for emergency first aid information.

# PESTICIDE GUIDE 7

## HEALTH HAZARDS

***Hazard Indicator: Danger***

Danger: contact may cause irreversible eye damage.

Caution: ingestion may be harmful.

Caution: inhalation of dry material or mists may cause respiratory irritation.

Caution: absorption through skin may be harmful.

Caution: contact may moderate cause skin irritation.

Fire may produce irritating or poisonous gases.

Runoff from fire or spills may cause pollution.

## SAFETY GUIDELINES

***Restricted Use Pesticide:*** No

***Solubility:*** Insoluble in water

***Toxicity Category:*** Class I Eye Hazard, Class III Dermal Hazard, Class IV Oral Hazard

The products are considered flammable; the formulations may contain cyclohexanone, ethylene chloride, monochlorobenzene, petroleum oil, and/or xylene.

***Restricted-entry Interval (REI):*** 48 hours

## HANDLING AND STORAGE PRECAUTIONS

***Use Precautions***

Avoid contact with susceptible crops and other desirable broadleaf plants.

Avoid spray drift and select and wear the proper PPE.

Mix ingredients in areas of sufficient ventilation or outdoors to minimize exposure. Avoid spilling material while mixing.

Wash with soap and water immediately after use or exposure. Follow good personal hygiene practices.

Do not reuse spray equipment for any purpose unless thoroughly cleaned with a suitable cleaner.

Reentry—do not enter treated areas without protective clothing until sprays have dried or after 48 hrs.

Notification of a pesticide application must be given to all workers who will be within 1/4 mile of a field during the application or before the REI expires. This notification may be either oral or written, e.g., a posted sign at the en-

trances to treated sites. For some products, both oral and written notifications are necessary.

### *Storage and Disposal Precautions*

The reader should consult Chapter 6, Part V, Sections A and B, pages 251-265 for recommendations on the proper storage and disposal precautions.

## PROTECTIVE CLOTHING

Protective eyewear such as goggles or face shield is strongly recommended. Chemical-resistant gloves and boots, long-sleeved shirt, long pants, and shoes and socks are strongly recommended. When mixing, use a splash apron and, if dust or vapor is a problem, use approved air purifying respirators for pesticides.

## EMERGENCY GUIDELINES

### *Fire or Explosive Hazard*

Evacuate nonessential personnel from fire area.

The products are considered flammable. They will support combustion and may decompose under fire conditions and give off toxic products. The reader should review Chapter 6, Section V, for information on storage, disposal, and handling of flammable products.

Some of these materials may react violently with water.

Segregate from oxidizers and incompatible materials.

The use of self-contained breathing apparatus and chemical protective clothing (chemical-resistant gloves, boots, hood, and suit) is recommended.

The use of high-pressure water may cause spread of hazardous materials.

*Small fires:* Dry chemical, $CO_2$, Halon, water spray, or standard foam.

*Large fires:* Water spray, fog, alcohol, or standard foam is recommended.

Move containers of chemicals from fire area if you can do so without risk.

Cool containers that are exposed to flames with water until fire is out and heat of container is reduced.

Stay upwind of fire.

### *Spills*

The reader should consult Chapter 6, Part V, Section C, pages 265-276 for the proper methods, adsorbent recommendations and directions to control, contain and clean-up spills, and for disposal of pesticide wastes.

### *First Aid*

The reader should consult Chapter 7, Section E, pages 312-314 for emergency first aid information.

# PESTICIDE GUIDE 8

## HEALTH HAZARDS

***Hazard Indicator: Danger***

Danger: corrosive; may cause severe eye irritation and corneal injury.

Caution: ingestion may be harmful.

Caution: inhalation of dry material or mists may cause respiratory irritation.

Caution: absorption through skin may be harmful.

Caution: contact may cause moderate skin irritation.

Fire may produce irritating or poisonous gases.

Runoff from fire or spills may cause pollution.

## SAFETY GUIDELINES

***Restricted Use Pesticide:*** No

***Solubility:*** Insoluble in water

***Toxicity Category:*** Class I Eye Hazard, Class III Dermal Hazard, Class III Inhalation Hazard, Class IV Oral Hazard

***Restricted-entry Interval (REI):*** 48 hours

## HANDLING AND STORAGE PRECAUTIONS

***Use Precautions***

Avoid contact with susceptible crops and beneficial insects.

Avoid spray drift.

Select and wear the proper PPE.

Mix ingredients in areas of sufficient ventilation or outdoors to minimize exposure. Avoid spilling material while mixing.

Wash with soap and water immediately after use or exposure. Follow good personal hygiene practices.

Do not reuse spray equipment for any purpose unless thoroughly cleaned with a suitable cleaner.

Reentry—do not enter treated areas without protective clothing until sprays have dried or after 48 hrs.

Notification of a pesticide application must be given to all workers who will be within 1/4 mile of a field during the application or before the REI expires. This notification may be either oral or written, e.g., a posted sign at the entrances to treated sites. For some products, both oral and written notifications are necessary.

### *Storage and Disposal Precautions*

The reader should consult Chapter 6, Part V, Sections A and B, pages 251-265 for recommendations on the proper storage and disposal precautions.

## PROTECTIVE CLOTHING

Protective eyewear such as goggles or face shield is strongly recommended. Chemical-resistant gloves and boots, long-sleeved shirt, long pants, and shoes and socks are strongly recommended. When mixing, use a splash apron and, if dust or vapor is a problem, use approved air purifying respirators for pesticides.

## EMERGENCY GUIDELINES

### *Fire or Explosive Hazard*

Evacuate nonessential personnel from fire area.

The products are not considered flammable. They will support combustion and may decompose under fire conditions and give off toxic products.

Some of these materials may react violently with water.

Segregate from oxidizers and incompatible materials.

The use of self-contained breathing apparatus and chemical protective clothing (chemical-resistant gloves, boots, hood, and suit) is recommended.

The use of high-pressure water may cause spread of hazardous materials.

*Small fires:* Dry chemical, $CO_2$, Halon, water spray, or standard foam.

*Large fires:* Water spray, fog, or standard foam is recommended.

Move containers of chemicals from fire area if you can do so without risk.

Cool containers that are exposed to flames with water until fire is out and heat of container is reduced.

Stay upwind of fire.

### *Spills*

The reader should consult Chapter 6, Part V, Section C, pages 265-276 for the proper methods, adsorbent recommendations and directions to control, contain and clean-up spills, and for disposal of pesticide wastes.

### *First Aid*

The reader should consult Chapter 7, Section E, pages 312-314 for emergency first aid information.

# PESTICIDE GUIDE 9

## HEALTH HAZARDS

***Hazard Indicator: Danger***

Danger: corrosive, may cause severe eye irritation and irreversible eye damage.

Caution: contact may cause skin irritation.

Caution: ingestion may be harmful.

Caution: absorption through skin may be harmful.

Caution: inhalation of dry material or mists may cause respiratory irritation.

Fire may produce irritating or poisonous gases.

Runoff from fire or spills may cause pollution.

## SAFETY GUIDELINES

***Restricted Use Pesticide:*** No

***Solubility:*** Slightly soluble in water

***Toxicity Category:*** Class I Eye Hazard, Class IV Oral Hazard, Class III Dermal Hazard

***Restricted-entry Interval (REI):*** 48 hours

## HANDLING AND STORAGE PRECAUTIONS

***Use Precautions***

Avoid contact with susceptible crops and beneficial insects.

Avoid spray drift.

Select and wear the proper PPE.

Mix ingredients in areas of sufficient ventilation or outdoors to minimize exposure. Avoid spilling material while mixing.

Wash with soap and water immediately after use or exposure. Follow good personal hygiene practices.

Do not reuse spray equipment for any purpose unless thoroughly cleaned with a suitable cleaner.

Reentry—do not enter treated areas without protective clothing until sprays have dried or after 48 hrs.

Notification of a pesticide application must be given to all workers who will be within 1/4 mile of a field during the application or before the REI expires. This notification may be either oral or written, e.g., a posted sign at the entrances to treated sites. For some products, both oral and written notifications are necessary.

### *Storage and Disposal Precautions*

The reader should consult Chapter 6, Part V, Sections A and B, pages 251-265 for recommendations on the proper storage and disposal precautions.

## PROTECTIVE CLOTHING

Protective eyewear such as goggles or face shield is strongly recommended. Chemical-resistant gloves and boots, long-sleeved shirt, long pants, and shoes and socks are strongly recommended. When handling undiluted product use face shield or goggles, impermeable gloves, splash apron, protective clothing and boots and, if dust or vapor is a problem, use approved air purifying respirators for pesticides.

## EMERGENCY GUIDELINES

### *Fire or Explosive Hazard*

Evacuate nonessential personnel from fire area.

The product is not considered flammable, but will both support combustion and may decompose under fire conditions and give off toxic products.

Some of these materials may react violently with water.

Segregate from oxidizers and incompatible materials.

The use of self-contained breathing apparatus and chemical protective clothing (chemical-resistant gloves, boots, hood, and suit) is recommended.

The use of high-pressure water may cause spread of hazardous materials.

*Small fires:* Dry chemical, $CO_2$, Halon, water spray, or standard foam.

*Large fires:* Water spray or fog is recommended.

Move containers of chemicals from fire area if you can do so without risk.

Cool containers that are exposed to flames with water until fire is out and heat of container is reduced.

Stay upwind of fire.

### *Spills*

The reader should consult Chapter 6, Part V, Section C, pages 265-276 for the proper methods, adsorbent recommendations and directions to control, contain and clean-up spills, and for disposal of pesticide wastes.

### *First Aid*

The reader should consult Chapter 7, Section E, pages 312-314 for emergency first aid information.

# PESTICIDE GUIDE 10

## HEALTH HAZARDS

***Hazard Indicator: Danger***

Danger: poisonous; ingestion may be fatal.

Warning: corrosive; contact may cause irreversible eye damage.

Caution: absorption through skin may be harmful.

Caution: inhalation of dry material or mists may cause respiratory irritation.

Caution: contact may cause skin irritation.

Fire may produce irritating or poisonous gases.

Runoff from fire or spills may cause pollution.

## SAFETY GUIDELINES

***Restricted Use Pesticide:*** No

***Solubility:*** Soluble in water

***Toxicity Category:*** Class I Oral Hazard, Class II Eye Hazard, Class III Dermal Hazard

***Restricted-entry Interval (REI):*** 48 hours

## HANDLING AND STORAGE PRECAUTIONS

***Use Precautions***

Avoid contact with susceptible crops and beneficial insects.

Avoid spray drift.

Select and wear the proper PPE.

Mix ingredients in areas of sufficient ventilation or outdoors to minimize exposure. Avoid spilling material while mixing.

Wash with soap and water immediately after use or exposure. Follow good personal hygiene practices.

Do not reuse spray equipment for any purpose unless thoroughly cleaned with a suitable cleaner.

Reentry—do not enter treated areas without protective clothing until sprays have dried or after 48 hrs.

Notification of a pesticide application must be given to all workers who will be within 1/4 mile of a field during the application or before the REI expires. This notification may be either oral or written, e.g., a posted sign at the entrances to treated sites. For some products, both oral and written notifications are necessary.

### *Storage and Disposal Precautions*

The reader should consult Chapter 6, Part V, Sections A and B, pages 251-265 for recommendations on the proper storage and disposal precautions.

## PROTECTIVE CLOTHING

Protective eyewear such as goggles and face shield, chemical-resistant gloves and boots, long-sleeved shirt, long pants, and shoes and socks are strongly recommended. When mixing, use a splash apron and, if dust or vapor is a problem, use approved air purifying respirators for pesticides.

## EMERGENCY GUIDELINES

### *Fire or Explosive Hazard*

Evacuate nonessential personnel from fire area.

The products are not considered flammable. They will support combustion and may decompose under fire conditions and give off toxic products.

Some of these materials may react violently with water.

Segregate from oxidizers and incompatible materials.

The use of self-contained breathing apparatus and chemical protective clothing (chemical-resistant gloves, boots, hood, and suit) is recommended.

The use of high-pressure water may cause spread of hazardous materials.

*Small fires:* Dry chemical, $CO_2$, Halon, water spray, or standard foam.

*Large fires:* Water spray, fog, or standard foam is recommended.

Move containers of chemicals from fire area if you can do so without risk.

Cool containers that are exposed to flames with water until fire is out and heat of container is reduced.

Stay upwind of fire.

### *Spills*

The reader should consult Chapter 6, Part V, Section C, pages 265-276 for the proper methods, adsorbent recommendations and directions to control, contain and clean-up spills, and for disposal of pesticide wastes.

### *First Aid*

The reader should consult Chapter 7, Section E, pages 312-314 for emergency first aid information.

# PESTICIDE GUIDE 11

## HEALTH HAZARDS

***Hazard Indicator: Danger***

Danger: poisonous, ingestion may be fatal.

Caution: can cause eye irritation.

Caution: inhalation of dry material or mists may cause respiratory irritation.

Caution: contact will irritate skin.

Fire may produce irritating or poisonous gases.

Runoff from fire or spills may cause pollution.

## SAFETY GUIDELINES

***Restricted Use Pesticide:*** Yes

***Solubility:*** Soluble in water

***Toxicity Category:*** Class I Oral Hazard

***Restricted-entry Interval (REI):*** 48 hours

## HANDLING AND STORAGE PRECAUTIONS

***Use Precautions***

Avoid contact with susceptible animals and crops.

Avoid spray drift.

Select and wear the proper PPE.

Mix ingredients in areas of sufficient ventilation or outdoors to minimize exposure. Avoid spilling material while mixing.

Wash with soap and water immediately after use or exposure. Follow good personal hygiene practices.

Do not reuse spray equipment for any purpose unless thoroughly cleaned with a suitable cleaner.

Reentry—do not enter treated areas without protective clothing until sprays have dried or after 48 hrs.

Notification of a pesticide application must be given to all workers who will be within 1/4 mile of a field during the application or before the REI expires. This notification may be either oral or written, e.g., a posted sign at the entrances to treated sites. For some products, both oral and written notifications are necessary.

### *Storage and Disposal Precautions*

The reader should consult Chapter 6, Part V, Sections A and B, pages 251-265 for recommendations on the proper storage and disposal precautions.

Caution should be taken; these pesticides are acutely hazardous.

## PROTECTIVE CLOTHING

Protective eyewear such as goggles or face shield, chemical-resistant gloves and boots, long-sleeved shirt, long pants, and shoes and socks are strongly recommended. When mixing, use a splash apron and, if dust or vapor is a problem, use approved air purifying respirators for pesticides.

## EMERGENCY GUIDELINES

### *Fire or Explosive Hazard*

Evacuate nonessential personnel from fire area.

The products are considered not flammable and may support combustion and may decompose under fire conditions and give off toxic products.

Some of these materials may react violently with water.

Segregate from oxidizers and incompatible materials.

The use of self-contained breathing apparatus and chemical protective clothing (chemical-resistant gloves, boots, hood, and suit) is recommended.

The use of high-pressure water may cause spread of hazardous materials.

*Small fires:* Dry chemical, $CO_2$, Halon, water spray, or alcohol type foam.

*Large fires:* Water spray, fog, or alcohol type foam is recommended.

Move containers of chemicals from fire area if you can do so without risk.

Cool containers that are exposed to flames with water until fire is out and heat of container is reduced.

Stay upwind of fire.

### *Spills*

The reader should consult Chapter 6, Part V, Section C, pages 265-276 for the proper methods, adsorbent recommendations and directions to control, contain and clean-up spills, and for disposal of pesticide wastes.

### *First Aid*

The reader should consult Chapter 7, Section E, pages 312-314 for emergency first aid information.

# PESTICIDE GUIDE 12

## HEALTH HAZARDS

***Hazard Indicator: Danger***

Danger: poisonous; ingestion may be fatal.

Danger: absorption through skin may be harmful.

Caution: inhalation of dry material or mists may cause respiratory irritation.

Caution: contact may cause skin irritation.

Caution: contact may cause eye irritation.

Fire may produce irritating or poisonous gases.

Runoff from fire or spills may cause pollution.

## SAFETY GUIDELINES

***Restricted Use Pesticide:*** Yes

***Solubility:*** Very slightly soluble in water

***Toxicity Category:*** Class I Oral Hazard, Class I Dermal Hazard

***Restricted-entry Interval (REI):*** 48 hours

## HANDLING AND STORAGE PRECAUTIONS

***Use Precautions***

Avoid contact with susceptible crops and beneficial insects.

Avoid spray drift.

Select and wear the proper PPE.

Mix ingredients in areas of sufficient ventilation or outdoors to minimize exposure. Avoid spilling material while mixing.

Wash with soap and water immediately after use or exposure. Follow good personal hygiene practices.

Do not reuse spray equipment for any purpose unless thoroughly cleaned with a suitable cleaner.

Reentry—do not enter treated areas without protective clothing until sprays have dried or after 48 hrs.

Notification of a pesticide application must be given to all workers who will be within 1/4 mile of a field during the application or before the REI expires. This notification may be either oral or written, e.g., a posted sign at the entrances to treated sites. For some products, both oral and written notifications are necessary.

### *Storage and Disposal Precautions*

The reader should consult Chapter 6, Part V, Sections A and B, pages 251-265 for recommendations on the proper storage and disposal precautions.

## PROTECTIVE CLOTHING

Protective eyewear, chemical-resistant gloves and boots, long-sleeved shirt, long pants, and shoes and socks are strongly recommended. When mixing, use a splash apron and, if dust or vapor is a problem, use approved air purifying respirators for pesticides.

## EMERGENCY GUIDELINES

### *Fire or Explosive Hazard*

Evacuate nonessential personnel from fire area.

The products are not considered flammable. They will support combustion and may decompose under fire conditions and give off toxic products.

Some of these materials may react violently with water.

Segregate from oxidizers and incompatible materials.

The use of self-contained breathing apparatus and chemical protective clothing (chemical-resistant gloves, boots, hood, and suit) is recommended.

The use of high-pressure water may cause spread of hazardous materials.

*Small fires:* Dry chemical, $CO_2$, Halon, water spray, or standard foam.

*Large fires:* Water spray, fog, or standard foam is recommended.

Move containers of chemicals from fire area if you can do so without risk.

Cool containers that are exposed to flames with water until fire is out and heat of container is reduced.

Stay upwind of fire.

### *Spills*

The reader should consult Chapter 6, Part V, Section C, pages 265-276 for the proper methods, adsorbent recommendations and directions to control, contain and clean-up spills, and for disposal of pesticide wastes.

### *First Aid*

The reader should consult Chapter 7, Section E, pages 312-314 for emergency first aid information.

# PESTICIDE GUIDE 13

## HEALTH HAZARDS

***Hazard Indicator: Danger***

Danger: poisonous; ingestion may be fatal.

Danger: absorption through skin or eyes may be fatal.

Danger: inhalation of the dry material or mists is highly toxic and may cause respiratory irritation.

Caution: contact may cause mild skin and eye irritation.

Fire may produce irritating or poisonous gases.

Runoff from fire or spills may cause pollution.

## SAFETY GUIDELINES

***Restricted Use Pesticide:*** Yes

***Solubility:*** Practically insoluble in water

***Toxicity Category:*** Class I Oral Hazard, Class I Dermal Hazard, Class I Inhalation Hazard

Formulated products may contain ethylbenzene, trimethylbenzene or xylene.

***Restricted-entry Interval (REI):*** 48 hours

## HANDLING AND STORAGE PRECAUTIONS

***Use Precautions***

Avoid contact with susceptible crops and beneficial insects.

Avoid spray drift.

Select and wear the proper PPE.

Mix ingredients in areas of sufficient ventilation or outdoors to minimize exposure. Avoid spilling material while mixing.

Wash with soap and water immediately after use or exposure. Follow good personal hygiene practices.

Do not reuse spray equipment for any purpose unless thoroughly cleaned with a suitable cleaner.

Reentry—do not enter treated areas without protective clothing until sprays have dried or after 48 hrs.

Notification of a pesticide application must be given to all workers who will be within 1/4 mile of a field during the application or before the REI expires. This notification may be either oral or written, e.g., a posted sign at the entrances to treated sites. For some products, both oral and written notifications are necessary.

### *Storage and Disposal Precautions*

The reader should consult Chapter 6, Part V, Sections A and B, pages 251-265 for recommendations on the proper storage and disposal precautions.

## PROTECTIVE CLOTHING

Protective eyewear such as goggles or face shield is strongly recommended; chemical-resistant gloves and boots, long-sleeved shirt, long pants, and shoes and socks are strongly recommended. When mixing, use a splash apron and, if dust or vapor is a problem, use approved air purifying respirators for pesticides.

## EMERGENCY GUIDELINES

### *Fire or Explosive Hazard*

Evacuate nonessential personnel from fire area.

The products are not considered flammable. They will support combustion and may decompose under fire conditions and give off toxic products.

Some of these materials may react violently with water.

Segregate from oxidizers and incompatible materials.

The use of self-contained breathing apparatus and chemical protective clothing (chemical-resistant gloves, boots, hood, and suit) is recommended.

The use of high-pressure water may cause spread of hazardous materials.

*Small fires:* Dry chemical, $CO_2$, Halon, water spray, or standard foam.

*Large fires:* Water spray, fog, or standard foam is recommended.

Move containers of chemicals from fire area if you can do so without risk.

Cool containers that are exposed to flames with water until fire is out and heat of container is reduced.

Stay upwind of fire.

### *Spills*

The reader should consult Chapter 6, Part V, Section C, pages 265-276 for the proper methods, adsorbent recommendations and directions to control, contain and clean-up spills, and for disposal of pesticide wastes.

### *First Aid*

The reader should consult Chapter 7, Section E, pages 312-314 for emergency first aid information.

# PESTICIDE GUIDE 14

## HEALTH HAZARDS

***Hazard Indicator: Danger***

Danger: poisonous, ingestion may be fatal.

Warning: absorption through skin may be fatal.

Caution: contact may cause skin irritation.

Caution: inhalation of dry material or mists may cause respiratory irritation.

Caution: contact may cause mild eye irritation.

Fire may produce irritating or poisonous gases.

Runoff from fire or spills may cause pollution.

## SAFETY GUIDELINES

***Restricted Use Pesticide:*** Yes

***Solubility:*** Practically insoluble in water

***Toxicity Category:*** Class I Oral Hazard, Class II Dermal Hazard

Formulations may contain low levels of kerosene

***Restricted-entry Interval (REI):*** 48 hours

## HANDLING AND STORAGE PRECAUTIONS

***Use Precautions***

Avoid contact with susceptible crops and other desirable broadleaf plants.

Avoid spray drift.

Select and wear the proper PPE.

Mix ingredients in areas of sufficient ventilation or outdoors to minimize exposure. Avoid spilling material while mixing.

Wash with soap and water immediately after use or exposure. Follow good personal hygiene practices.

Do not reuse spray equipment for any purpose unless thoroughly cleaned with a suitable cleaner.

Reentry—do not enter treated areas without protective clothing until sprays have dried or after 48 hrs.

Notification of a pesticide application must be given to all workers who will be within 1/4 mile of a field during the application or before the REI expires. This notification may be either oral or written, e.g., a posted sign at the entrances to treated sites. For some products, both oral and written notifications are necessary.

### *Storage and Disposal Precautions*

The reader should consult Chapter 6, Part V, Sections A and B, pages 251-265 for recommendations on the proper storage and disposal precautions.

## PROTECTIVE CLOTHING

Protective eyewear such as goggles or face shield, chemical-resistant gloves and boots, long-sleeved shirt, long pants, and shoes and socks are strongly recommended. When mixing, use a splash apron and, if dust or vapor is a problem, use approved air purifying respirators for pesticides.

## EMERGENCY GUIDELINES

### *Fire or Explosive Hazard*

Evacuate nonessential personnel from fire area.

The products are not considered flammable. They will support combustion and may decompose under fire conditions and give off toxic products.

Some of these materials may react violently with water.

Segregate from oxidizers and incompatible materials.

The use of self-contained breathing apparatus and chemical protective clothing (chemical-resistant gloves, boots, hood, and suit) is recommended.

The use of high-pressure water may cause spread of hazardous materials.

*Small fires:* Dry chemical, $CO_2$, Halon, water spray, or standard foam.

*Large fires:* Water spray, fog, or standard foam is recommended.

Move containers of chemicals from fire area if you can do so without risk.

Cool containers that are exposed to flames with water until fire is out and heat of container is reduced.

Stay upwind of fire.

### *Spills*

The reader should consult Chapter 6, Part V, Section C, pages 265-276 for the proper methods, adsorbent recommendations and directions to control, contain and clean-up spills, and for disposal of pesticide wastes.

### *First Aid*

The reader should consult Chapter 7, Section E, pages 312-314 for emergency first aid information.

# PESTICIDE GUIDE 15

## HEALTH HAZARDS

***Hazard Indicator: Danger***

Danger: poisonous; ingestion may be fatal.

Danger: highly toxic; and inhalation of dry material or mists may cause respiratory irritation.

Warning: absorption through skin may be fatal.

Caution: may cause moderate eye and skin irritation.

Fire may produce irritating or poisonous gases.

Runoff from fire or spills may cause pollution.

## SAFETY GUIDELINES

***Restricted Use Pesticide:*** Yes

***Solubility:*** Soluble/miscible to insoluble in water

***Toxicity Category:*** Class I Oral Hazard, Class I Inhalation Hazard, Class II Dermal Hazard

Formulations may contain xylene and/or ethyl benzene.

***Restricted-entry Interval (REI):*** 48 hours

## HANDLING AND STORAGE PRECAUTIONS

***Use Precautions***

Avoid contact with susceptible crops and beneficial insects.

Avoid spray drift.

Select and wear the proper PPE.

Mix ingredients in areas of sufficient ventilation or outdoors to minimize exposure. Avoid spilling material while mixing.

Wash with soap and water immediately after use or exposure. Follow good personal hygiene practices.

Do not reuse spray equipment for any purpose unless thoroughly cleaned with a suitable cleaner.

Reentry—do not enter treated areas without protective clothing until sprays have dried or after 48 hrs.

Notification of a pesticide application must be given to all workers who will be within 1/4 mile of a field during the application or before the REI expires. This notification may be either oral or written, e.g., a posted sign at the entrances to treated sites. For some products, both oral and written notifications are necessary.

### *Storage and Disposal Precautions*

The reader should consult Chapter 6, Part V, Sections A and B, pages 251-265 for recommendations on the proper storage and disposal precautions.

## PROTECTIVE CLOTHING

Protective eyewear, chemical-resistant gloves and boots, long-sleeved shirt, long pants, and shoes and socks are strongly recommended. When mixing, use a splash apron and, if dust or vapor is a problem, use approved air purifying respirators for pesticides.

## EMERGENCY GUIDELINES

### *Fire or Explosive Hazard*

Evacuate nonessential personnel from fire area.

The products are not considered flammable. They will support combustion and may decompose under fire conditions and give off toxic products.

Some of these materials may react violently with water.

Segregate from oxidizers and incompatible materials.

The use of self-contained breathing apparatus and chemical protective clothing (chemical-resistant gloves, boots, hood, and suit) is recommended.

The use of high-pressure water may cause spread of hazardous materials.

*Small fires:* Dry chemical, $CO_2$, Halon, water spray, or standard foam.

*Large fires:* Water spray, fog, or standard foam is recommended.

Move containers of chemicals from fire area if you can do so without risk.

Cool containers that are exposed to flames with water until fire is out and heat of container is reduced.

Stay upwind of fire.

### *Spills*

The reader should consult Chapter 6, Part V, Section C, pages 265-276 for the proper methods, adsorbent recommendations and directions to control, contain and clean-up spills, and for disposal of pesticide wastes.

### *First Aid*

The reader should consult Chapter 7, Section E, pages 312-314 for emergency first aid information.

# PESTICIDE GUIDE 16

## HEALTH HAZARDS

***Hazard Indicator: Danger***

Danger: poisonous; ingestion may be fatal.

Warning: absorption through skin may be harmful.

Caution: contact may cause eye and skin irritation.

Caution: inhalation of dry material or mists may cause respiratory irritation.

Fire may produce irritating or poisonous gases.

Runoff from fire or spills may cause pollution.

## SAFETY GUIDELINES

***Restricted Use Pesticide:*** **Yes**

***Solubility:*** Very slightly soluble in water

***Toxicity Category:*** Class I Oral Hazard, Class II Dermal Hazard, Class III Inhalation Hazard

***Restricted-entry Interval (REI):*** 48 hours

## HANDLING AND STORAGE PRECAUTIONS

***Use Precautions***

Avoid contact with susceptible crops and beneficial insects.

Avoid spray drift.

Select and wear the proper PPE.

Mix ingredients in areas of sufficient ventilation or outdoors to minimize exposure. Avoid spilling material while mixing.

Wash with soap and water immediately after use or exposure. Follow good personal hygiene practices.

Do not reuse spray equipment for any purpose unless thoroughly cleaned with a suitable cleaner.

Reentry—do not enter treated areas without protective clothing until sprays have dried or after 48 hrs.

Notification of a pesticide application must be given to all workers who will be within 1/4 mile of a field during the application or before the REI expires. This notification may be either oral or written, e.g., a posted sign at the entrances to treated sites. For some products, both oral and written notifications are necessary.

### *Storage and Disposal Precautions*

The reader should consult Chapter 6, Part V, Sections A and B, pages 251-265 for recommendations on the proper storage and disposal precautions.

## PROTECTIVE CLOTHING

Protective eyewear, chemical-resistant gloves and boots, long-sleeved shirt, long pants, and shoes and socks are strongly recommended. When mixing, use a splash apron and, if dust or vapor is a problem, use approved air purifying respirators for pesticides.

## EMERGENCY GUIDELINES

### *Fire or Explosive Hazard*

Evacuate nonessential personnel from fire area.

The products are not considered flammable. They will support combustion and may decompose under fire conditions and give off toxic products.

Some of these materials may react violently with water.

Segregate from oxidizers and incompatible materials.

The use of self-contained breathing apparatus and chemical protective clothing (chemical-resistant gloves, boots, hood, and suit) is recommended.

The use of high-pressure water may cause spread of hazardous materials.

*Small fires:* Dry chemical, $CO_2$, Halon, water spray, or foam.

*Large fires:* Water spray, fog, or standard or alcohol foam is recommended.

Move containers of chemicals from fire area if you can do so without risk.

Cool containers that are exposed to flames with water until fire is out and heat of container is reduced.

Stay upwind of fire.

### *Spills*

The reader should consult Chapter 6, Part V, Section C, pages 265-276 for the proper methods, adsorbent recommendations and directions to control, contain and clean-up spills, and for disposal of pesticide wastes.

### *First Aid*

The reader should consult Chapter 7, Section E, pages 312-314 for emergency first aid information.

# PESTICIDE GUIDE 17

## HEALTH HAZARDS

***Hazard Indicator: Danger***

Danger: poisonous—ingestion may be fatal.

Danger: toxic, inhalation of dry material or mists may cause severe respiratory irritation.

Warning: absorption through skin may be fatal.

Caution: corrosive, may cause severe eye irritation and damage.

Caution: contact may cause moderate skin irritation.

Fire may produce irritating or poisonous gases.

Runoff from fire or spills may cause pollution.

## SAFETY GUIDELINES

***Restricted Use Pesticide:*** Yes

***Solubility:*** Soluble to insoluble in water

***Toxicity Category:*** Class I Oral Hazard, Class I Inhalation Hazard, Class III Dermal Hazard

Several products are considered flammable; the formulations may contain ethylbenzene, methylene chloride, methyl alcohol. cyclohexanone, petroleum solvent, trimethylbenzene, and xylene.

***Restricted-entry Interval (REI):*** 48 hours

## HANDLING AND STORAGE PRECAUTIONS

***Use Precautions***

Avoid contact with susceptible crops and beneficial insects.

Avoid spray drift.

Select and wear the proper PPE.

Mix ingredients in areas of sufficient ventilation or outdoors to minimize exposure. Avoid spilling material while mixing.

Wash with soap and water immediately after use or exposure. Follow good personal hygiene practices.

Do not reuse spray equipment for any purpose unless thoroughly cleaned with a suitable cleaner.

Reentry—do not enter treated areas without protective clothing until sprays have dried or after 48 hrs.

Notification of a pesticide application must be given to all workers who will be within 1/4 mile of a field during the application or before the REI expires. This notification may be either oral or written, e.g., a posted sign at the

entrances to treated sites. For some products, both oral and written notifications are necessary.

### *Storage and Disposal Precautions*

The reader should consult Chapter 6, Part V, Sections A and B, pages 251-265 for recommendations on the proper storage and disposal precautions.

## PROTECTIVE CLOTHING

Protective eyewear such as goggles or face shield is strongly recommended. Chemical-resistant gloves and boots, long-sleeved shirt, long pants, and shoes and socks are strongly recommended. When handling undiluted product use goggles, impermeable gloves, splash apron, protective clothing and boots and, if dust or vapor is a problem, use approved air purifying respirators for pesticides.

## EMERGENCY GUIDELINES

### *Fire or Explosive Hazard*

Evacuate nonessential personnel from fire area.

Most products are not considered flammable. They will support combustion and may decompose under fire conditions and give off toxic products. However, several products are flammable. For information on product flammability the reader should consult Table 1 under formulations (see footnotes). The reader should review Chapter 6, Section V, for information on storage, disposal, and handling of flammable products.

Some of these materials may react violently with water.

Segregate from oxidizers and incompatible materials.

The use of self-contained breathing apparatus and chemical protective clothing (chemical-resistant gloves, boots, hood, and suit) is recommended.

The use of high-pressure water may cause spread of hazardous materials.

*Small fires:* Dry chemical, $CO_2$, Halon, water spray, or standard foam.

*Large fires:* Water spray, fog, or alcohol foam is recommended.

Move containers of chemicals from fire area if you can do so without risk.

Cool containers that are exposed to flames with water until fire is out and heat of container is reduced.

### *Spills*

The reader should consult Chapter 6, Part V, Section C, pages 265-276 for the proper methods, adsorbent recommendations and directions to control, contain and clean-up spills, and for disposal of pesticide wastes.

### *First Aid*

The reader should consult Chapter 7, Section E, pages 312-314 for emergency first aid information.

# PESTICIDE GUIDE 18

## HEALTH HAZARDS

***Hazard Indicator: Danger***

Danger: poisonous, ingestion may be fatal.

Danger: may cause severe eye irritation and corneal damage.

Danger: contact may cause severe skin irritation and skin damage.

Danger: absorption through skin may be fatal.

Caution: inhalation of dry material or mists may cause respiratory irritation.

Fire may produce irritating or poisonous gases.

Runoff from fire or spills may cause pollution.

## SAFETY GUIDELINES

***Restricted Use Pesticide:*** Yes

***Solubility:*** Soluble/miscible in water

***Toxicity Category:*** Class I Oral Hazard, Class I Dermal Hazard, Class I Eye Hazard

***Restricted-entry Interval (REI):*** 48 hours

## HANDLING AND STORAGE PRECAUTIONS

***Use Precautions***

Avoid contact with susceptible crops and other desirable broadleaf plants.

Avoid spray drift.

Select and wear the proper PPE.

Mix ingredients in areas of sufficient ventilation or outdoors to minimize exposure. Avoid spilling material while mixing.

Wash with soap and water immediately after use or exposure. Follow good personal hygiene practices.

Do not reuse spray equipment for any purpose unless thoroughly cleaned with a suitable cleaner.

Reentry—do not enter treated areas without protective clothing until sprays have dried or after 48 hrs.

Notification of a pesticide application must be given to all workers who will be within 1/4 mile of a field during the application or before the REI expires. This notification may be either oral or written, e.g., a posted sign at the entrances to treated sites. For some products, both oral and written notifications are necessary.

### *Storage and Disposal Precautions*

The reader should consult Chapter 6, Part V, Sections A and B, pages 251-265 for recommendations on the proper storage and disposal precautions.

## PROTECTIVE CLOTHING

Protective eyewear such as goggles or face shield is strongly recommended; chemical-resistant gloves and boots, long-sleeved shirt, long pants, and shoes and socks are strongly recommended. When handling undiluted product use face shield or goggles, impermeable gloves, splash apron, protective clothing and boots and, if dust or vapor is a problem, use approved air purifying respirators for pesticides.

## EMERGENCY GUIDELINES

### *Fire or Explosive Hazard*

Evacuate nonessential personnel from fire area.

The product is not considered flammable, but will both support combustion and may decompose under fire conditions and give off toxic products.

Some of these materials may react violently with water.

Segregate from oxidizers and incompatible materials.

The use of self-contained breathing apparatus and chemical protective clothing (chemical-resistant gloves, boots, hood, and suit) is recommended.

The use of high-pressure water may cause spread of hazardous materials.

*Small fires:* Dry chemical, $CO_2$, Halon, water spray, or alcohol foam.

*Large fires:* Water spray or fog is recommended.

Move containers of chemicals from fire area if you can do so without risk.

Cool containers that are exposed to flames with water until fire is out and heat of container is reduced.

Stay upwind of fire.

### *Spills*

The reader should consult Chapter 6, Part V, Section C, pages 265-276 for the proper methods, adsorbent recommendations and directions to control, contain and clean-up spills, and for disposal of pesticide wastes.

### *First Aid*

The reader should consult Chapter 7, Section E, pages 312-314 for emergency first aid information.

# PESTICIDE GUIDE 19

## HEALTH HAZARDS

***Hazard Indicator: Danger***

Danger: may cause severe eye irritation and irreversible eye damage.

Danger: poisonous, ingestion may be fatal.

Danger: inhalation of dry material or mists may cause respiratory irritation and may be fatal.

Warning: absorption through skin may be harmful.

Warning: contact may cause severe skin irritation and burns.

Fire may produce irritating or poisonous gases.

Runoff from fire or spills may cause pollution.

## SAFETY GUIDELINES

***Restricted Use Pesticide:*** Yes

***Solubility:*** Slightly soluble in water

***Toxicity Category:*** Class I Eye Hazard, Class I Oral Hazard, Class I Inhalation Hazard, Class II Dermal Hazard

The products are considered flammable; the formulations may contain xylene, cumene, trimethylbenzene, and 1,3-dichloropropene. 1,3-Dichloro-propene is listed as a potential carcinogen.

***Restricted-entry Interval (REI):*** 48 hours

## HANDLING AND STORAGE PRECAUTIONS

***Use Precautions***

Avoid contact with susceptible crops and beneficial insects.

Avoid spray drift and select and wear the proper PPE.

Mix ingredients in areas of sufficient ventilation or outdoors to minimize exposure. Avoid spilling material while mixing.

Wash with soap and water immediately after use or exposure. Follow good personal hygiene practices.

Do not reuse spray equipment for any purpose unless thoroughly cleaned with a suitable cleaner.

Reentry—do not enter treated areas without protective clothing until sprays have dried or after 48 hrs.

Notification of a pesticide application must be given to all workers who will be within 1/4 mile of a field during the application or before the REI expires. This notification may be either oral or written, e.g., a posted sign at the

entrances to treated sites. For some products, both oral and written notifications are necessary.

### *Storage and Disposal Precautions*

The reader should consult Chapter 6, Part V, Sections A and B, pages 251-265 for recommendations on the proper storage and disposal precautions.

## PROTECTIVE CLOTHING

Protective eyewear such as goggles or face shield is strongly recommended; chemical-resistant gloves and boots, long-sleeved shirt, long pants, and shoes and socks are strongly recommended. When mixing, use a splash apron and, if dust or vapor is a problem, use approved air purifying respirators for pesticides.

## EMERGENCY GUIDELINES

### *Fire or Explosive Hazard*

Evacuate nonessential personnel from fire area.

The products are considered flammable. They will support combustion and may decompose under fire conditions and give off toxic products. The reader should review Chapter 6, Section V, for information on storage, disposal, and handling of flammable products.

Some of these materials may react violently with water.

Segregate from oxidizers and incompatible materials.

The use of self-contained breathing apparatus and chemical protective clothing (chemical-resistant gloves, boots, hood, and suit) is recommended.

The use of high-pressure water may cause spread of hazardous materials.

*Small fires:* Dry chemical, $CO_2$, Halon, water spray, or standard foam.

*Large fires:* Water spray, fog, or standard or alcohol foam is recommended.

Move containers of chemicals from fire area if you can do so without risk.

Cool containers that are exposed to flames with water until fire is out and heat of container is reduced.

Stay upwind of fire.

### *Spills*

The reader should consult Chapter 6, Part V, Section C, pages 265-276 for the proper methods, adsorbent recommendations and directions to control, contain and clean-up spills, and for disposal of pesticide wastes.

### *First Aid*

The reader should consult Chapter 7, Section E, pages 312-314 for emergency first aid information.

# PESTICIDE GUIDE 20

## HEALTH HAZARDS

***Hazard Indicator: Danger***

Danger: poisonous, ingestion may be fatal.

Danger: may cause severe eye irritation and injury.

Danger: contact may cause severe skin irritation.

Caution: inhalation of dry material or mists may cause respiratory irritation.

Caution: absorption through skin may be harmful.

Fire may produce irritating or poisonous gases.

Runoff from fire or spills may cause pollution.

## SAFETY GUIDELINES

***Restricted Use Pesticide:*** Yes

***Solubility:*** Insoluble in water

***Toxicity Category:*** Class I Oral Hazard, Class I Eye Hazard, Class III Dermal Hazard

Formulations may contain xylene, light paraffinic petroleum distillates, and trimethylbenezene.

***Restricted-entry Interval (REI):*** 48 hours

## HANDLING AND STORAGE PRECAUTIONS

***Use Precautions***

Avoid contact with susceptible crops and beneficial insects.

Avoid spray drift and select and wear the proper PPE.

Mix ingredients in areas of sufficient ventilation or outdoors to minimize exposure. Avoid spilling material while mixing.

Wash with soap and water immediately after use or exposure. Follow good personal hygiene practices.

Do not reuse spray equipment for any purpose unless thoroughly cleaned with a suitable cleaner.

Reentry—do not enter treated areas without protective clothing until sprays have dried or after 48 hrs.

Notification of a pesticide application must be given to all workers who will be within 1/4 mile of a field during the application or before the REI expires. This notification may be either oral or written, e.g., a posted sign at the entrances to treated sites. For some products, both oral and written notifications are necessary.

### *Storage and Disposal Precautions*

The reader should consult Chapter 6, Part V, Sections A and B, pages 251-265 for recommendations on the proper storage and disposal precautions.

## PROTECTIVE CLOTHING

Protective eyewear such as goggles or face shield is strongly recommended. Chemical-resistant gloves and boots, long-sleeved shirt, long pants, and shoes and socks are strongly recommended. When mixing, use a splash apron and, if dust or vapor is a problem, use approved air purifying respirators for pesticides.

## EMERGENCY GUIDELINES

### *Fire or Explosive Hazard*

Evacuate nonessential personnel from fire area.

The products are not considered flammable. They will support combustion and may decompose under fire conditions and give off toxic products.

Some of these materials may react violently with water.

Segregate from oxidizers and incompatible materials.

The use of self-contained breathing apparatus and chemical protective clothing (chemical-resistant gloves, boots, hood, and suit) is recommended.

The use of high-pressure water may cause spread of hazardous materials.

*Small fires:* Dry chemical, $CO_2$, Halon, water spray, or standard foam.

*Large fires:* Water spray, fog, or alcohol foam is recommended.

Move containers of chemicals from fire area if you can do so without risk.

Cool containers that are exposed to flames with water until fire is out and heat of container is reduced.

Stay upwind of fire.

### *Spills*

The reader should consult Chapter 6, Part V, Section C, pages 265-276 for the proper methods, adsorbent recommendations and directions to control, contain and clean-up spills, and for disposal of pesticide wastes.

### *First Aid*

The reader should consult Chapter 7, Section E, pages 312-314 for emergency first aid information.

# PESTICIDE GUIDE 21

## HEALTH HAZARDS

***Hazard Indicator: Danger***

Danger: may cause severe eye irritation and irreversible burns.

Warning: ingestion may be fatal.

Warning: contact may cause severe skin irritation.

Warning: absorption through skin may be harmful.

Caution: inhalation of dry material or mists may cause respiratory irritation.

Fire may produce irritating or poisonous gases.

Runoff from fire or spills may cause pollution.

## SAFETY GUIDELINES

***Restricted Use Pesticide:*** Yes

***Solubility:*** Insoluble in water

***Toxicity Category:*** Class I Eye Hazard, Class II Oral Hazard, Class II Dermal Hazard

The products are considered flammable, the formulations may contain xylenes, trimethylbenzenes and emulsifiers.

***Restricted-entry Interval (REI):*** 48 hours

## HANDLING AND STORAGE PRECAUTIONS

***Use Precautions***

Avoid contact with susceptible crops and beneficial insects.

Avoid spray drift.

Select and wear the proper PPE.

Mix ingredients in areas of sufficient ventilation or outdoors to minimize exposure. Avoid spilling material while mixing.

Wash with soap and water immediately after use or exposure. Follow good personal hygiene practices.

Do not reuse spray equipment for any purpose unless thoroughly cleaned with a suitable cleaner.

Reentry—do not enter treated areas without protective clothing until sprays have dried or after 48 hrs.

Notification of a pesticide application must be given to all workers who will be within 1/4 mile of a field during the application or before the REI expires. This notification may be either oral or written, e.g., a posted sign at the

entrances to treated sites. For some products, both oral and written notifications are necessary.

### *Storage and Disposal Precautions*

The reader should consult Chapter 6, Part V, Sections A and B, pages 251-265 for recommendations on the proper storage and disposal precautions.

## PROTECTIVE CLOTHING

Protective eyewear such as goggles and face shield, chemical-resistant gloves and boots, long-sleeved shirt, long pants, and shoes and socks are strongly recommended. When mixing, use a splash apron and, if dust or vapor is a problem, use approved air purifying respirators for pesticides.

## EMERGENCY GUIDELINES

### *Fire or Explosive Hazard*

Evacuate nonessential personnel from fire area.

The products are considered flammable. They will support combustion and may decompose under fire conditions and give off toxic products. The reader should review Chapter 6, Section V, for additional information on storage, disposal, and handling of flammable products.

Some of these materials may react violently with water.

Segregate from oxidizers and incompatible materials.

The use of self-contained breathing apparatus and chemical protective clothing (chemical-resistant gloves, boots, hood, and suit) is recommended.

The use of high-pressure water may cause spread of hazardous materials.

*Small fires:* Dry chemical, $CO_2$, Halon, water spray, or standard foam.

*Large fires:* Water spray, fog, or standard foam is recommended.

Move containers of chemicals from fire area if you can do so without risk.

Cool containers that are exposed to flames with water until fire is out and heat of container is reduced.

Stay upwind of fire.

### *Spills*

The reader should consult Chapter 6, Part V, Section C, pages 265-276 for the proper methods, adsorbent recommendations and directions to control, contain and clean-up spills, and for disposal of pesticide wastes.

### *First Aid*

The reader should consult Chapter 7, Section E, pages 312-314 for emergency first aid information.

# PESTICIDE GUIDE 22

## HEALTH HAZARDS

***Hazard Indicator: Danger***

Danger: may cause severe eye irritation and irrevesible eye damage.

Warning: ingestion may be fatal.

Warning: absorption through skin may be harmful.

Warning: inhalation of dry material or mists may cause respiratory irritation and may be fatal.

Warning: contact may cause moderate skin irritation and burns.

Fire may produce irritating or poisonous gases.

Runoff from fire or spills may cause pollution.

## SAFETY GUIDELINES

***Restricted Use Pesticide:*** Yes

***Solubility:*** Very slightly soluble in water

***Toxicity Category:*** Class I Eye Hazard, Class II Oral Hazard, Class II Dermal Hazard, Class II Inhalation Hazard

The products are considered flammable; the formulations may contain a mixture of chlorinated hexenes, hexanes, and hexadienes, and 1,3-dichloropropene which is listed as a potential carcinogen.

***Restricted-entry Interval (REI):*** 48 hours

## HANDLING AND STORAGE PRECAUTIONS

***Use Precautions***

Avoid contact with susceptible crops and beneficial insects.

Avoid spray drift and select and wear the proper PPE.

Mix ingredients in areas of sufficient ventilation or outdoors to minimize exposure. Avoid spilling material while mixing.

Wash with soap and water immediately after use or exposure. Follow good personal hygiene practices.

Do not reuse spray equipment for any purpose unless thoroughly cleaned with a suitable cleaner.

Reentry—do not enter treated areas without protective clothing until sprays have dried or after 48 hrs.

Notification of a pesticide application must be given to all workers who will be within 1/4 mile of a field during the application or before the REI expires. This notification may be either oral or written, e.g., a posted sign at the

entrances to treated sites. For some products, both oral and written notifications are necessary.

### *Storage and Disposal Precautions*

The reader should consult Chapter 6, Part V, Sections A and B, pages 251-265 for recommendations on the proper storage and disposal precautions.

## PROTECTIVE CLOTHING

Protective eyewear such as goggles or face shield is strongly recommended; chemical-resistant gloves and boots, long-sleeved shirt, long pants, and shoes and socks are strongly recommended. When mixing, use a splash apron and, if dust or vapor is a problem, use approved air purifying respirators for pesticides.

## EMERGENCY GUIDELINES

### *Fire or Explosive Hazard*

Evacuate nonessential personnel from fire area.

The products are considered flammable. They will support combustion and may decompose under fire conditions and give off toxic products. The reader should review Chapter 6, Section V, for additional information on storage, disposal, and handling of flammable products.

Some of these materials may react violently with water.

Segregate from oxidizers and incompatible materials.

The use of self-contained breathing apparatus and chemical protective clothing (chemical-resistant gloves, boots, hood, and suit) is recommended.

The use of high-pressure water may cause spread of hazardous materials.

*Small fires:* Dry chemical, $CO_2$, Halon, water spray, or standard foam.

*Large fires:* Water spray, fog, or standard or alcohol foam is recommended.

Move containers of chemicals from fire area if you can do so without risk.

Cool containers that are exposed to flames with water until fire is out and heat of container is reduced.

Stay upwind of fire.

### *Spills*

The reader should consult Chapter 6, Part V, Section C, pages 265-276 for the proper methods, adsorbent recommendations and directions to control, contain and clean-up spills, and for disposal of pesticide wastes.

### *First Aid*

The reader should consult Chapter 7, Section E, pages 312-314 for emergency first aid information.

# PESTICIDE GUIDE 23

## HEALTH HAZARDS

***Hazard Indicator: Danger***

Danger: corrosive; contact may cause irreversible eye damage.

Warning: ingestion may be fatal.

Caution: inhalation of dry material or mists may cause respiratory irritation.

Caution: absorption through skin may be harmful.

Caution: contact may cause moderate skin and eye irritation.

Fire may produce irritating or poisonous gases.

Runoff from fire or spills may cause pollution.

## SAFETY GUIDELINES

***Restricted Use Pesticide:*** Yes

***Solubility:*** Soluble in water

***Toxicity Category Catregory:*** Class I Eye Hazard, Class II Oral Hazard, Class III Dermal Hazard

***Restricted-entry Interval (REI):*** 48 hours

## HANDLING AND STORAGE PRECAUTIONS

***Use Precautions***

Avoid contact with susceptible crops and beneficial insects.

Avoid spray drift.

Select and wear the proper PPE.

Mix ingredients in areas of sufficient ventilation or outdoors to minimize exposure. Avoid spilling material while mixing.

Wash with soap and water immediately after use or exposure. Follow good personal hygiene practices.

Do not reuse spray equipment for any purpose unless thoroughly cleaned with a suitable cleaner.

Reentry—do not enter treated areas without protective clothing until sprays have dried or after 48 hrs.

Notification of a pesticide application must be given to all workers who will be within 1/4 mile of a field during the application or before the REI expires. This notification may be either oral or written, e.g., a posted sign at the entrances to treated sites. For some products, both oral and written notifications are necessary.

### *Storage and Disposal Precautions*

The reader should consult Chapter 6, Part V, Sections A and B, pages 251-265 for recommendations on the proper storage and disposal precautions.

## PROTECTIVE CLOTHING

Protective eyewear such as goggles or face shield is strongly recommended. Chemical-resistant gloves and boots, long-sleeved shirt, long pants, and shoes and socks are also strongly recommended. When handling undiluted product use face shield or goggles, impermeable gloves, splash apron, protective clothing and boots and, if dust or vapor is a problem, use approved air purifying respirators for pesticides.

## EMERGENCY GUIDELINES

### *Fire or Explosive Hazard*

Evacuate nonessential personnel from fire area.

The product is not considered flammable, but will both support combustion and may decompose under fire conditions and give off toxic products.

Some of these materials may react violently with water.

Segregate from oxidizers and incompatible materials.

The use of self-contained breathing apparatus and chemical protective clothing (chemical-resistant gloves, boots, hood, and suit) is recommended.

The use of high-pressure water may cause spread of hazardous materials.

*Small fires:* Dry chemical, $CO_2$, Halon, water spray, or standard foam.

*Large fires:* Water spray or fog is recommended.

Move containers of chemicals from fire area if you can do so without risk.

Cool containers that are exposed to flames with water until fire is out and heat of container is reduced.

Stay upwind of fire.

### *Spills*

The reader should consult Chapter 6, Part V, Section C, pages 265-276 for the proper methods, adsorbent recommendations and directions to control, contain and clean-up spills, and for disposal of pesticide wastes.

### *First Aid*

The reader should consult Chapter 7, Section E, pages 312-314 for emergency first aid information.

# PESTICIDE GUIDE 24

## HEALTH HAZARDS

***Hazard Indicator: Danger***

Danger: corrosive, contact may cause irreversible eye damage.

Warning: ingestion may be fatal.

Caution: inhalation of dry material or mists may cause respiratory irritation.

Caution: absorption through skin may be harmful.

Caution: contact may cause moderate skin and eye irritation.

Fire may produce irritating or poisonous gases.

Runoff from fire or spills may cause pollution.

## SAFETY GUIDELINES

***Restricted Use Pesticide:*** Yes

***Solubility:*** Practically insoluble in water

***Toxicity Category:*** Class I Eye Hazard, Class II Oral Hazard, Class III Dermal Hazard, Class III Inhalation Hazard

The products are considered flammable; the formulations may contain xylene.

***Restricted-entry Interval (REI):*** 48 hours

## HANDLING AND STORAGE PRECAUTIONS

***Use Precautions***

Avoid contact with susceptible crops and beneficial insects.

Avoid spray drift.

Select and wear the proper PPE.

Mix ingredients in areas of sufficient ventilation or outdoors to minimize exposure. Avoid spilling material while mixing.

Wash with soap and water immediately after use or exposure. Follow good personal hygiene practices.

Do not reuse spray equipment for any purpose unless thoroughly cleaned with a suitable cleaner.

Reentry—do not enter treated areas without protective clothing until sprays have dried or after 48 hrs.

Notification of a pesticide application must be given to all workers who will be within 1/4 mile of a field during the application or before the REI expires. This notification may be either oral or written, e.g., a posted sign at the

entrances to treated sites. For some products, both oral and written notifications are necessary.

### *Storage and Disposal Precautions*

The reader should consult Chapter 6, Part V, Sections A and B, pages 251-265 for recommendations on the proper storage and disposal precautions.

## PROTECTIVE CLOTHING

Protective eyewear such as goggles or face shield is strongly recommended. Chemical-resistant gloves and boots, long-sleeved shirt, long pants, and shoes and socks are also strongly recommended. When handling undiluted product use face shield or goggles, impermeable gloves, splash apron, protective clothing and boots and, if dust or vapor is a problem, use approved air purifying respirators for pesticides.

## EMERGENCY GUIDELINES

### *Fire or Explosive Hazard*

Evacuate nonessential personnel from fire area.

The products are considered flammable. They will support combustion and may decompose under fire conditions and give off toxic products. The reader should review Chapter 6, Section V, for additional information on storage, disposal, and handling of flammable products.

Some of these materials may react violently with water.

Segregate from oxidizers and incompatible materials.

The use of self-contained breathing apparatus and chemical protective clothing (chemical-resistant gloves, boots, hood, and suit) is recommended.

The use of high-pressure water may cause spread of hazardous materials.

*Small fires:* Dry chemical, $CO_2$, Halon, water spray, or standard foam.

*Large fires:* Water spray or fog is recommended.

Move containers of chemicals from fire area if you can do so without risk.

Cool containers with water until fire is out and heat of container is reduced.

Stay upwind of fire.

### *Spills*

The reader should consult Chapter 6, Part V, Section C, pages 265-276 for the proper methods, adsorbent recommendations and directions to control, contain and clean-up spills, and for disposal of pesticide wastes.

### *First Aid*

The reader should consult Chapter 7, Section E, pages 312-314 for emergency first aid information.

# PESTICIDE GUIDE 25

## HEALTH HAZARDS

***Hazard Indicator: Danger***

Danger: avoid eye contact; can cause irreversible damage to eyes.

Warning: ingestion may be fatal.

Caution: inhalation of the dry material or mists may cause respiratory irritation.

Caution: could be harmful if absorbed through skin.

Caution: contact may irritate skin.

Fire may produce irritating or poisonous gases.

Runoff from fire or spills may cause pollution.

## SAFETY GUIDELINES

***Restricted Use Pesticide:*** Yes

***Solubility:*** Insoluble in water

***Toxicity Category:*** Class I Eye Hazard, Class II Oral Hazard, Class IV Dermal Hazard

***Restricted-entry Interval (REI):*** 48 hours

## HANDLING AND STORAGE PRECAUTIONS

***Use Precautions***

Avoid contact with 2,4-D susceptible crops and beneficial insects.

Avoid spray drift.

Select and wear the proper PPE.

Mix ingredients in areas of sufficient ventilation or outdoors to minimize exposure.

Wash with soap and water immediately after use or exposure. Follow good personal hygiene practices.

Do not reuse spray equipment for any purpose unless thoroughly cleaned with a suitable cleaner.

Reentry—do not enter treated areas without protective clothing until sprays have dried or after 48 hrs.

Notification of a pesticide application must be given to all workers who will be within 1/4 mile of a field during the application or before the REI expires. This notification may be either oral or written, e.g., a posted sign at the entrances to treated sites. For some products, both oral and written notifications are necessary.

### *Storage and Disposal Precautions*

The reader should consult Chapter 6, Part V, Sections A and B, pages 251-265 for recommendations on the proper storage and disposal precautions.

## PROTECTIVE CLOTHING

Protective eyewear such as goggles and face shield, chemical-resistant gloves and boots, long-sleeved shirt, long pants, and shoes and socks are strongly recommended. When mixing use a splash apron and use only MSHA/NIOSH approved air purifying respirators for pesticides.

## EMERGENCY GUIDELINES

### *Fire or Explosive Hazard*

Evacuate nonessential personnel from fire area.

The products are considered not flammable. They may support combustion and may decompose under fire conditions and give off toxic products.

Some of these materials may react violently with water.

Segregate from oxidizers and incompatible materials.

The use of self-contained breathing apparatus and chemical protective clothing (chemical-resistant gloves, boots, hood, and suit) is recommended.

The use of high-pressure water may cause spread of hazardous materials.

*Small fires:* Dry chemical, $CO_2$, Halon, water spray, or standard foam.

*Large fires:* Water spray, fog, or standard foam is recommended.

Move containers of chemicals from fire area if you can do so without risk.

Cool containers that are exposed to flames with water until fire is out and heat of container is reduced.

Stay upwind of fire.

### *Spills*

The reader should consult Chapter 6, Part V, Section C, pages 265-276 for the proper methods, adsorbent recommendations and directions to control, contain and clean-up spills, and for disposal of pesticide wastes.

### *First Aid*

The reader should consult Chapter 7, Section E, pages 312-314 for emergency first aid information.

# PESTICIDE GUIDE 26

## HEALTH HAZARDS

***Hazard Indicator: Danger***

Danger: corrosive, eye contact may cause irreversible eye damage.

Caution: ingestion may be harmful.

Caution: contact may cause skin absorption and/or irritation.

Caution: inhalation of the dry material or mists may cause respiratory irritation.

Fire may produce irritating or poisonous gases.

Runoff from fire or spills may cause pollution.

## SAFETY GUIDELINES

***Restricted Use Pesticide:*** Yes

***Solubility:*** Practically insoluble in water

***Toxicity Category:*** Class I Eye Hazard, Class III Oral Hazard, Class III Dermal Hazard

***Restricted-entry Interval (REI):*** 48 hours

## HANDLING AND STORAGE PRECAUTIONS

***Use Precautions***

Avoid contact with susceptible crops and other desirable broadleaf plants.

Avoid spray drift.

Select and wear the proper PPE.

Mix ingredients in areas of sufficient ventilation or outdoors to minimize exposure. Avoid spilling material while mixing.

Wash with soap and water immediately after use or exposure. Follow good personal hygiene practices.

Do not reuse spray equipment for any purpose unless thoroughly cleaned with a suitable cleaner.

Reentry—do not enter treated areas without protective clothing until sprays have dried or after 48 hrs.

Notification of a pesticide application must be given to all workers who will be within 1/4 mile of a field during the application or before the REI expires. This notification may be either oral or written, e.g., a posted sign at the entrances to treated sites. For some products, both oral and written notifications are necessary.

### *Storage and Disposal Precautions*

The reader should consult Chapter 6, Part V, Sections A and B, pages 251-265 for recommendations on the proper storage and disposal precautions.

## PROTECTIVE CLOTHING

Protective eyewear such as goggles or face shield, chemical-resistant gloves and boots, long-sleeved shirt, long pants, and shoes and socks are strongly recommended. When mixing, use a splash apron and, if dust or vapor is a problem, use approved air purifying respirators for pesticides.

## EMERGENCY GUIDELINES

### *Fire or Explosive Hazard*

Evacuate nonessential personnel from fire area.

The products are not considered flammable. They will support combustion and may decompose under fire conditions and give off toxic products.

Some of these materials may react violently with water.

Segregate from oxidizers and incompatible materials.

The use of self-contained breathing apparatus and chemical protective clothing (chemical-resistant gloves, boots, hood, and suit) is recommended.

The use of high-pressure water may cause spread of hazardous materials.

*Small fires:* Dry chemical, $CO_2$, Halon, water spray, or standard foam.

*Large fires:* Water spray, fog, or standard foam is recommended.

Move containers of chemicals from fire area if you can do so without risk.

Cool containers that are exposed to flames with water until fire is out and heat of container is reduced.

Stay upwind of fire.

### *Spills*

The reader should consult Chapter 6, Part V, Section C, pages 265-276 for the proper methods, adsorbent recommendations and directions to control, contain and clean-up spills, and for disposal of pesticide wastes.

### *First Aid*

The reader should consult Chapter 7, Section E, pages 312-314 for emergency first aid information.

# PESTICIDE GUIDE 27

## HEALTH HAZARDS

***Hazard Indicator: Danger***

Danger: may cause severe eye irritation and irreversible eye damage.

Warning: absorption through skin may be harmful.

Warning: contact may cause moderate skin irritation.

Caution: inhalation of dry material or mists may cause respiratory irritation.

Caution: ingestion may be harmful.

Fire may produce irritating or poisonous gases.

Runoff from fire or spills may cause pollution.

## SAFETY GUIDELINES

***Restricted Use Pesticide:*** Yes

***Solubility:*** Soluble/miscible in water

***Toxicity Category:*** Class I Eye Hazard, Class II Dermal Hazard, Class III Oral Hazard, Class III Inhalation Hazard

Formulations may contain mixed xylenes, ethylbenzene, cyclohexanone.

***Restricted-entry Interval (REI):*** 48 hours

## HANDLING AND STORAGE PRECAUTIONS

***Use Precautions***

Avoid contact with susceptible crops and beneficial insects.

Avoid spray drift.

Select and wear the proper PPE.

Mix ingredients in areas of sufficient ventilation or outdoors to minimize exposure. Avoid spilling material while mixing.

Wash with soap and water immediately after use or exposure. Follow good personal hygiene practices.

Do not reuse spray equipment for any purpose unless thoroughly cleaned with a suitable cleaner.

Reentry—do not enter treated areas without protective clothing until sprays have dried or after 48 hrs.

Notification of a pesticide application must be given to all workers who will be within 1/4 mile of a field during the application or before the REI expires. This notification may be either oral or written, e.g., a posted sign at the

entrances to treated sites. For some products, both oral and written notifications are necessary.

### *Storage and Disposal Precautions*

The reader should consult Chapter 6, Part V, Sections A and B, pages 251-265 for recommendations on the proper storage and disposal precautions.

## PROTECTIVE CLOTHING

Protective eyewear such as goggles or face shield, chemical-resistant gloves and boots, long-sleeved shirt, long pants, and shoes and socks are strongly recommended. When mixing, use a splash apron and, if dust or vapor is a problem, use approved air purifying respirators for pesticides.

## EMERGENCY GUIDELINES

### *Fire or Explosive Hazard*

Evacuate nonessential personnel from fire area.

The products are not considered flammable. They will support combustion and may decompose under fire conditions and give off toxic products.

Some of these materials may react violently with water.

Segregate from oxidizers and incompatible materials.

The use of self-contained breathing apparatus and chemical protective clothing (chemical-resistant gloves, boots, hood, and suit) is recommended.

The use of high-pressure water may cause spread of hazardous materials.

*Small fires:* Dry chemical, $CO_2$, Halon, water spray, or foam.

*Large fires:* Water spray, fog, or standard or alcohol foam is recommended.

Move containers of chemicals from fire area if you can do so without risk.

Cool containers that are exposed to flames with water until fire is out and heat of container is reduced.

Stay upwind of fire.

### *Spills*

The reader should consult Chapter 6, Part V, Section C, pages 265-276 for the proper methods, adsorbent recommendations and directions to control, contain and clean-up spills, and for disposal of pesticide wastes.

### *First Aid*

The reader should consult Chapter 7, Section E, pages 312-314 for emergency first aid information.

# PESTICIDE GUIDE 28

## HEALTH HAZARDS

***Hazard Indicator: Danger***

Danger: corrosive, contact may cause irreversible eye damage.

Warning: moderately toxic, inhalation of dry material or mists may cause respiratory irritation.

Caution: ingestion may be harmful.

Caution: absorption through skin may be harmful.

Caution: contact may cause moderate skin irritation.

Fire may produce irritating or poisonous gases.

Runoff from fire or spills may cause pollution.

## SAFETY GUIDELINES

***Restricted Use Pesticide***: No

***Solubility:*** Practically insoluble in water

***Toxicity Category:*** Class I Eye Hazard, Class II Inhalation Hazard, Class III Oral Hazard, Class III Dermal Hazard

***Restricted-entry Interval (REI):*** 48 hours

## HANDLING AND STORAGE PRECAUTIONS

***Use Precautions***

Avoid contact with susceptible crops and other desirable broadleaf plants.

Avoid spray drift.

Select and wear the proper PPE.

Mix ingredients in areas of sufficient ventilation or outdoors to minimize exposure. Avoid spilling material while mixing.

Wash with soap and water immediately after use or exposure. Follow good personal hygiene practices.

Do not reuse spray equipment for any purpose unless thoroughly cleaned with a suitable cleaner.

Reentry—do not enter treated areas without protective clothing until sprays have dried or after 48 hrs.

Notification of a pesticide application must be given to all workers who will be within 1/4 mile of a field during the application or before the REI expires. This notification may be either oral or written, e.g., a posted sign at the entrances to treated sites. For some products, both oral and written notifications are necessary.

### *Storage and Disposal Precautions*

The reader should consult Chapter 6, Part V, Sections A and B, pages 251-265 for recommendations on the proper storage and disposal precautions.

## PROTECTIVE CLOTHING

Protective eyewear such as goggles or face shield is strongly recommended. Chemical-resistant gloves and boots, long-sleeved shirt, long pants, and shoes and socks are also strongly recommended. When handling undiluted product use face shield or goggles, impermeable gloves, splash apron, protective clothing and boots and, if dust or vapor is a problem, use approved air purifying respirators for pesticides.

## EMERGENCY GUIDELINES

### *Fire or Explosive Hazard*

Evacuate nonessential personnel from fire area.

The product is not considered flammable, but will both support combustion and may decompose under fire conditions and give off toxic products.

Some of these materials may react violently with water.

Segregate from oxidizers and incompatible materials.

The use of self-contained breathing apparatus and chemical protective clothing (chemical-resistant gloves, boots, hood, and suit) is recommended.

The use of high-pressure water may cause spread of hazardous materials.

*Small fires:* Dry chemical, $CO_2$, Halon, water spray, or standard foam.

*Large fires:* Water spray or fog is recommended.

Move containers of chemicals from fire area if you can do so without risk.

Cool containers that are exposed to flames with water until fire is out and heat of container is reduced.

Stay upwind of fire.

### *Spills*

The reader should consult Chapter 6, Part V, Section C, pages 265-276 for the proper methods, adsorbent recommendations and directions to control, contain and clean-up spills, and for disposal of pesticide wastes.

### *First Aid*

The reader should consult Chapter 7, Section E, pages 312-314 for emergency first aid information.

# PESTICIDE GUIDE 29

## HEALTH HAZARDS

***Hazard Indicator: Danger***

Danger: poisonous, ingestion may be fatal.

Danger: product has been shown to be a cancer hazard.

Caution: contact may cause skin irritation.

Caution: inhalation of the dry material or mists may cause respiratory irritation.

Caution: can cause eye irritation.

Fire may produce irritating or poisonous gases.

Runoff from fire or spills may cause pollution.

## SAFETY GUIDELINES

***Restricted Use Pesticide:*** Yes

***Solubility:*** Completely soluble in water

***Toxicity Category:*** Class I Oral Hazard, Class III Dermal Hazard, Class IV Eye Hazard

Several formulated products may contain small amounts of dichloromethane which has been determined to be carcinogenic in laboratory animals.

***Restricted-entry Interval (REI):*** 48 hours

## HANDLING AND STORAGE PRECAUTIONS

***Use Precautions***

Avoid contact with susceptible crops and other desirable broadleaf plants.

Avoid spray drift.

Select and wear the proper PPE.

Mix ingredients in areas of sufficient ventilation or outdoors to minimize exposure. Avoid spilling material while mixing.

Wash with soap and water immediately after use or exposure. Follow good personal hygiene practices.

Do not reuse spray equipment for any purpose unless thoroughly cleaned with a suitable cleaner.

Reentry—do not enter treated areas without protective clothing until sprays have dried or after 48 hrs.

Notification of a pesticide application must be given to all workers who will be within 1/4 mile of a field during the application or before the REI expires.

This notification may be either oral or written, e.g., a posted sign at the entrances to treated sites. For some products, both oral and written notifications are necessary.

### *Storage and Disposal Precautions*

These pesticides are acutely hazardous and the reader should consult Chapter 6, Part V, Sections A and B, pages 251-265 for recommendations on the proper storage and disposal precautions.

## PROTECTIVE CLOTHING

Protective eyewear, chemical-resistant gloves and boots, long-sleeved shirt, long pants, and shoes and socks are strongly recommended. When mixing, use a splash apron and, if dust or vapor is a problem, use approved air purifying respirators for pesticides.

## EMERGENCY GUIDELINES

### *Fire or Explosive Hazard*

Evacuate nonessential personnel from fire area.

The products are not considered flammable. They will support combustion and may decompose under fire conditions and give off toxic products.

Some of these materials may react violently with water.

Segregate from oxidizers and incompatible materials.

The use of self-contained breathing apparatus and chemical protective clothing (chemical-resistant gloves, boots, hood, and suit) is recommended.

The use of high-pressure water may cause spread of hazardous materials.

*Small fires:* Dry chemical, $CO_2$, Halon, water spray, or standard foam.

*Large fires:* Water spray, fog, or standard foam is recommended.

Move containers of chemicals from fire area if you can do so without risk.

Cool containers that are exposed to flames with water until fire is out and heat of container is reduced.

Stay upwind of fire.

### *Spills*

The reader should consult Chapter 6, Part V, Section C, pages 265-276 for the proper methods, adsorbent recommendations and directions to control, contain and clean-up spills, and for disposal of pesticide wastes.

### *First Aid*

The reader should consult Chapter 7, Section E, pages 312-314 for emergency first aid information.

# PESTICIDE GUIDE 29

## HEALTH HAZARDS

***Hazard Indicator: Danger***

Danger: poisonous, ingestion may be fatal.

Danger: product has been shown to be a cancer hazard.

Caution: contact may cause skin irritation.

Caution: inhalation of the dry material or mists may cause respiratory irritation.

Caution: can cause eye irritation.

Fire may produce irritating or poisonous gases.

Runoff from fire or spills may cause pollution.

## SAFETY GUIDELINES

***Restricted Use Pesticide:*** Yes

***Solubility:*** Completely soluble in water

***Toxicity Category:*** Class I Oral Hazard, Class III Dermal Hazard, Class IV Eye Hazard

Several formulated products may contain small amounts of dichloromethane which has been determined to be carcinogenic in laboratory animals.

***Restricted-entry Interval (REI):*** 48 hours

## HANDLING AND STORAGE PRECAUTIONS

***Use Precautions***

Avoid contact with susceptible crops and other desirable broadleaf plants.

Avoid spray drift.

Select and wear the proper PPE.

Mix ingredients in areas of sufficient ventilation or outdoors to minimize exposure. Avoid spilling material while mixing.

Wash with soap and water immediately after use or exposure. Follow good personal hygiene practices.

Do not reuse spray equipment for any purpose unless thoroughly cleaned with a suitable cleaner.

Reentry—do not enter treated areas without protective clothing until sprays have dried or after 48 hrs.

Notification of a pesticide application must be given to all workers who will be within 1/4 mile of a field during the application or before the REI expires.

This notification may be either oral or written, e.g., a posted sign at the entrances to treated sites. For some products, both oral and written notifications are necessary.

### *Storage and Disposal Precautions*

These pesticides are acutely hazardous and the reader should consult Chapter 6, Part V, Sections A and B, pages 251-265 for recommendations on the proper storage and disposal precautions.

## PROTECTIVE CLOTHING

Protective eyewear, chemical-resistant gloves and boots, long-sleeved shirt, long pants, and shoes and socks are strongly recommended. When mixing, use a splash apron and, if dust or vapor is a problem, use approved air purifying respirators for pesticides.

## EMERGENCY GUIDELINES

### *Fire or Explosive Hazard*

Evacuate nonessential personnel from fire area.

The products are not considered flammable. They will support combustion and may decompose under fire conditions and give off toxic products.

Some of these materials may react violently with water.

Segregate from oxidizers and incompatible materials.

The use of self-contained breathing apparatus and chemical protective clothing (chemical-resistant gloves, boots, hood, and suit) is recommended.

The use of high-pressure water may cause spread of hazardous materials.

*Small fires:* Dry chemical, $CO_2$, Halon, water spray, or standard foam.

*Large fires:* Water spray, fog, or standard foam is recommended.

Move containers of chemicals from fire area if you can do so without risk.

Cool containers that are exposed to flames with water until fire is out and heat of container is reduced.

Stay upwind of fire.

### *Spills*

The reader should consult Chapter 6, Part V, Section C, pages 265-276 for the proper methods, adsorbent recommendations and directions to control, contain and clean-up spills, and for disposal of pesticide wastes.

### *First Aid*

The reader should consult Chapter 7, Section E, pages 312-314 for emergency first aid information.

# PESTICIDE GUIDE 30

## HEALTH HAZARDS

***Hazard Indicator:* *Warning***

Warning: ingestion may be fatal.

Caution: may cause eye irritation.

Caution: contact may cause skin irritation.

Caution: inhalation of the dry material or mists may cause respiratory irritation.

Caution: absorption through skin may be harmful.

Fire may produce irritating or poisonous gases.

Runoff from fire or spills may cause pollution.

## SAFETY GUIDELINES

***Restricted Use Pesticide:*** No

***Solubility:*** Practically insoluble in water

***Toxicity Category:*** Class II Oral Hazard

***Restricted-entry Interval (REI):*** 24 hours

## HANDLING AND STORAGE PRECAUTIONS

***Use Precautions***

Avoid contact with susceptible crops and beneficial insects.

Avoid spray drift.

Select and wear the proper PPE.

Mix ingredients in areas of sufficient ventilation or outdoors to minimize exposure. Avoid spilling material while mixing.

Wash with soap and water immediately after use or exposure. Follow good personal hygiene practices.

Do not reuse spray equipment for any purpose unless thoroughly cleaned with a suitable cleaner.

Reentry—do not enter treated areas without protective clothing until sprays have dried or after 24-hrs.

Notification of a pesticide application must be given to all workers who will be within 1/4 mile of a field during the application or before the REI expires. This notification may be either oral or written, e.g., a posted sign at the entrances to treated sites. For some products, both oral and written notifications are necessary.

### *Storage and Disposal Precautions*

The reader should consult Chapter 6, Part V, Sections A and B, pages 251-265 for recommendations on the proper storage and disposal precautions.

## PROTECTIVE CLOTHING

Protective eyewear, chemical-resistant gloves and boots, long-sleeved shirt, long pants, and shoes and socks are strongly recommended. When handling undiluted product use goggles, impermeable gloves, splash apron, protective clothing and boots and, if dust or vapor is a problem, use approved air purifying respirators for pesticides.

## EMERGENCY GUIDELINES

### *Fire or Explosive Hazard*

Evacuate nonessential personnel from fire area.

The products are not considered flammable. They will support combustion and may decompose under fire conditions and give off toxic products.

Some of these materials may react violently with water.

Segregate from oxidizers and incompatible materials.

The use of self-contained breathing apparatus and chemical protective clothing (chemical-resistant gloves, boots, hood, and suit) is recommended.

The use of high-pressure water may cause spread of hazardous materials.

*Small fires:* Dry chemical, $CO_2$, Halon, water spray, or alcohol foam.

*Large fires:* Water spray, fog, or alcohol foam is recommended.

Move containers of chemicals from fire area if you can do so without risk.

Cool containers that are exposed to flames with water until fire is out and heat of container is reduced.

Stay upwind of fire.

### *Spills*

The reader should consult Chapter 6, Part V, Section C, pages 265-276 for the proper methods, adsorbent recommendations and directions to control, contain and clean-up spills, and for disposal of pesticide wastes.

### *First Aid*

The reader should consult Chapter 7, Section E, pages 312-314 for emergency first aid information.

# PESTICIDE GUIDE 31

## HEALTH HAZARDS

***Hazard Indicator: Warning***

Warning: ingestion may be fatal.

Warning: inhalation of dry material or mists may cause respiratory irritation and tissue damage.

No information available on skin absorption, or skin and eye irritation.

Fire may produce irritating or poisonous gases.

Runoff from fire or spills may cause pollution.

## SAFETY GUIDELINES

***Restricted Use Pesticide:*** No

***Solubility:*** Insoluble in water

***Toxicity Category:*** Class II Oral Hazard, Class II Inhalation Hazard

***Restricted-entry Interval (REI):*** 24 hours

## HANDLING AND STORAGE PRECAUTIONS

***Use Precautions***

Avoid contact with susceptible animals and beneficial insects.

Avoid spray drift.

Select and wear the proper PPE.

Mix ingredients in areas of sufficient ventilation or outdoors to minimize exposure. Avoid spilling material while mixing.

Wash with soap and water immediately after use or exposure. Follow good personal hygiene practices.

Do not reuse spray equipment for any purpose unless thoroughly cleaned with a suitable cleaner.

Reentry—do not enter treated areas without protective clothing until sprays have dried or after 24 hrs.

Notification of a pesticide application must be given to all workers who will be within 1/4 mile of a field during the application or before the REI expires. This notification may be either oral or written, e.g., a posted sign at the entrances to treated sites. For some products, both oral and written notifications are necessary.

### *Storage and Disposal Precautions*

These pesticides are acutely hazardous, therefore, the reader should consult Chapter 6, Part V, Sections A and B, pages 251-265 for recommendations on the proper storage and disposal precautions.

## PROTECTIVE CLOTHING

Protective eyewear, chemical-resistant gloves and boots, long-sleeved shirt, long pants, and shoes and socks are strongly recommended. When mixing, use a splash apron and, if dust or vapor is a problem, use approved air purifying respirators for pesticides.

## EMERGENCY GUIDELINES

### *Fire or Explosive Hazard*

Evacuate nonessential personnel from fire area.

The products are considered not flammable and may support combustion and may decompose under fire conditions and give off toxic products.

Some of these materials may react violently with water.

Segregate from oxidizers and incompatible materials.

The use of self-contained breathing apparatus and chemical protective clothing (chemical-resistant gloves, boots, hood, and suit) is recommended.

The use of high-pressure water may cause spread of hazardous materials.

*Small fires*: Dry chemical, $CO_2$, Halon, water spray, or alcohol type foam.

*Large fires:* Water spray, fog, or alcohol type foam is recommended.

Move containers of chemicals from fire area if you can do so without risk.

Cool containers that are exposed to flames with water until fire is out and heat of container is reduced.

Stay upwind of fire.

### *Spills*

The reader should consult Chapter 6, Part V, Section C, pages 265-276 for the proper methods, adsorbent recommendations and directions to control, contain and clean-up spills, and for disposal of pesticide wastes.

### *First Aid*

The reader should consult Chapter 7, Section E, pages 312-314 for emergency first aid information.

# PESTICIDE GUIDE 32

## HEALTH HAZARDS

***Hazard Indicator: Warning***

Warning: ingestion may be fatal.

Warning: absorption through skin may be fatal.

Caution: contact may cause eye irritation.

Caution: inhalation of dry material or mists may cause respiratory irritation.

Caution: contact may cause moderate skin irritation.

Fire may produce irritating or poisonous gases.

Runoff from fire or spills may cause pollution.

## SAFETY GUIDELINES

***Restricted Use Pesticide:*** No

***Solubility:*** Very slightly soluble to insoluble in water

***Toxicity Category:*** Class II Dermal Hazard, Class II Oral Hazard

***Restricted-entry Interval (REI):*** 24 hours

## HANDLING AND STORAGE PRECAUTIONS

***Use Precautions***

Avoid contact with susceptible crops and beneficial insects..

Avoid spray drift.

Select and wear the proper PPE.

Mix ingredients in areas of sufficient ventilation or outdoors to minimize exposure. Avoid spilling material while mixing.

Wash with soap and water immediately after use or exposure. Follow good personal hygiene practices.

Do not reuse spray equipment for any purpose unless thoroughly cleaned with a suitable cleaner.

Reentry—do not enter treated areas without protective clothing until sprays have dried or after 24 hrs.

Notification of a pesticide application must be given to all workers who will be within 1/4 mile of a field during the application or before the REI expires. This notification may be either oral or written, e.g., a posted sign at the entrances to treated sites. For some products, both oral and written notifications are necessary.

### *Storage and Disposal Precautions*

The reader should consult Chapter 6, Part V, Sections A and B, pages 251-265 for recommendations on the proper storage and disposal precautions.

## PROTECTIVE CLOTHING

Protective eyewear, chemical-resistant gloves and boots, long-sleeved shirt, long pants, and shoes and socks are strongly recommended. When mixing, use a splash apron and, if dust or vapor is a problem, use approved air purifying respirators for pesticides.

## EMERGENCY GUIDELINES

### *Fire or Explosive Hazard*

Evacuate nonessential personnel from fire area.

The products are not considered flammable. They will support combustion and may decompose under fire conditions and give off toxic products.

Some of these materials may react violently with water.

Segregate from oxidizers and incompatible materials.

The use of self-contained breathing apparatus and chemical protective clothing (chemical-resistant gloves, boots, hood, and suit) is recommended.

The use of high-pressure water may cause spread of hazardous materials.

*Small fires:* Dry chemical, $CO_2$, Halon, water spray ,or standard foam.

*Large fires:* Water spray, fog, or standard foam is recommended.

Move containers of chemicals from fire area if you can do so without risk.

Cool containers that are exposed to flames with water until fire is out and heat of container is reduced.

Stay upwind of fire.

### *Spills*

The reader should consult Chapter 6, Part V, Section C, pages 265-276 for the proper methods, adsorbent recommendations and directions to control, contain and clean-up spills, and for disposal of pesticide wastes.

### *First Aid*

The reader should consult Chapter 7, Section E, pages 312-314 for emergency first aid information.

# PESTICIDE GUIDE 33

## HEALTH HAZARDS

*Hazard Indicator: Warning*

Warning: ingestion may be fatal.

Warning: absorption through skin may be harmful.

Warning: may be toxic, inhalation of dry material or mists may cause respiratory irritation.

Caution: may cause moderate eye irritation.

Caution: contact may cause moderate skin irritation.

Fire may produce irritating or poisonous gases.

Runoff from fire or spills may cause pollution.

## SAFETY GUIDELINES

*Restricted Use Pesticide:* No

*Solubility:* Miscible in water

*Toxicity Category:* Class II Dermal Hazard, Class II Oral Hazard, Class II Inhalation Hazard

*Restricted-entry Interval (REI):* 24 hours

## HANDLING AND STORAGE PRECAUTIONS

*Use Precautions*

Avoid contact with susceptible crops and other desirable broadleaf plants.

Avoid spray drift.

Select and wear the proper PPE.

Mix ingredients in areas of sufficient ventilation or outdoors to minimize exposure. Avoid spilling material while mixing.

Wash with soap and water immediately after use or exposure. Follow good personal hygiene practices.

Do not reuse spray equipment for any purpose unless thoroughly cleaned with a suitable cleaner.

Reentry—do not enter treated areas without protective clothing until sprays have dried or after 24 hrs.

Notification of a pesticide application must be given to all workers who will be within 1/4 mile of a field during the application or before the REI expires. This notification may be either oral or written, e.g., a posted sign at the entrances to treated sites. For some products, both oral and written notifications are necessary.

### *Storage and Disposal Precautions*

The reader should consult Chapter 6, Part V, Sections A and B, pages 251-265 for recommendations on the proper storage and disposal precautions.

## PROTECTIVE CLOTHING

Protective eyewear, chemical-resistant gloves and boots, long-sleeved shirt, long pants, and shoes and socks are strongly recommended. When mixing, use a splash apron and, if dust or vapor is a problem, use approved air purifying respirators for pesticides.

## EMERGENCY GUIDELINES

### *Fire or Explosive Hazard*

Evacuate nonessential personnel from fire area.

The products are not considered flammable. They will support combustion and may decompose under fire conditions and give off toxic products.

Some of these materials may react violently with water.

Segregate from oxidizers and incompatible materials.

The use of self-contained breathing apparatus and chemical protective clothing (chemical-resistant gloves, boots, hood, and suit) is recommended.

The use of high-pressure water may cause spread of hazardous materials.

*Small fires:* Dry chemical, $CO_2$, Halon, water spray, or standard foam.

*Large fires:* Water spray, fog, or standard foam is recommended.

Move containers of chemicals from fire area if you can do so without risk.

Cool containers that are exposed to flames with water until fire is out and heat of container is reduced.

Stay upwind of fire.

### *Spills*

The reader should consult Chapter 6, Part V, Section C, pages 265-276 for the proper methods, adsorbent recommendations and directions to control, contain and clean-up spills, and for disposal of pesticide wastes.

### *First Aid*

The reader should consult Chapter 7, Section E, pages 312-314 for emergency first aid information.

# PESTICIDE GUIDE 34

## HEALTH HAZARDS

***Hazard Indicator: Warning***

Warning: ingestion may be fatal.

Caution: contact may cause skin irritation.

Caution: may cause moderate eye irritation.

Caution: absorption through skin may be harmful.

Caution: inhalation of the dry material or mists may cause respiratory irritation.

Fire may produce irritating or poisonous gases.

Runoff from fire or spills may cause pollution.

## SAFETY GUIDELINES

***Restricted Use Pesticide:*** No

***Solubility:*** Slightly soluble to practically insoluble in water

**Toxicity Category:** Class II Oral Hazard, Class III Dermal Hazard

Several formulated products may contain ethylbenzene, xylenes, naphthalene, trimethylbenzenes, and/or emulsifiers.

***Restricted-entry Interval (REI):*** 24 hours

## HANDLING AND STORAGE PRECAUTIONS

***Use Precautions***

Avoid contact with susceptible crops and beneficial insects.

Avoid spray drift.

Select and wear the proper PPE.

Mix ingredients in areas of sufficient ventilation or outdoors to minimize exposure. Avoid spilling material while mixing.

Wash with soap and water immediately after use or exposure. Follow good personal hygiene practices.

Do not reuse spray equipment for any purpose unless thoroughly cleaned with a suitable cleaner.

Reentry—do not enter treated areas without protective clothing until sprays have dried or after 24 hrs.

Notification of a pesticide application must be given to all workers who will be within 1/4 mile of a field during the application or before the REI expires. This notification may be either oral or written, e.g., a posted sign at the

entrances to treated sites. For some products, both oral and written notifications are necessary.

### *Storage and Disposal Precautions*

The reader should consult Chapter 6, Part V, Sections A and B, pages 251-265 for recommendations on the proper storage and disposal precautions.

## PROTECTIVE CLOTHING

Protective eyewear, chemical-resistant gloves and boots, long-sleeved shirt, long pants, and shoes and socks are strongly recommended. When mixing, use a splash apron and, if dust or vapor is a problem, use approved air purifying respirators for pesticides.

## EMERGENCY GUIDELINES

### *Fire or Explosive Hazard*

Evacuate nonessential personnel from fire area.

The products are not considered flammable. They will support combustion and may decompose under fire conditions and give off toxic products.

Some of these materials may react violently with water.

Segregate from oxidizers and incompatible materials.

The use of self-contained breathing apparatus and chemical protective clothing (chemical-resistant gloves, boots, hood, and suit) is recommended.

The use of high-pressure water may cause spread of hazardous materials.

*Small fires:* Dry chemical, $CO_2$, Halon, water spray, or standard foam.

*Large fires:* Water spray, fog, or standard foam is recommended.

Move containers of chemicals from fire area if you can do so without risk.

Cool containers that are exposed to flames with water until fire is out and heat of container is reduced.

Stay upwind of fire.

### *Spills*

The reader should consult Chapter 6, Part V, Section C, pages 265-276 for the proper methods, adsorbent recommendations and directions to control, contain and clean-up spills, and for disposal of pesticide wastes.

### *First Aid*

The reader should consult Chapter 7, Section E, pages 312-314 for emergency first aid information.

# PESTICIDE GUIDE 35

## HEALTH HAZARDS

***Hazard Indicator: Warning***

Warning: ingestion may be fatal.

Warning: may be toxic, inhalation of dry material or mists may cause respiratory irritation.

Caution: may cause moderate eye irritation.

Caution: absorption through skin may be harmful.

Caution: contact may cause moderate skin irritation.

Fire may produce irritating or poisonous gases.

Runoff from fire or spills may cause pollution.

## SAFETY GUIDELINES

***Restricted Use Pesticide:*** No

***Solubility:*** Practically insoluble in water

***Toxicity Category:*** Class II Oral Hazard, Class II Inhalation Hazard, Class III Dermal Hazard

***Restricted-entry Interval (REI):*** 24 hours

## HANDLING AND STORAGE PRECAUTIONS

***Use Precautions***

Avoid contact with susceptible crops and beneficial insects.

Avoid spray drift.

Select and wear the proper PPE.

Mix ingredients in areas of sufficient ventilation or outdoors to minimize exposure. Avoid spilling material while mixing.

Wash with soap and water immediately after use or exposure. Follow good personal hygiene practices.

Do not reuse spray equipment for any purpose unless thoroughly cleaned with a suitable cleaner.

Reentry—do not enter treated areas without protective clothing until sprays have dried or after 24 hrs.

Notification of a pesticide application must be given to all workers who will be within 1/4 mile of a field during the application or before the REI expires. This notification may be either oral or written, e.g., a posted sign at the entrances to treated sites. For some products, both oral and written notifications are necessary.

### *Storage and Disposal Precautions*

The reader should consult Chapter 6, Part V, Sections A and B, pages 251-265 for recommendations on the proper storage and disposal precautions.

## PROTECTIVE CLOTHING

Protective eyewear such as goggles or face shield, chemical-resistant gloves and boots, long-sleeved shirt, long pants, and shoes and socks are strongly recommended. When mixing, use a splash apron and, if dust or vapor is a problem, use approved air purifying respirators for pesticides.

## EMERGENCY GUIDELINES

### *Fire or Explosive Hazard*

Evacuate nonessential personnel from fire area.

The products are not considered flammable. They will support combustion and may decompose under fire conditions and give off toxic products.

Some of these materials may react violently with water.

Segregate from oxidizers and incompatible materials.

The use of self-contained breathing apparatus and chemical protective clothing (chemical-resistant gloves, boots, hood, and suit) is recommended.

The use of high-pressure water may cause spread of hazardous materials.

*Small fires:* Dry chemical, $CO_2$, Halon, water spray, or standard foam.

*Large fires:* Water spray, fog, or standard foam is recommended.

Move containers of chemicals from fire area if you can do so without risk.

Cool containers that are exposed to flames with water until fire is out and heat of container is reduced.

Stay upwind of fire.

### *Spills*

The reader should consult Chapter 6, Part V, Section C, pages 265-276 for the proper methods, adsorbent recommendations and directions to control, contain and clean-up spills, and for disposal of pesticide wastes.

### *First Aid*

The reader should consult Chapter 7, Section E, pages 312-314 for emergency first aid information.

# PESTICIDE GUIDE 36

## HEALTH HAZARDS

***Hazard Indicator:* *Warning***

Warning: absorption through skin may be fatal.

Caution: ingestion may be harmful.

Caution: inhalation of the dry material or mists may cause respiratory irritation.

Caution: skin contact may cause moderate irritation.

Caution: contact may cause moderate eye irritation.

Fire may produce irritating or poisonous gases.

Runoff from fire or spills may cause pollution.

## SAFETY GUIDELINES

***Restricted Use Pesticide:*** No

***Solubility:*** Insoluble in water

***Toxicity Category:*** Class III Oral Hazard, Class II Dermal Hazard

***Restricted-entry Interval (REI):*** 24 hours

Some formulated products may contain petroleum distillates, xylene, ethylbenzene, and other inert ingredients (trade secrets).

## HANDLING AND STORAGE PRECAUTIONS

***Use Precautions***

Avoid contact with susceptible crops and other desirable broadleaf plants.

Avoid spray drift.

Select and wear the proper PPE.

Mix ingredients in areas of sufficient ventilation or outdoors to minimize exposure. Avoid spilling material while mixing.

Wash with soap and water immediately after use or exposure. Follow good personal hygiene practices.

Do not reuse spray equipment for any purpose unless thoroughly cleaned with a suitable cleaner.

Reentry—do not enter treated areas without protective clothing until sprays have dried or after 24 hrs.

Notification of a pesticide application must be given to all workers who will be within 1/4 mile of a field during the application or before the REI expires. This notification may be either oral or written, e.g., a posted sign at the

entrances to treated sites. For some products, both oral and written notifications are necessary.

### *Storage and Disposal Precautions*

The reader should consult Chapter 6, Part V, Sections A and B, pages 251-265 for recommendations on the proper storage and disposal precautions.

## PROTECTIVE CLOTHING

Protective eyewear, chemical-resistant gloves and boots, long-sleeved shirt, long pants, and shoes and socks are strongly recommended. When mixing, use a splash apron and, if dust or vapor is a problem, use approved air purifying respirators for pesticides.

## EMERGENCY GUIDELINES

### *Fire or Explosive Hazard*

Evacuate nonessential personnel from fire area.

The products are not considered flammable. They will support combustion and may decompose under fire conditions and give off toxic products.

Some of these materials may react violently with water.

Segregate from oxidizers and incompatible materials.

The use of self-contained breathing apparatus and chemical protective clothing (chemical-resistant gloves, boots, hood, and suit) is recommended.

The use of high-pressure water may cause spread of hazardous materials.

*Small fires:* Dry chemical, $CO_2$, Halon, water spray, or standard foam.

*Large fires:* Water spray, fog, or standard foam is recommended.

Move containers of chemicals from fire area if you can do so without risk.

Cool containers that are exposed to flames with water until fire is out and heat of container is reduced.

Stay upwind of fire.

### *Spills*

The reader should consult Chapter 6, Part V, Section C, pages 265-276 for the proper methods, adsorbent recommendations and directions to control, contain and clean-up spills, and for disposal of pesticide wastes.

### *First Aid*

The reader should consult Chapter 7, Section E, pages 312-314 for emergency first aid information.

# PESTICIDE GUIDE 37

## HEALTH HAZARDS

***Hazard Indicator: Warning***

Warning: absorption through skin may be fatal.

Warning: moderately toxic, inhalation of dry material or mists may cause respiratory irritation.

Warning: may cause moderate eye irritation.

Warning: contact may cause moderate skin irritation.

Caution: ingestion may be harmful.

Fire may produce irritating or poisonous gases.

Runoff from fire or spills may cause pollution.

## SAFETY GUIDELINES

***Restricted Use Pesticide:*** No

***Solubility:*** Practically insoluble in water

***Toxicity Category:*** Class II Dermal Hazard, Class II Inhalation Hazard, Class III Oral Hazard

***Restricted-entry Interval (REI):*** 24 hours

## HANDLING AND STORAGE PRECAUTIONS

***Use Precautions***

Avoid contact with susceptible crops and other desirable broadleaf plants.

Avoid spray drift.

Select and wear the proper PPE.

Mix ingredients in areas of sufficient ventilation or outdoors to minimize exposure. Avoid spilling material while mixing.

Wash with soap and water immediately after use or exposure. Follow good personal hygiene practices.

Do not reuse spray equipment for any purpose unless thoroughly cleaned with a suitable cleaner.

Reentry—do not enter treated areas without protective clothing until sprays have dried or after 24 hrs.

Notification of a pesticide application must be given to all workers who will be within 1/4 mile of a field during the application or before the REI expires. This notification may be either oral or written, e.g., a posted sign at the entrances to treated sites. For some products, both oral and written notifications are necessary.

### *Storage and Disposal Precautions*

The reader should consult Chapter 6, Part V, Sections A and B, pages 251-265 for recommendations on the proper storage and disposal precautions.

## PROTECTIVE CLOTHING

Protective eyewear such as goggles or face shield is strongly recommended. Chemical-resistant gloves and boots, long-sleeved shirt, long pants, and shoes and socks are also strongly recommended. When handling undiluted product use goggles, impermeable gloves, splash apron, protective clothing, and boots and, if dust or vapor is a problem, use approved air purifying respirators for pesticides.

## EMERGENCY GUIDELINES

### *Fire or Explosive Hazard*

Evacuate nonessential personnel from fire area.

The products are not considered flammable. They will support combustion and may decompose under fire conditions and give off toxic products.

Some of these materials may react violently with water.

Segregate from oxidizers and incompatible materials.

The use of self-contained breathing apparatus and chemical protective clothing (chemical-resistant gloves, boots, hood, and suit) is recommended.

The use of high-pressure water may cause spread of hazardous materials.

*Small fires:* Dry chemical, $CO_2$, Halon, water spray, or alcohol foam.

*Large fires:* Water spray, fog, or alcohol foam is recommended.

Move containers of chemicals from fire area if you can do so without risk.

Cool containers that are exposed to flames with water until fire is out and heat of container is reduced.

Stay upwind of fire.

### *Spills*

The reader should consult Chapter 6, Part V, Section C, pages 265-276 for the proper methods, adsorbent recommendations and directions to control, contain and clean-up spills, and for disposal of pesticide wastes.

### *First Aid*

The reader should consult Chapter 7, Section E, pages 312-314 for emergency first aid information.

# PESTICIDE GUIDE 38

## HEALTH HAZARDS

***Hazard Indicator: Warning***

Warning: product contains paraformaldehyde, a probable human carcinogen.

Caution: absorption through skin may be harmful.

Caution: ingestion may be harmful.

Caution: contact may cause skin irritation.

Caution: can cause eye irritation.

Caution: inhalation of dry material or mists may cause respiratory irritation.

Fire may produce irritating or poisonous gases.

Runoff from fire or spills may cause pollution.

## SAFETY GUIDELINES

***Restricted Use Pesticide:*** No

***Solubility:*** Insoluble in water

***Toxicity Category:*** Class III Oral Hazard, Class III Dermal Hazard

***Restricted-entry Interval (REI):*** 24 hours

## HANDLING AND STORAGE PRECAUTIONS

***Use Precautions***

Avoid contact with susceptible animals and crops.

Avoid spray drift.

Select and wear the proper PPE.

Mix ingredients in areas of sufficient ventilation or outdoors to minimize exposure. Avoid spilling material while mixing.

Wash with soap and water immediately after use or exposure. Follow good personal hygiene practices.

Do not reuse spray equipment for any purpose unless thoroughly cleaned with a suitable cleaner.

Reentry—do not enter treated areas without protective clothing until sprays have dried or after 24 hrs.

Notification of a pesticide application must be given to all workers who will be within 1/4 mile of a field during the application or before the REI expires. This notification may be either oral or written, e.g., a posted sign at the entrances to treated sites. For some products, both oral and written notifications are necessary.

### *Storage and Disposal Precautions*

These pesticides are acutely hazardous, therefore, the reader should consult Chapter 6, Part V, Sections A and B, pages 251-265 for recommendations on the proper storage and disposal precautions.

## PROTECTIVE CLOTHING

Protective eyewear, chemical-resistant gloves and boots, long-sleeved shirt, long pants, and shoes and socks are strongly recommended. When mixing, use a splash apron and, if dust or vapor is a problem, use approved air purifying respirators for pesticides.

## EMERGENCY GUIDELINES

### *Fire or Explosive Hazard*

Evacuate nonessential personnel from fire area.

The products are not considered flammable. They will support combustion and may decompose under fire conditions and give off toxic products.

Some of these materials may react violently with water.

Segregate from oxidizers and incompatible materials.

The use of self-contained breathing apparatus and chemical protective clothing (chemical-resistant gloves, boots, hood, and suit) is recommended.

The use of high-pressure water may cause spread of hazardous materials.

*Small fires:* Dry chemical, $CO_2$, Halon, water spray, or standard foam.

*Large fires:* Water spray, fog, or standard foam is recommended.

Move containers of chemicals from fire area if you can do so without risk.

Cool containers that are exposed to flames with water until fire is out and heat of container is reduced.

Stay upwind of fire.

### *Spills*

The reader should consult Chapter 6, Part V, Section C, pages 265-276 for the proper methods, adsorbent recommendations and directions to control, contain and clean-up spills, and for disposal of pesticide wastes.

### *First Aid*

The reader should consult Chapter 7, Section E, pages 312-314 for emergency first aid information.

# PESTICIDE GUIDE 39

## HEALTH HAZARDS

***Hazard Indicator: Warning***

Warning: highly toxic, inhalation of dry material or mists may cause severe respiratory irritation.

Caution: ingestion may be harmful.

Caution: absorption through skin may be harmful.

Caution: may cause moderate eye irritation.

Caution: contact may cause moderate skin irritation.

Fire may produce irritating or poisonous gases.

Runoff from fire or spills may cause pollution.

## SAFETY GUIDELINES

***Restricted Use Pesticide:*** No

***Solubility:*** Insoluble in water

***Toxicity Category:*** Class II Inhalation Hazard, Class III Oral Hazard, Class III Dermal Hazard

Products that contain chlorophenoxyacetic acid herbicides should be handled with care. Chlorophenoxyacetic acid herbicides are considered by the IARC (International Agency for Research on Cancer) as class 2B carcinogens.

***Restricted-entry Interval (REI):*** 24 hours

## HANDLING AND STORAGE PRECAUTIONS

***Use Precautions***

Avoid contact with susceptible crops and other desirable broadleaf plants.

Avoid spray drift.

Select and wear the proper PPE.

Mix ingredients in areas of sufficient ventilation or outdoors to minimize exposure. Avoid spilling material while mixing.

Wash with soap and water immediately after use or exposure. Follow good personal hygiene practices.

Do not reuse spray equipment for any purpose unless thoroughly cleaned with a suitable cleaner.

Reentry—do not enter treated areas without protective clothing until sprays have dried or after 24 hrs.

Notification of a pesticide application must be given to all workers who will be within 1/4 mile of a field during the application or before the REI expires. This notification may be either oral or written, e.g., a posted sign at the entrances to treated sites. For some products, both oral and written notifications are necessary.

### *Storage and Disposal Precautions*

The reader should consult Chapter 6, Part V, Sections A and B, pages 251-265 for recommendations on the proper storage and disposal precautions.

## PROTECTIVE CLOTHING

Protective eyewear such as goggles or face shield, chemical-resistant gloves and boots, long-sleeved shirt, long pants, and shoes and socks are strongly recommended. When mixing, use a splash apron and, if dust or vapor is a problem, use approved air purifying respirators for pesticides.

## EMERGENCY GUIDELINES

### *Fire or Explosive Hazard*

Evacuate nonessential personnel from fire area.

The products are not considered flammable. They will support combustion and may decompose under fire conditions and give off toxic products.

Some of these materials may react violently with water.

Segregate from oxidizers and incompatible materials.

The use of self-contained breathing apparatus and chemical protective clothing (chemical-resistant gloves, boots, hood, and suit) is recommended.

The use of high pressure water may cause spread of hazardous materials.

*Small fires:* Dry chemical, $CO_2$, Halon, water spray, or standard foam.

*Large fires:* Water spray, fog, or alcohol foam is recommended.

Move containers of chemicals from fire area if you can do so without risk.

Cool containers that are exposed to flames with water until fire is out and heat of container is reduced.

Stay upwind of fire.

### *Spills*

The reader should consult Chapter 6, Part V, Section C, pages 265-276 for the proper methods, adsorbent recommendations and directions to control, contain and clean-up spills, and for disposal of pesticide wastes.

### *First Aid*

The reader should consult Chapter 7, Section E, pages 312-314 for emergency first aid information.

# PESTICIDE GUIDE 40

## HEALTH HAZARDS

***Hazard Indicator: Warning***

Warning: contact may cause substantial but temporary eye damage.

Warning: ingestion may be fatal.

Caution: inhalation of the dry material or mists may cause respiratory irritation.

Caution: absorption through skin may be harmful.

Caution: contact may cause moderate skin irritation.

Fire may produce irritating or poisonous gases.

Runoff from fire or spills may cause pollution.

## SAFETY GUIDELINES

***Restricted Use Pesticide:*** No

***Solubility:*** Practically insoluble in water

***Toxicity Category:*** Class II Eye Hazard, Class II Oral Hazard, Class III Dermal Hazard

Formulations may contain petroleum distillates.

***Restricted-entry Interval (REI):*** 24 hours

## HANDLING AND STORAGE PRECAUTIONS

***Use Precautions***

Avoid contact with susceptible crops, other desirable broadleaf plants, and beneficial insects.

Avoid spray drift.

Select and wear the proper PPE.

Mix ingredients in areas of sufficient ventilation or outdoors to minimize exposure. Avoid spilling material while mixing.

Wash with soap and water immediately after use or exposure. Follow good personal hygiene practices.

Do not reuse spray equipment for any purpose unless thoroughly cleaned with a suitable cleaner.

Reentry—do not enter treated areas without protective clothing until sprays have dried or after 24 hrs.

Notification of a pesticide application must be given to all workers who will be within 1/4 mile of a field during the application or before the REI expires.

This notification may be either oral or written, e.g., a posted sign at the entrances to treated sites. For some products, both oral and written notifications are necessary.

### *Storage and Disposal Precautions*

The reader should consult Chapter 6, Part V, Sections A and B, pages 251-265 for recommendations on the proper storage and disposal precautions.

## PROTECTIVE CLOTHING

Protective eyewear such as goggles or face shield is strongly recommended. Chemical-resistant gloves and boots, long-sleeved shirt, long pants, and shoes and socks are also strongly recommended. When handling undiluted product use face shield or goggles, impermeable gloves, splash apron, protective clothing and boots and, if dust or vapor is a problem, use approved air purifying respirators for pesticides.

## EMERGENCY GUIDELINES

### *Fire or Explosive Hazard*

Evacuate nonessential personnel from fire area.

The product is not considered flammable, but will both support combustion and may decompose under fire conditions and give off toxic products.

Some of these materials may react violently with water.

Segregate from oxidizers and incompatible materials.

The use of self-contained breathing apparatus and chemical protective clothing (chemical-resistant gloves, boots, hood, and suit) is recommended.

The use of high-pressure water may cause spread of hazardous materials.

*Small fires:* Dry chemical, $CO_2$, Halon, water spray, or standard foam.

*Large fires:* Water spray or fog is recommended.

Move containers of chemicals from fire area if you can do so without risk.

Cool containers that are exposed to flames with water until fire is out and heat of container is reduced.

Stay upwind of fire.

### *Spills*

The reader should consult Chapter 6, Part V, Section C, pages 265-276 for the proper methods, adsorbent recommendations and directions to control, contain and clean-up spills, and for disposal of pesticide wastes.

### *First Aid*

The reader should consult Chapter 7, Section E, pages 312-314 for emergency first aid information.

# PESTICIDE GUIDE 41

## HEALTH HAZARDS

***Hazard Indicator: Warning***

Warning: absorption through skin may be fatal.

Warning: can cause moderate eye irritation and eye damage.

Caution: ingestion may be harmful.

Caution: contact will irritate skin.

Caution: inhalation of dry material or mists may cause respiratory irritation.

Fire may produce irritating or poisonous gases.

Runoff from fire or spills may cause pollution.

## SAFETY GUIDELINES

***Restricted Use Pesticide:*** No

***Solubility:*** Practically insoluble in water

***Toxicity Caterory:*** Class II Eye Hazard, Class II Dermal Hazard, Class III Oral Hazard

***Restricted-entry Interval (REI):*** 24 hours

## HANDLING AND STORAGE PRECAUTIONS

***Use Precautions***

Avoid contact with susceptible crops and other desirable broadleaf plants.

Avoid spray drift.

Select and wear the proper PPE.

Mix ingredients in areas of sufficient ventilation or outdoors to minimize exposure. Avoid spilling material while mixing.

Wash with soap and water immediately after use or exposure. Follow good personal hygiene practices.

Do not reuse spray equipment for any purpose unless thoroughly cleaned with a suitable cleaner.

Reentry—do not enter treated areas without protective clothing until sprays have dried or after 48 hrs.

Notification of a pesticide application must be given to all workers who will be within 1/4 mile of a field during the application or before the REI expires. This notification may be either oral or written, e.g., a posted sign at the entrances to treated sites. For some products, both oral and written notifications are necessary.

### *Storage and Disposal Precautions*

The reader should consult Chapter 6, Part V, Sections A and B, pages 251-265 for recommendations on the proper storage and disposal precautions.

## PROTECTIVE CLOTHING

Protective eyewear, chemical-resistant gloves and boots, long-sleeved shirt, long pants, and shoes and socks are strongly recommended. When mixing, use a splash apron and, if dust or vapor is a problem, use approved air purifying respirators for pesticides.

## EMERGENCY GUIDELINES

### *Fire or Explosive Hazard*

Evacuate nonessential personnel from fire area.

The products are not considered flammable. They will support combustion and may decompose under fire conditions and give off toxic products.

Some of these materials may react violently with water.

Segregate from oxidizers and incompatible materials.

The use of self-contained breathing apparatus and chemical protective clothing (chemical-resistant gloves, boots, hood, and suit) is recommended.

The use of high-pressure water may cause spread of hazardous materials.

*Small fires:* Dry chemical, $CO_2$, Halon, water spray, or standard foam.

*Large fires:* Water spray, fog, or standard foam is recommended.

Move containers of chemicals from fire area if you can do so without risk.

Cool containers that are exposed to flames with water until fire is out and heat of container is reduced.

Stay upwind of fire.

### *Spills*

The reader should consult Chapter 6, Part V, Section C, pages 265-276 for the proper methods, adsorbent recommendations and directions to control, contain and clean-up spills, and for disposal of pesticide wastes.

### *First Aid*

The reader should consult Chapter 7, Section E, pages 312-314 for emergency first aid information.

# PESTICIDE GUIDE 42

## HEALTH HAZARDS

***Hazard Indicator: Warning***

Warning: can cause moderate eye irritation and eye damage.

Warning: corrosive, contact may cause severe skin irritation and scarring.

Warning: slightly toxic, inhalation of dry material or mists may cause respiratory irritation.

Caution: ingestion may be harmful.

Caution: absorption through skin may be harmful.

Fire may produce irritating or poisonous gases.

Runoff from fire or spills may cause pollution.

## SAFETY GUIDELINES

***Restricted Use Pesticide:*** No

***Solubility:*** Very slightly soluble in water

***Toxicity Category:*** Class II Eye Hazard, Class II Dermal Hazard, Class II Inhalation Hazard, Class III Oral Hazard

***Restricted-entry Interval (REI):*** 24 hours

## HANDLING AND STORAGE PRECAUTIONS

***Use Precautions***

Avoid contact with susceptible crops and other desirable broadleaf plants.

Avoid spray drift.

Select and wear the proper PPE.

Mix ingredients in areas of sufficient ventilation or outdoors to minimize exposure. Avoid spilling material while mixing.

Wash with soap and water immediately after use or exposure. Follow good personal hygiene practices.

Do not reuse spray equipment for any purpose unless thoroughly cleaned with a suitable cleaner.

Reentry—do not enter treated areas without protective clothing until sprays have dried or after 24 hrs.

Notification of a pesticide application must be given to all workers who will be within 1/4 mile of a field during the application or before the REI expires. This notification may be either oral or written, e.g., a posted sign at the entrances to treated sites. For some products, both oral and written notifications are necessary.

### *Storage and Disposal Precautions*

The reader should consult Chapter 6, Part V, Sections A and B, pages 251-265 for recommendations on the proper storage and disposal precautions.

## PROTECTIVE CLOTHING

Protective eyewear such as goggles or face shield is strongly recommended. Chemical-resistant gloves and boots, long-sleeved shirt, long pants, and shoes and socks are strongly recommended. When handling undiluted product use goggles, impermeable gloves, splash apron, protective clothing and boots and, if dust or vapor is a problem, use approved air purifying respirators for pesticides.

## EMERGENCY GUIDELINES

### *Fire or Explosive Hazard*

Evacuate nonessential personnel from fire area.

The products are not considered flammable. They will support combustion and may decompose under fire conditions and give off toxic products.

Some of these materials may react violently with water.

Segregate from oxidizers and incompatible materials.

The use of self-contained breathing apparatus and chemical protective clothing (chemical-resistant gloves, boots, hood, and suit) is recommended.

The use of high-pressure water may cause spread of hazardous materials.

*Small fires:* Dry chemical, $CO_2$, Halon, water spray, or alcohol foam.

*Large fires:* Water spray, fog, or alcohol foam is recommended.

Move containers of chemicals from fire area if you can do so without risk.

Cool containers that are exposed to flames with water until fire is out and heat of container is reduced.

Stay upwind of fire.

### *Spills*

The reader should consult Chapter 6, Part V, Section C, pages 265-276 for the proper methods, adsorbent recommendations and directions to control, contain and clean-up spills, and for disposal of pesticide wastes.

### *First Aid*

The reader should consult Chapter 7, Section E, pages 312-314 for emergency first aid information.

# PESTICIDE GUIDE 43

## HEALTH HAZARDS

***Hazard Indicator: Warning***

Warning: may cause severe eye irritation and irreversible eye damage.

Warning: contact may cause severe skin irritation.

Caution: ingestion may be harmful.

Caution: inhalation of the dry material or mists may cause respiratory irritation.

Caution: absorption through skin may be harmful.

Fire may produce irritating or poisonous gases.

Runoff from fire or spills may cause pollution.

## SAFETY GUIDELINES

***Restricted Use Pesticide:*** No

***Solubility:*** Practically insoluble in water

***Toxicity Category:*** Class II Eye Hazard, Class II Dermal Hazard, Class III Oral Hazard, Class III Inhalation Hazard

Formulations may contain xylene, isobutanol, and alkyl phenol condensate.

***Restricted-entry Interval (REI):*** 24 hours

## HANDLING AND STORAGE PRECAUTIONS

***Use Precautions***

Avoid contact with susceptible crops and beneficial insects.

Avoid spray drift.

Select and wear the proper PPE.

Mix ingredients in areas of sufficient ventilation or outdoors to minimize exposure. Avoid spilling material while mixing.

Wash with soap and water immediately after use or exposure. Follow good personal hygiene practices.

Do not reuse spray equipment for any purpose unless thoroughly cleaned with a suitable cleaner.

Reentry—do not enter treated areas without protective clothing until sprays have dried or after 24 hrs.

Notification of a pesticide application must be given to all workers who will be within 1/4 mile of a field during the application or before the REI expires. This notification may be either oral or written, e.g., a posted sign at the en-

trances to treated sites. For some products, both oral and written notifications are necessary.

### *Storage and Disposal Precautions*

The reader should consult Chapter 6, Part V, Sections A and B, pages 251-265 for recommendations on the proper storage and disposal precautions.

## PROTECTIVE CLOTHING

Protective eyewear such as goggles or face shield is strongly recommended. Chemical-resistant gloves and boots, long-sleeved shirt, long pants, and shoes and socks are strongly recommended. When mixing, use a splash apron and, if dust or vapor is a problem, use approved air purifying respirators for pesticides.

## EMERGENCY GUIDELINES

### *Fire or Explosive Hazard*

Evacuate nonessential personnel from fire area.

The products are considered not flammable. They will support combustion and may decompose under fire conditions and give off toxic products.

Some of these materials may react violently with water.

Segregate from oxidizers and incompatible materials.

The use of self-contained breathing apparatus and chemical protective clothing (chemical-resistant gloves, boots, hood, and suit) is recommended.

The use of high-pressure water may cause spread of hazardous materials.

*Small fires:* Dry chemical, $CO_2$, Halon, water spray, or alcohol foam.

*Large fires:* Water spray, fog, or standard or alcohol foam is recommended.

Move containers of chemicals from fire area if you can do so without risk.

Cool containers that are exposed to flames with water until fire is out and heat of container is reduced.

Stay upwind of fire.

### *Spills*

The reader should consult Chapter 6, Part V, Section C, pages 265-276 for the proper methods, adsorbent recommendations and directions to control, contain and clean-up spills, and for disposal of pesticide wastes.

### *First Aid*

The reader should consult Chapter 7, Section E, pages 312-314 for emergency first aid information.

# PESTICIDE GUIDE 44

## HEALTH HAZARDS

***Hazard Indicator: Warning***

Warning: may cause severe eye irritation and tempory eye damage.

Caution: contact may cause moderate skin irritation.

Caution: ingestion may be harmful.

Caution: absorption through skin may be harmful.

Caution: inhalation of the dry material or mists may cause respiratory irritation.

Fire may produce irritating or poisonous gases.

Runoff from fire or spills may cause pollution.

## SAFETY GUIDELINES

***Restricted Use Pesticide:*** No

***Solubility:*** Insoluble in water

***Toxicity Category:*** Class II Eye Hazard, Class III Oral Hazard, Class III Dermal Hazard

Formulations may contain one or more of the following: 1-butanol, cyclohexanone, 2,6-dichloro-2-trichloromethyl pyridine, N,N dimethylformamide, ethylbenzene, ethylene dichloride, hydrotreated heavy naphtha, isobutanol, isophorone (isophorone has shown some level of carcinogincity in rodents), kerosene, methyl isobutyl ketone, monochlorobenzene, naphtha, naphthalene, petroleum distillates, petroleum oil, propyl glycol monomethyl ether, surfactants, trimethylbenzene, and/or xylene.

***Restricted-entry Interval (REI):*** 24 hours

## HANDLING AND STORAGE PRECAUTIONS

### *Use Precautions*

Avoid contact with susceptible crops, other desirable broadleaf plants, beneficial insects and/or susceptible animals.

Avoid spray drift.

Select and wear the proper PPE.

Mix ingredients in areas of sufficient ventilation or outdoors to minimize exposure. Avoid spilling material while mixing.

Wash with soap and water immediately after use or exposure. Follow good personal hygiene practices.

Do not reuse spray equipment for any purpose unless thoroughly cleaned with a suitable cleaner.

Reentry—do not enter treated areas without protective clothing until sprays have dried or after 24 hrs.

Notification of a pesticide application must be given to all workers who will be within 1/4 mile of a field during the application or before the REI expires. This notification may be either oral or written, e.g., a posted sign at the entrances to treated sites. For some products, both oral and written notifications are necessary.

### *Storage and Disposal Precautions*

The reader should consult Chapter 6, Part V, Sections A and B, pages 251-265 for recommendations on the proper storage and disposal precautions.

## PROTECTIVE CLOTHING

Protective eyewear such as goggles or face shield is strongly recommended; chemical-resistant gloves and boots, long-sleeved shirt, long pants, and shoes and socks are strongly recommended. When handling undiluted product use goggles, impermeable gloves, splash apron, protective clothing and boots and, if dust or vapor is a problem, use approved air purifying respirators for pesticides.

## EMERGENCY GUIDELINES

### *Fire or Explosive Hazard*

Evacuate nonessential personnel from fire area.

Most products are not considered flammable. They will support combustion and may decompose under fire conditions and give off toxic products. However, several products are flammable. For information on product flammability the reader should consult Table 1 Pesticides Listed Alphabetically by Trade Name under formulations (see footnotes). The reader should review Chapter 6, Section V, for information on storage, disposal and handling of flammable products.

Some of these materials may react violently with water.

Segregate from oxidizers and incompatible materials.

The use of self-contained breathing apparatus and chemical protective clothing (chemical-resistant gloves, boots, hood, and suit) is recommended.

The use of high-pressure water may cause spread of hazardous materials.

*Small fires:* Dry chemical, $CO_2$, Halon, water spray, or standard foam.

*Large fires:* Water spray, fog, or foam (read label for information on foam selection—standard or alcohol foam) is recommended.

Move containers of chemicals from fire area if you can do so without risk.

Cool containers that are exposed to flames with water until fire is out and heat of container is reduced.

Stay upwind of fire.

### *Spills*

The reader should consult Chapter 6, Part V, Section C, pages 265-276 for the proper methods, adsorbent recommendations and directions to control, contain and clean-up spills, and for disposal of pesticide wastes.

### *First Aid*

The reader should consult Chapter 7, Section E, pages 312-314 for emergency first aid information.

# PESTICIDE GUIDE 45

## HEALTH HAZARDS

***Hazard Indicator: Warning***

Warning: contact may cause severe but temporary eye injury.

Warning: inhalation of the dry material or mists can cause respiratory irritation and may be fatal.

Caution: ingestion may be harmful.

Caution: absorption through skin may be harmful.

Fire may produce irritating or poisonous gases.

Runoff from fire or spills may cause pollution.

## SAFETY GUIDELINES

***Restricted Use Pesticide:*** No

***Solubility:*** Practically insoluble in water

***Toxicity Category***: Class II Inhalation Hazard, Class II Eye Hazard, Class III Oral Hazard, Class III Dermal Hazard

***Restricted-entry Interval (REI):*** 24 hours

## HANDLING AND STORAGE PRECAUTIONS

***Use Precautions***

Avoid contact with susceptible crops and beneficial insects.

Avoid spray drift.

Select and wear the proper PPE.

Mix ingredients in areas of sufficient ventilation or outdoors to minimize exposure. Avoid spilling material while mixing.

Wash with soap and water immediately after use or exposure. Follow good personal hygiene practices.

Do not reuse spray equipment for any purpose unless thoroughly cleaned with a suitable cleaner.

Reentry—do not enter treated areas without protective clothing until sprays have dried or after 12 hrs.

Notification of a pesticide application must be given to all workers who will be within 1/4 mile of a field during the application or before the REI expires. This notification may be either oral or written, e.g., a posted sign at the entrances to treated sites. For some products, both oral and written notifications are necessary.

### *Storage and Disposal Precautions*

The reader should consult Chapter 6, Part V, Sections A and B, pages 251-265 for recommendations on the proper storage and disposal precautions.

## PROTECTIVE CLOTHING

Protective eyewear, chemical-resistant gloves and boots, long-sleeved shirt, long pants, and shoes and socks are strongly recommended. When mixing, use a splash apron and, if dust or vapor is a problem, use approved air purifying respirators for pesticides.

## EMERGENCY GUIDELINES

### *Fire or Explosive Hazard*

Evacuate nonessential personnel from fire area.

The products are not considered flammable. They will support combustion and may decompose under fire conditions and give off toxic products.

Some of these materials may react violently with water.

Segregate from oxidizers and incompatible materials.

The use of self-contained breathing apparatus and chemical protective clothing (chemical-resistant gloves, boots, hood, and suit) is recommended.

The use of high-pressure water may cause spread of hazardous materials.

*Small fires:* Dry chemical, $CO_2$, Halon, water spray, or standard foam.

*Large fires:* Water spray, fog, or standard foam is recommended.

Move containers of chemicals from fire area if you can do so without risk.

Cool containers that are exposed to flames with water until fire is out and heat of container is reduced.

Stay upwind of fire.

### *Spills*

The reader should consult Chapter 6, Part V, Section C, pages 265-276 for the proper methods, adsorbent recommendations and directions to control, contain and clean-up spills, and for disposal of pesticide wastes.

### *First Aid*

The reader should consult Chapter 7, Section E, pages 312-314 for emergency first aid information.

# PESTICIDE GUIDE 46

## HEALTH HAZARDS

***Hazard Indicator: Warning***

Warning: may cause substantial but temporary eye damage.

Caution: contact may cause moderate skin irritation and sensitization.

Caution: ingestion may be harmful.

Caution: inhalation of the dry material or mists may cause respiratory irritation.

Caution: absorption through skin may be harmful.

Fire may produce irritating or poisonous gases.

Runoff from fire or spills may cause pollution.

## SAFETY GUIDELINES

***Restricted Use Pesticide:*** No

***Solubility:*** Practically insoluble in water

***Toxicity Category:*** Class II Eye Hazard, Class III Oral Hazard, Class III Dermal Hazard, Class III Inhalation Hazard

Some of these products are considered flammable; the formulations may contain aromatic hydrocarbons, naphtha, glycol, naphthalene, petroleum solvent.

***Restricted-entry Interval (REI):*** 24 hours

## HANDLING AND STORAGE PRECAUTIONS

### *Use Precautions*

Avoid contact with susceptible crops, desirable broadleaf plants, beneficial insects, and/or susceptible animals.

Avoid spray drift.

Select and wear the proper PPE.

Mix ingredients in areas of sufficient ventilation or outdoors to minimize exposure. Avoid spilling material while mixing.

Wash with soap and water immediately after use or exposure. Follow good personal hygiene practices.

Do not reuse spray equipment for any purpose unless thoroughly cleaned with a suitable cleaner.

Reentry—do not enter treated areas without protective clothing until sprays have dried or after 24 hrs.

Notification of a pesticide application must be given to all workers who will be within 1/4 mile of a field during the application or before the REI expires. This notification may be either oral or written, e.g., a posted sign at the entrances to treated sites. For some products, both oral and written notifications are necessary.

### *Storage and Disposal Precautions*

The reader should consult Chapter 6, Part V, Sections A and B, pages 251-265 for recommendations on the proper storage and disposal precautions.

## PROTECTIVE CLOTHING

Protective eyewear such as goggles or face shield is strongly recommended. Chemical-resistant gloves and boots, long-sleeved shirt, long pants, and shoes and socks are strongly recommended. When handling undiluted product use goggles, impermeable gloves, splash apron, protective clothing and boots and, if dust or vapor is a problem, use approved air purifying respirators for pesticides.

## EMERGENCY GUIDELINES

### *Fire or Explosive Hazard*

Evacuate nonessential personnel from fire area.

Most products are not considered flammable. They will support combustion and may decompose under fire conditions and give off toxic products. However, several products are flammable. For information on product flammability the reader should consult Table 1 Pesticides Listed Alphabetically by Trade Name under formulations (see footnotes). The reader should review Chapter 6, Section V, for additional information on storage, disposal, and handling of flammable products.

Some of these materials may react violently with water.

Segregate from oxidizers and incompatible materials.

The use of self-contained breathing apparatus and chemical protective clothing (chemical-resistant gloves, boots, hood, and suit) is recommended.

The use of high-pressure water may cause spread of hazardous materials.

*Small fires:* Dry chemical, $CO_2$, Halon, water spray, or alcohol foam.

*Large fires:* Water spray, fog, or alcohol foam is recommended.

Move containers of chemicals from fire area if you can do so without risk.

Cool containers that are exposed to flames with water until fire is out and heat of container is reduced.

Stay upwind of fire.

### *Spills*

The reader should consult Chapter 6, Part V, Section C, pages 265-276 for the proper methods, adsorbent recommendations and directions to control, contain and clean-up spills, and for disposal of pesticide wastes.

### *First Aid*

The reader should consult Chapter 7, Section E, pages 312-314 for emergency first aid information.

# PESTICIDE GUIDE 47

## HEALTH HAZARDS

***Hazard Indicator: Warning***

Warning: avoid eye contact; can cause moderate irritation with slight corneal injury.

Caution: ingestion may be harmful.

Caution: may be absorbed through skin with prolonged contact.

Caution: prolonged or repeated contact may irritate skin.

Caution: inhalation of dry material or mists may cause respiratory irritation.

Fire may produce irritating or poisonous gases.

Runoff from fire or spills may cause pollution.

## SAFETY GUIDELINES

***Restricted Use Pesticide:*** No

***Solubility:*** Miscible in water

***Toxicity Category:*** Class II Eye Hazard, Class III or IV Oral Hazard, Class III or IV Dermal Hazard

Several formulations may contain monochlorotoluene.

***Restricted-entry Interval (REI):*** 24 hours

## HANDLING AND STORAGE PRECAUTIONS

***Use Precautions***

Avoid contact with susceptible crops and other desirable broadleaf plants.

Avoid spray drift.

Select and wear the proper PPE.

Mix ingredients in areas of sufficient ventilation or outdoors to minimize exposure. Avoid spilling material while mixing.

Wash with soap and water immediately after use or exposure. Follow good personal hygiene practices.

Do not reuse spray equipment for any purpose unless thoroughly cleaned with a suitable cleaner.

Reentry—do not enter treated areas without protective clothing until sprays have dried or after 48 hrs.

Notification of a pesticide application must be given to all workers who will be within 1/4 mile of a field during the application or before the REI expires. This notification may be either oral or written, e.g., a posted sign at the

entrances to treated sites. For some products, both oral and written notifications are necessary.

### *Storage and Disposal Precautions*

The reader should consult Chapter 6, Part V, Sections A and B, pages 251-265 for recommendations on the proper storage and disposal precautions.

## PROTECTIVE CLOTHING

Protective eyewear such as, goggles or face shield, chemical-resistant gloves and boots, long-sleeved shirt, long pants, and shoes and socks are strongly recommended. When mixing, use a splash apron and, if dust or vapor is a problem, use approved air purifying respirators for pesticides.

## EMERGENCY GUIDELINES

### *Fire or Explosive Hazard*

Evacuate nonessential personnel from fire area.

The products are not considered flammable. They will support combustion and may decompose under fire conditions and give off toxic products.

Some of these materials may react violently with water.

Segregate from oxidizers and incompatible materials.

The use of self-contained breathing apparatus and chemical protective clothing (chemical-resistant gloves, boots, hood, and suit) is recommended.

The use of high-pressure water may cause spread of hazardous materials.

*Small fires:* Dry chemical, $CO_2$, Halon, water spray, or standard foam.

*Large fires:* Water spray, fog, or standard foam is recommended.

Move containers of chemicals from fire area if you can do so without risk.

Cool containers that are exposed to flames with water until fire is out and heat of container is reduced.

Stay upwind of fire

### *Spills*

The reader should consult Chapter 6, Part V, Section C, pages 265-276 for the proper methods, adsorbent recommendations and directions to control, contain and clean-up spills, and for disposal of pesticide wastes.

### *First Aid*

The reader should consult Chapter 7, Section E, pages 312-314 for emergency first aid information.

# PESTICIDE GUIDE 48

## HEALTH HAZARDS

***Hazard Indicator: Warning***

Warning: contact may cause severe eye irritation and temporary damage.

Warning: moderately toxic, inhalation of the dry material or mists may cause respiratory irritation.

Caution: ingestion may be harmful.

Caution: absorption through skin may be harmful.

Caution: contact may cause skin irritation.

Fire may produce irritating or poisonous gases.

Runoff from fire or spills may cause pollution.

## SAFETY GUIDELINES

***Restricted Use Pesticide:*** No

***Solubility:*** Soluble in water

***Toxicity Category:*** Class II Eye Hazard, Class II Inhalation Hazard, Class III Dermal Hazard, Class IV Oral Hazard

***Restricted-entry Interval (REI):*** 24 hours

## HANDLING AND STORAGE PRECAUTIONS

***Use Precautions***

Avoid contact with susceptible crops and other desirable broadleaf plants.

Avoid spray drift.

Select and wear the proper PPE.

Mix ingredients in areas of sufficient ventilation or outdoors to minimize exposure. Avoid spilling material while mixing.

Wash with soap and water immediately after use or exposure. Follow good personal hygiene practices.

Do not reuse spray equipment for any purpose unless thoroughly cleaned with a suitable cleaner.

Reentry—do not enter treated areas without protective clothing until sprays have dried or after 24 hrs.

Notification of a pesticide application must be given to all workers who will be within 1/4 mile of a field during the application or before the REI expires. This notification may be either oral or written, e.g., a posted sign at the entrances to treated sites. For some products, both oral and written notifications are necessary.

### *Storage and Disposal Precautions*

The reader should consult Chapter 6, Part V, Sections A and B, pages 251-265 for recommendations on the proper storage and disposal precautions.

## PROTECTIVE CLOTHING

Protective eyewear such as goggles and face shield is strongly recommended. Chemical-resistant gloves and boots, long-sleeved shirt, long pants, and shoes and socks are also strongly recommended. When mixing, use a splash apron and, if dust or vapor is a problem, use approved air purifying respirators for pesticides.

## EMERGENCY GUIDELINES

### *Fire or Explosive Hazard*

Evacuate nonessential personnel from fire area.

The products are not considered flammable. They will support combustion and may decompose under fire conditions and give off toxic products.

Some of these materials may react violently with water.

Segregate from oxidizers and incompatible materials.

The use of self-contained breathing apparatus and chemical protective clothing (chemical-resistant gloves, boots, hood, and suit) is recommended.

The use of high-pressure water may cause spread of hazardous materials.

*Small fires:* Dry chemical, $CO_2$, Halon, water spray, or standard foam.

*Large fires:* Water spray, fog, or standard foam is recommended.

Move containers of chemicals from fire area if you can do so without risk.

Cool containers that are exposed to flames with water until fire is out and heat of container is reduced.

Stay upwind of fire.

### *Spills*

The reader should consult Chapter 6, Part V, Section C, pages 265-276 for the proper methods, adsorbent recommendations and directions to control, contain and clean-up spills, and for disposal of pesticide wastes.

### *First Aid*

The reader should consult Chapter 7, Section E, pages 312-314 for emergency first aid information.

# PESTICIDE GUIDE 49

## HEALTH HAZARDS

***Hazard Indicator: Warning***

Warning: ingestion may be fatal.

Warning: absorption through skin may be harmful.

Caution: may cause eye irritation.

Caution: contact may cause skin irritation.

Caution: inhalation of the dry material or mists may cause severe respiratory irritation.

Fire may produce irritating or poisonous gases.

Runoff from fire or spills may cause pollution.

## SAFETY GUIDELINES

***Restricted Use Pesticide:*** Yes

***Solubility:*** Very slightly soluble in water

***Toxicity Category:*** Class II Oral Hazard, Class II Dermal Hazard

***Restricted-entry Interval (REI):*** 24 hours

## HANDLING AND STORAGE PRECAUTIONS

***Use Precautions***

Avoid contact with susceptible crops and beneficial insects.

Avoid spray drift.

Select and wear the proper PPE.

Mix ingredients in areas of sufficient ventilation or outdoors to minimize exposure. Avoid spilling material while mixing.

Wash with soap and water immediately after use or exposure. Follow good personal hygiene practices.

Do not reuse spray equipment for any purpose unless thoroughly cleaned with a suitable cleaner.

Reentry—do not enter treated areas without protective clothing until sprays have dried or after 24 hrs.

Notification of a pesticide application must be given to all workers who will be within 1/4 mile of a field during the application or before the REI expires. This notification may be either oral or written, e.g., a posted sign at the entrances to treated sites. For some products, both oral and written notifications are necessary.

### *Storage and Disposal Precautions*

The reader should consult Chapter 6, Part V, Sections A and B, pages 251-265 for recommendations on the proper storage and disposal precautions.

## PROTECTIVE CLOTHING

Protective eyewear such as goggles or face shield, chemical-resistant gloves and boots, long-sleeved shirt, long pants, and shoes and socks are strongly recommended. When mixing, use a splash apron and, if dust or vapor is a problem, use approved air purifying respirators for pesticides.

## EMERGENCY GUIDELINES

### *Fire or Explosive Hazard*

Evacuate nonessential personnel from fire area.

The products are not considered flammable. They will support combustion and may decompose under fire conditions and give off toxic products.

Some of these materials may react violently with water.

Segregate from oxidizers and incompatible materials.

The use of self-contained breathing apparatus and chemical protective clothing (chemical-resistant gloves, boots, hood, and suit) is recommended.

The use of high-pressure water may cause spread of hazardous materials.

*Small fires:* Dry chemical, $CO_2$, Halon, water spray, or standard foam.

*Large fires:* Water spray, fog, or alcohol foam is recommended.

Move containers of chemicals from fire area if you can do so without risk.

Cool containers that are exposed to flames with water until fire is out and heat of container is reduced.

Stay upwind of fire.

### *Spills*

The reader should consult Chapter 6, Part V, Section C, pages 265-276 for the proper methods, adsorbent recommendations and directions to control, contain and clean-up spills, and for disposal of pesticide wastes.

### *First Aid*

The reader should consult Chapter 7, Section E, pages 312-314 for emergency first aid information.

# PESTICIDE GUIDE 50

## HEALTH HAZARDS

***Hazard Indicator: Warning***

Warning: ingestion may be fatal.

Warning: toxic, inhalation of the dry material or mists may cause severe respiratory irritation.

Warning: absorption through skin may be fatal.

Fire may produce irritating or poisonous gases.

Runoff from fire or spills may cause pollution.

## SAFETY GUIDELINES

***Restricted Use Pesticide:*** Yes

***Solubility:*** Soluble to insoluble in water

***Toxicity Category:*** Class II Oral Hazard, Class II Inhalation Hazard, Class II Dermal Hazard

Several products are considered flammable; the formulations may contain carbon tetrachloride, 2,2-dichloroethyl dimethyl ester, ethylbenzene, methylbromide, methyl isobutyl ketone, petroleum distillates, phosphoric acid, trimethybenzenes, xylene, and other inert ingredients (trade secrets).

***Restricted-entry Interval (REI):*** 24 hours

## HANDLING AND STORAGE PRECAUTIONS

***Use Precautions***

Avoid contact with susceptible crops, other desirable broadleaf plants, and beneficial insects.

Avoid spray drift.

Select and wear the proper PPE.

Mix ingredients in areas of sufficient ventilation or outdoors to minimize exposure. Avoid spilling material while mixing.

Wash with soap and water immediately after use or exposure. Follow good personal hygiene practices.

Do not reuse spray equipment for any purpose unless thoroughly cleaned with a suitable cleaner.

Reentry—do not enter treated areas without protective clothing until sprays have dried or after 24 hrs.

Notification of a pesticide application must be given to all workers who will be within 1/4 mile of a field during the application or before the REI expires. This notification may be either oral or written, e.g., a posted sign at the

entrances to treated sites. For some products, both oral and written notifications are necessary.

### *Storage and Disposal Precautions*

The reader should consult Chapter 6, Part V, Sections A and B, pages 251-265 for recommendations on the proper storage and disposal precautions.

## PROTECTIVE CLOTHING

Protective eyewear such as goggles or face shield is strongly recommended. Chemical-resistant gloves and boots, long-sleeved shirt, long pants, and shoes and socks are strongly recommended. When handling undiluted product use goggles, impermeable gloves, splash apron, protective clothing and boots and, if dust or vapor is a problem, use approved air purifying respirators for pesticides.

## EMERGENCY GUIDELINES

### *Fire or Explosive Hazard*

Evacuate nonessential personnel from fire area.

Most products are not considered flammable. They will support combustion and may decompose under fire conditions and give off toxic products. However, several products are flammable. For information on product flammability the reader should consult Table 1 Pesticides Listed Alphabetically by Trade Name under formulations (see footnotes). The reader should review Chapter 6, Section V, for information on storage, disposal, and handling of flammable products.

Some of these materials may react violently with water.

Segregate from oxidizers and incompatible materials.

The use of self-contained breathing apparatus and chemical protective clothing (chemical-resistant gloves, boots, hood, and suit) is recommended.

The use of high-pressure water may cause spread of hazardous materials.

*Small fires:* Dry chemical, $CO_2$, Halon, water spray, or standard foam.

*Large fires:* Water spray, fog, or alcohol foam is recommended.

Move containers of chemicals from fire area if you can do so without risk.

Cool containers that are exposed to flames with water until fire is out and heat of container is reduced.

Stay upwind of fire.

### *Spills*

The reader should consult Chapter 6, Part V, Section C, pages 265-276 for the proper methods, adsorbent recommendations and directions to control, contain and clean-up spills, and for disposal of pesticide wastes.

### *First Aid*

The reader should consult Chapter 7, Section E, pages 312-314 for emergency first aid information.

# PESTICIDE GUIDE 51

## HEALTH HAZARDS

***Hazard Indicator: Warning***

Warning: ingestion may be fatal.

Warning: absorption through skin may be fatal.

Caution: may cause mild eye irritation.

Caution: contact may cause mild skin irritation and weak sensitizer.

Caution: inhalation of the dry material or mists may cause respiratory irritation.

Fire may produce irritating or poisonous gases.

Runoff from fire or spills may cause pollution.

## SAFETY GUIDELINES

***Restricted Use Pesticide:*** Yes

***Solubility:*** Practically insoluble in water

***Toxicity Category:*** Class II Oral Hazard, Class II Dermal Hazard, Class III Inhalation Hazard

Formulations may contain methyl alcohol, xylenes, trimethylbenzene, and light aromatic petroleum solvent.

***Restricted-entry Interval (REI):*** 24 hours

## HANDLING AND STORAGE PRECAUTIONS

***Use Precautions***

Avoid contact with susceptible crops and beneficial insects.

Avoid spray drift.

Select and wear the proper PPE.

Mix ingredients in areas of sufficient ventilation or outdoors to minimize exposure. Avoid spilling material while mixing.

Wash with soap and water immediately after use or exposure. Follow good personal hygiene practices.

Do not reuse spray equipment for any purpose unless thoroughly cleaned with a suitable cleaner.

Reentry—do not enter treated areas without protective clothing until sprays have dried or after 24 hrs.

Notification of a pesticide application must be given to all workers who will be within 1/4 mile of a field during the application or before the REI expires.

This notification may be either oral or written, e.g., a posted sign at the entrances to treated sites. For some products, both oral and written notifications are necessary.

### *Storage and Disposal Precautions*

The reader should consult Chapter 6, Part V, Sections A and B, pages 251-265 for recommendations on the proper storage and disposal precautions.

## PROTECTIVE CLOTHING

Protective eyewear such as goggles or face shield is strongly recommended. Chemical-resistant gloves and boots, long-sleeved shirt, long pants, and shoes and socks are strongly recommended. When handling undiluted product use goggles, impermeable gloves, splash apron, protective clothing and boots and, if dust or vapor is a problem, use approved air purifying respirators for pesticides.

## EMERGENCY GUIDELINES

### *Fire or Explosive Hazard*

Evacuate nonessential personnel from fire area.

The products are not considered flammable. They will support combustion and may decompose under fire conditions and give off toxic products.

Some of these materials may react violently with water.

Segregate from oxidizers and incompatible materials.

The use of self-contained breathing apparatus and chemical protective clothing (chemical-resistant gloves, boots, hood, and suit) is recommended.

The use of high-pressure water may cause spread of hazardous materials.

*Small fires:* Dry chemical, $CO_2$, Halon, water spray, or alcohol foam.

*Large fires:* Water spray, fog, or alcohol foam is recommended.

Move containers of chemicals from fire area if you can do so without risk.

Cool containers that are exposed to flames with water until fire is out and heat of container is reduced.

Stay upwind of fire.

### *Spills*

The reader should consult Chapter 6, Part V, Section C, pages 265-276 for the proper methods, adsorbent recommendations and directions to control, contain and clean-up spills, and for disposal of pesticide wastes.

### *First Aid*

The reader should consult Chapter 7, Section E, pages 312-314 for emergency first aid information.

# PESTICIDE GUIDE 52

## HEALTH HAZARDS

***Hazard Indicator: Warning***

Warning: ingestion may be fatal.

Caution: may cause eye irritation.

Caution: contact may cause skin irritation.

Caution: inhalation of the dry material or mists may cause respiratory irritation.

Caution: absorption through skin may be harmful.

Fire may produce irritating or poisonous gases.

Runoff from fire or spills may cause pollution.

## SAFETY GUIDELINES

***Restricted Use Pesticide:*** Yes

***Solubility:*** Practically insoluble in water

***Toxicity Category:*** Class II Oral Hazard, Class III Dermal Hazard

***Restricted-entry Interval (REI):*** 24 hours

## HANDLING AND STORAGE PRECAUTIONS

***Use Precautions***

Avoid contact with susceptible crops, other desirable broadleaf plants, beneficial insects, and/or susceptible animals.

Avoid spray drift.

Select and wear the proper PPE.

Mix ingredients in areas of sufficient ventilation or outdoors to minimize exposure. Avoid spilling material while mixing.

Wash with soap and water immediately after use or exposure. Follow good personal hygiene practices.

Do not reuse spray equipment for any purpose unless thoroughly cleaned with a suitable cleaner.

Reentry—do not enter treated areas without protective clothing until sprays have dried or after 24 hrs.

Notification of a pesticide application must be given to all workers who will be within 1/4 mile of a field during the application or before the REI expires. This notification may be either oral or written, e.g., a posted sign at the entrances to treated sites. For some products, both oral and written notifications are necessary.

### *Storage and Disposal Precautions*

The reader should consult Chapter 6, Part V, Sections A and B, pages 251-265 for recommendations on the proper storage and disposal precautions.

## PROTECTIVE CLOTHING

Protective eyewear, chemical-resistant gloves and boots, long-sleeved shirt, long pants, and shoes and socks are strongly recommended. When handling undiluted product use goggles, impermeable gloves, splash apron, protective clothing and boots and, if dust or vapor is a problem, use approved air purifying respirators for pesticides.

## EMERGENCY GUIDELINES

### *Fire or Explosive Hazard*

Evacuate nonessential personnel from fire area.

The products are not considered flammable. They will support combustion and may decompose under fire conditions and give off toxic products.

Some of these materials may react violently with water.

Segregate from oxidizers and incompatible materials.

The use of self-contained breathing apparatus and chemical protective clothing (chemical-resistant gloves, boots, hood, and suit) is recommended.

The use of high-pressure water may cause spread of hazardous materials.

*Small fires:* Dry chemical, $CO_2$, Halon, water spray, or standard foam.

*Large fires:* Water spray, fog, or alcohol foam is recommended.

Move containers of chemicals from fire area if you can do so without risk.

Cool containers that are exposed to flames with water until fire is out and heat of container is reduced.

Stay upwind of fire.

### *Spills*

The reader should consult Chapter 6, Part V, Section C, pages 265-276 for the proper methods, adsorbent recommendations and directions to control, contain and clean-up spills, and for disposal of pesticide wastes.

### *First Aid*

The reader should consult Chapter 7, Section E, pages 312-314 for emergency first aid information.

# PESTICIDE GUIDE 53

## HEALTH HAZARDS

***Hazard Indicator: Warning***

Warning: ingestion may be fatal.

Caution: may cause eye irritation.

Caution: inhalation of the dry material or mists may cause respiratory irritation.

Caution: absorption through skin may be harmful.

Caution: contact may cause moderate skin irritation.

Fire may produce irritating or poisonous gases.

Runoff from fire or spills may cause pollution.

## SAFETY GUIDELINES

***Restricted Use Pesticide:*** Yes

***Solubility:*** Insoluble in water

***Toxicity Category:*** Class II Oral Hazard, Class III Dermal Hazard, Class III Inhalation Hazard

Formulations may contain ethylbenzene and/or xylene.

***Restricted-entry Interval (REI):*** 24 hours

## HANDLING AND STORAGE PRECAUTIONS

***Use Precautions***

Avoid contact with susceptible crops and beneficial insects.

Avoid spray drift.

Select and wear the proper PPE.

Mix ingredients in areas of sufficient ventilation or outdoors to minimize exposure. Avoid spilling material while mixing.

Wash with soap and water immediately after use or exposure. Follow good personal hygiene practices.

Do not reuse spray equipment for any purpose unless thoroughly cleaned with a suitable cleaner.

Reentry—do not enter treated areas without protective clothing until sprays have dried or after 24 hrs.

Notification of a pesticide application must be given to all workers who will be within 1/4 mile of a field during the application or before the REI expires. This notification may be either oral or written, e.g., a posted sign at the

entrances to treated sites. For some products, both oral and written notifications are necessary.

### *Storage and Disposal Precautions*

The reader should consult Chapter 6, Part V, Sections A and B, pages 251-265 for recommendations on the proper storage and disposal precautions.

## PROTECTIVE CLOTHING

Protective eyewear, chemical-resistant gloves and boots, long-sleeved shirt, long pants, and shoes and socks are strongly recommended. When mixing, use a splash apron and, if dust or vapor is a problem, use approved air purifying respirators for pesticides.

## EMERGENCY GUIDELINES

### *Fire or Explosive Hazard*

Evacuate nonessential personnel from fire area.

The products are not considered flammable. They will support combustion and may decompose under fire conditions and give off toxic products.

Some of these materials may react violently with water.

Segregate from oxidizers and incompatible materials.

The use of self-contained breathing apparatus and chemical protective clothing (chemical-resistant gloves, boots, hood, and suit) is recommended.

The use of high-pressure water may cause spread of hazardous materials.

*Small fires:* Dry chemical, $CO_2$, Halon, water spray, or standard foam.

*Large fires:* Water spray, fog, or standard foam is recommended.

Move containers of chemicals from fire area if you can do so without risk.

Cool containers that are exposed to flames with water until fire is out and heat of container is reduced.

Stay upwind of fire.

### *Spills*

The reader should consult Chapter 6, Part V, Section C, pages 265-276 for the proper methods, adsorbent recommendations and directions to control, contain and clean-up spills, and for disposal of pesticide wastes.

### *First Aid*

The reader should consult Chapter 7, Section E, pages 312-314 for emergency first aid information.

# PESTICIDE GUIDE 54

## HEALTH HAZARDS

***Hazard Indicator:*** ***Warning***

Warning: absorption through skin may be fatal.

Warning: toxic, inhalation of dry material or mists may cause respiratory irritation.

Caution: may cause moderate eye irritation.

Caution: contact may cause slight skin irritation.

Caution: ingestion may be harmful.

Fire may produce irritating or poisonous gases.

Runoff from fire or spills may cause pollution.

## SAFETY GUIDELINES

***Restricted Use Pesticide:*** Yes

***Solubility:*** Practically insoluble in water

***Toxicity Category:*** Class II Dermal Hazard, Class II Inhalation Hazard, Class III Oral Hazard

Formulations may contain petroleum solvent consisting of ethylbenzene, naphtha, trimethylbenzene, and xylene.

***Restricted-entry Interval (REI):*** 24 hours

## HANDLING AND STORAGE PRECAUTIONS

***Use Precautions***

Avoid contact with susceptible crops and beneficial insects.

Avoid spray drift.

Select and wear the proper PPE.

Mix ingredients in areas of sufficient ventilation or outdoors to minimize exposure. Avoid spilling material while mixing.

Wash with soap and water immediately after use or exposure. Follow good personal hygiene practices.

Do not reuse spray equipment for any purpose unless thoroughly cleaned with a suitable cleaner.

Reentry—do not enter treated areas without protective clothing until sprays have dried or after 24 hrs.

Notification of a pesticide application must be given to all workers who will be within 1/4 mile of a field during the application or before the REI expires.

This notification may be either oral or written, e.g., a posted sign at the entrances to treated sites. For some products, both oral and written notifications are necessary.

### *Storage and Disposal Precautions*

The reader should consult Chapter 6, Part V, Sections A and B, pages 251-265 for recommendations on the proper storage and disposal precautions.

## PROTECTIVE CLOTHING

Protective eyewear such as goggles or face shield is strongly recommended. Chemical-resistant gloves and boots, long-sleeved shirt, long pants, and shoes and socks are strongly recommended. When handling undiluted product use goggles, impermeable gloves, splash apron, protective clothing and boots and, if dust or vapor is a problem, use approved air purifying respirators for pesticides.

## EMERGENCY GUIDELINES

### *Fire or Explosive Hazard*

Evacuate nonessential personnel from fire area.

The products are not considered flammable. They will support combustion and may decompose under fire conditions and give off toxic products.

Some of these materials may react violently with water.

Segregate from oxidizers and incompatible materials.

The use of self-contained breathing apparatus and chemical protective clothing (chemical-resistant gloves, boots, hood, and suit) is recommended.

The use of high-pressure water may cause spread of hazardous materials.

*Small fires:* Dry chemical, $CO_2$, Halon, water spray, or alcohol foam.

*Large fires:* Water spray, fog, or alcohol foam is recommended.

Move containers of chemicals from fire area if you can do so without risk.

Cool containers that are exposed to flames with water until fire is out and heat of container is reduced.

Stay upwind of fire.

### *Spills*

The reader should consult Chapter 6, Part V, Section C, pages 265-276 for the proper methods, adsorbent recommendations and directions to control, contain and clean-up spills, and for disposal of pesticide wastes.

### *First Aid*

The reader should consult Chapter 7, Section E, pages 312-314 for emergency first aid information.

# PESTICIDE GUIDE 55

## HEALTH HAZARDS

***Hazard Indicator: Warning***

Warning: contact will cause severe skin irritation.

Warning: this product contain alachlor which has been determined to cause tumors in laboratory animals.

Caution: can cause eye irritation.

Caution: ingestion may be harmful.

Caution: inhalation of dry material or mists may cause respiratory irritation.

Fire may produce irritating or poisonous gases.

Runoff from fire or spills may cause pollution.

## SAFETY GUIDELINES

***Restricted Use Pesticide:*** Yes

***Solubility:*** Insoluble in water

***Toxicity Category:*** Class II Dermal hazard, Class III Oral hazard, Class III or IV Inhalation hazard

Bromooxynil is classified as a developmental toxicant. MPCA shows limited evidence as a carcinogen.

Formulations may contain aromatic and chlorinated aromatic hydrocarbons and petroleum distillates.

***Restricted-entry Interval (REI):*** 24 hours

## HANDLING AND STORAGE PRECAUTIONS

***Use Precautions***

Avoid contact with susceptible crops and other desirable broadleaf plants.

Avoid spray drift.

Select and wear the proper PPE.

Mix ingredients in areas of sufficient ventilation or outdoors to minimize exposure. Avoid spilling material while mixing.

Wash with soap and water immediately after use or exposure. Follow good personal hygiene practices.

Do not reuse spray equipment for any purpose unless thoroughly cleaned with a suitable cleaner.

Reentry—do not enter treated areas without protective clothing until sprays have dried or after 24 hrs.

Notification of a pesticide application must be given to all workers who will be within 1/4 mile of a field during the application or before the REI expires. This notification may be either oral or written, e.g., a posted sign at the entrances to treated sites. For some products, both oral and written notifications are necessary.

### *Storage and Disposal Precautions*

The reader should consult Chapter 6, Part V, Sections A and B, pages 251-265 for recommendations on the proper storage and disposal precautions.

## PROTECTIVE CLOTHING

Protective eyewear, chemical-resistant gloves and boots, long-sleeved shirt, long pants, and shoes and socks are strongly recommended. When mixing, use a splash apron and, if dust or vapor is a problem, use approved air purifying respirators for pesticides.

## EMERGENCY GUIDELINES

### *Fire or Explosive Hazard*

Evacuate nonessential personnel from fire area.

The products are not considered flammable. They will support combustion and may decompose under fire conditions and give off toxic products.

Some of these materials may react violently with water.

Segregate from oxidizers and incompatible materials.

The use of self-contained breathing apparatus and chemical protective clothing (chemical-resistant gloves, boots, hood, and suit) is recommended.

The use of high-pressure water may cause spread of hazardous materials.

*Small fires:* Dry chemical, $CO_2$, Halon, water spray, or standard foam.

*Large fires:* Water spray, fog, or standard foam is recommended.

Move containers of chemicals from fire area if you can do so without risk.

Cool containers that are exposed to flames with water until fire is out and heat of container is reduced.

Stay upwind of fire.

### *Spills*

The reader should consult Chapter 6, Part V, Section C, pages 265-276 for the proper methods, adsorbent recommendations and directions to control, contain and clean-up spills, and for disposal of pesticide wastes.

### *First Aid*

The reader should consult Chapter 7, Section E, pages 312-314 for emergency first aid information.

# PESTICIDE GUIDE 56

## HEALTH HAZARDS

***Hazard Indicator: Warning***

Warning: this product contains alachlor which has been determined to cause tumors in laboratory animals.

Caution: inhalation of the dry material or mists may be harmful and may cause respiratory irritation.

Caution: contact may cause skin irritation.

Caution: ingestion may be harmful.

Caution: contact may cause eye irritation.

Fire may produce irritating or poisonous gases.

Runoff from fire or spills may cause pollution.

## SAFETY GUIDELINES

***Restricted Use Pesticide:*** Yes

***Solubility:*** Practically insoluble in water

***Toxicity Category:*** Class III Dermal Hazard, Class III Inhalation Hazard, Class III Oral, Hazards

***Restricted-entry Interval (REI):*** 12 hours

## HANDLING AND STORAGE PRECAUTIONS

***Use Precautions***

Avoid contact with susceptible crops and other desirable broadleaf plants.

Avoid spray drift.

Select and wear the proper PPE.

Mix ingredients in areas of sufficient ventilation or outdoors to minimize exposure. Avoid spilling material while mixing.

Wash with soap and water immediately after use or exposure. Follow good personal hygiene practices.

Do not reuse spray equipment for any purpose unless thoroughly cleaned with a suitable cleaner.

Reentry—do not enter treated areas without protective clothing until sprays have dried or after 12 hrs.

Notification of a pesticide application must be given to all workers who will be within 1/4 mile of a field during the application or before the REI expires. This notification may be either oral or written, e.g., a posted sign at the

entrances to treated sites. For some products, both oral and written notifications are necessary.

### *Storage and Disposal Precautions*

The reader should consult Chapter 6, Part V, Sections A and B, pages 251-265 for recommendations on the proper storage and disposal precautions.

## PROTECTIVE CLOTHING

Protective eyewear, chemical-resistant gloves and boots, long-sleeved shirt, long pants, and shoes and socks are strongly recommended. When mixing, use a splash apron and, if dust or vapor is a problem, use approved air purifying respirators for pesticides.

## EMERGENCY GUIDELINES

### *Fire or Explosive Hazard*

Evacuate nonessential personnel from fire area.

The products are not considered flammable. They will support combustion and may decompose under fire conditions and give off toxic products.

Some of these materials may react violently with water.

Segregate from oxidizers and incompatible materials.

The use of self-contained breathing apparatus and chemical protective clothing (chemical-resistant gloves, boots, hood, and suit) is recommended.

The use of high-pressure water may cause spread of hazardous materials.

*Small fires:* Dry chemical, $CO_2$, Halon, water spray, or standard foam.

*Large fires:* Water spray, fog, or standard foam is recommended.

Move containers of chemicals from fire area if you can do so without risk.

Cool containers that are exposed to flames with water until fire is out and heat of container is reduced.

Stay upwind of fire.

### *Spills*

The reader should consult Chapter 6, Part V, Section C, pages 265-276 for the proper methods, adsorbent recommendations and directions to control, contain and clean-up spills, and for disposal of pesticide wastes.

### *First Aid*

The reader should consult Chapter 7, Section E, pages 312-314 for emergency first aid information.

# PESTICIDE GUIDE 57

## HEALTH HAZARDS

***Hazard Indicator: Warning***

Warning: may cause moderate eye irritation.

Warning: absorption through skin may be harmful.

Caution: ingestion may be harmful.

Caution: contact may cause mild skin irritation and weak sensitizer.

Caution: inhalation of the dry material or mists may cause respiratory irritation.

Fire may produce irritating or poisonous gases.

Runoff from fire or spills may cause pollution.

## SAFETY GUIDELINES

***Restricted Use Pesticide:*** Yes

***Solubility:*** Practically insoluble in water

***Toxicity Category***: Class II Eye Hazard, Class III Oral Hazard, Class II Dermal Hazard

***Restricted-entry Interval (REI):*** 24 hours

## HANDLING AND STORAGE PRECAUTIONS

***Use Precautions***

Avoid contact with susceptible crops and beneficial insects.

Avoid spray drift.

Select and wear the proper PPE.

Mix ingredients in areas of sufficient ventilation or outdoors to minimize exposure. Avoid spilling material while mixing.

Wash with soap and water immediately after use or exposure. Follow good personal hygiene practices.

Do not reuse spray equipment for any purpose unless thoroughly cleaned with a suitable cleaner.

Reentry—do not enter treated areas without protective clothing until sprays have dried or after 24 hrs.

Notification of a pesticide application must be given to all workers who will be within 1/4 mile of a field during the application or before the REI expires. This notification may be either oral or written, e.g., a posted sign at the entrances to treated sites. For some products, both oral and written notifications are necessary.

### *Storage and Disposal Precautions*

The reader should consult Chapter 6, Part V, Sections A and B, pages 251-265 for recommendations on the proper storage and disposal precautions.

## PROTECTIVE CLOTHING

Protective eyewear such as goggles or face shield is strongly recommended. Chemical-resistant gloves and boots, long-sleeved shirt, long pants, and shoes and socks are strongly recommended. When handling undiluted product use goggles, impermeable gloves, splash apron, protective clothing and boots and, if dust or vapor is a problem, use approved air purifying respirators for pesticides.

## EMERGENCY GUIDELINES

### *Fire or Explosive Hazard*

Evacuate nonessential personnel from fire area.

The products are not considered flammable. They will support combustion and may decompose under fire conditions and give off toxic products.

Some of these materials may react violently with water.

Segregate from oxidizers and incompatible materials.

The use of self-contained breathing apparatus and chemical protective clothing (chemical-resistant gloves, boots, hood, and suit) is recommended.

The use of high-pressure water may cause spread of hazardous materials.

*Small fires:* Dry chemical, $CO_2$, Halon, water spray, or alcohol foam.

*Large fires:* Water spray, fog, or alcohol foam is recommended.

Move containers of chemicals from fire area if you can do so without risk.

Cool containers that are exposed to flames with water until fire is out and heat of container is reduced.

Stay upwind of fire.

### *Spills*

The reader should consult Chapter 6, Part V, Section C, pages 265-276 for the proper methods, adsorbent recommendations and directions to control, contain and clean-up spills, and for disposal of pesticide wastes.

### *First Aid*

The reader should consult Chapter 7, Section E, pages 312-314 for emergency first aid information.

# PESTICIDE GUIDE 58

## HEALTH HAZARDS

***Hazard Indicator: Warning***

Warning: may cause moderate eye irritation.

Caution: ingestion may be harmful.

Caution: absorption through skin may be harmful.

Caution: contact may cause mild skin irritation and weak sensitizer.

Caution: inhalation of the dry material or mists may cause respiratory irritation.

Fire may produce irritating or poisonous gases.

Runoff from fire or spills may cause pollution.

## SAFETY GUIDELINES

***Restricted Use Pesticide:*** Yes

***Solubility:*** Practically insoluble in water

***Toxicity Category:*** Class II Eye Hazard, Class III Oral Hazard, Class III Dermal Hazard

Formulations may contain petroleum distillates.

***Restricted-entry Interval (REI):*** 24 hours

## HANDLING AND STORAGE PRECAUTIONS

***Use Precautions***

Avoid contact with susceptible crops and beneficial insects.

Avoid spray drift.

Select and wear the proper PPE.

Mix ingredients in areas of sufficient ventilation or outdoors to minimize exposure. Avoid spilling material while mixing.

Wash with soap and water immediately after use or exposure. Follow good personal hygiene practices.

Do not reuse spray equipment for any purpose unless thoroughly cleaned with a suitable cleaner.

Reentry—do not enter treated areas without protective clothing until sprays have dried or after 24 hrs.

Notification of a pesticide application must be given to all workers who will be within 1/4 mile of a field during the application or before the REI expires. This notification may be either oral or written, e.g., a posted sign at the

entrances to treated sites. For some products, both oral and written notifications are necessary.

### *Storage and Disposal Precautions*

The reader should consult Chapter 6, Part V, Sections A and B, pages 251-265 for recommendations on the proper storage and disposal precautions.

## PROTECTIVE CLOTHING

Protective eyewear such as goggles or face shield is strongly recommended. Chemical-resistant gloves and boots, long-sleeved shirt, long pants, and shoes and socks are strongly recommended. When handling undiluted product use goggles, impermeable gloves, splash apron, protective clothing and boots and, if dust or vapor is a problem, use approved air purifying respirators for pesticides.

## EMERGENCY GUIDELINES

### *Fire or Explosive Hazard*

Evacuate nonessential personnel from fire area.

The products are not considered flammable. They will support combustion and may decompose under fire conditions and give off toxic products.

Some of these materials may react violently with water.

Segregate from oxidizers and incompatible materials.

The use of self-contained breathing apparatus and chemical protective clothing (chemical-resistant gloves, boots, hood, and suit) is recommended.

The use of high-pressure water may cause spread of hazardous materials.

*Small fires:* Dry chemical, $CO_2$, Halon, water spray, or alcohol foam.

*Large fires:* Water spray, fog, or alcohol foam is recommended.

Move containers of chemicals from fire area if you can do so without risk.

Cool containers that are exposed to flames with water until fire is out and heat of container is reduced.

Stay upwind of fire.

### *Spills*

The reader should consult Chapter 6, Part V, Section C, pages 265-276 for the proper methods, adsorbent recommendations and directions to control, contain and clean-up spills, and for disposal of pesticide wastes.

### *First Aid*

The reader should consult Chapter 7, Section E, pages 312-314 for emergency first aid information.

# PESTICIDE GUIDE 59

## HEALTH HAZARDS

***Hazard Indicator: Warning***

Warning: can cause severe eye irritation and eye damage.

Warning: toxic, inhalation of the dry material or mists may cause respiratory irritation.

Warning: this product contains alachlor which has been determined to cause tumors in laboratory animals.

Caution: ingestion may be harmful.

Caution: contact will irritate skin and cause an allergic reaction.

Fire may produce irritating or poisonous gases.

Runoff from fire or spills may cause pollution.

## SAFETY GUIDELINES

***Restricted Use Pesticide:*** Yes

***Solubility:*** Moderately soluble to insoluble in water

***Toxicity Category:*** Class II Eye Hazard, Class II Inhalation Hazard, Class III Oral Hazard, Class III Dermal Hazard

***Restricted-entry Interval (REI):*** 24 hours

Formulations may contain petroleum distillates with naphthalene.

## HANDLING AND STORAGE PRECAUTIONS

***Use Precautions***

Avoid contact with susceptible crops and other desirable broadleaf plants.

Avoid spray drift.

Select and wear the proper PPE.

Mix ingredients in areas of sufficient ventilation or outdoors to minimize exposure. Avoid spilling material while mixing.

Wash with soap and water immediately after use or exposure. Follow good personal hygiene practices.

Do not reuse spray equipment for any purpose unless thoroughly cleaned with a suitable cleaner.

Reentry—do not enter treated areas without protective clothing until sprays have dried or after 24 hrs.

Notification of a pesticide application must be given to all workers who will be within 1/4 mile of a field during the application or before the REI expires.

This notification may be either oral or written, e.g., a posted sign at the entrances to treated sites. For some products, both oral and written notifications are necessary.

### *Storage and Disposal Precautions*

The reader should consult Chapter 6, Part V, Sections A and B, pages 251-265 for recommendations on the proper storage and disposal precautions.

## PROTECTIVE CLOTHING

Protective eyewear such as goggles and face shield, chemical-resistant gloves and boots, long-sleeved shirt, long pants, and shoes and socks are strongly recommended. When mixing, use a splash apron and, if dust or vapor is a problem, use approved air purifying respirators for pesticides.

## EMERGENCY GUIDELINES

### *Fire or Explosive Hazard*

Evacuate nonessential personnel from fire area.

The products are not considered flammable. They will support combustion and may decompose under fire conditions and give off toxic products.

Some of these materials may react violently with water.

Segregate from oxidizers and incompatible materials.

The use of self-contained breathing apparatus and chemical protective clothing (chemical-resistant gloves, boots, hood, and suit) is recommended.

The use of high-pressure water may cause spread of hazardous materials.

*Small fires:* Dry chemical, $CO_2$, Halon, water spray, or alcohol type foam.

*Large fires:* Water spray, fog, or alcohol type foam is recommended.

Move containers of chemicals from fire area if you can do so without risk.

Cool containers that are exposed to flames with water until fire is out and heat of container is reduced.

Stay upwind of fire.

### *Spills*

The reader should consult Chapter 6, Part V, Section C, pages 265-276 for the proper methods, adsorbent recommendations and directions to control, contain and clean-up spills, and for disposal of pesticide wastes.

### *First Aid*

The reader should consult Chapter 7, Section E, pages 312-314 for emergency first aid information.

# PESTICIDE GUIDE 60

## HEALTH HAZARDS

***Hazard Indicator: Warning***

Warning: may cause moderate eye irritation.

Caution: contact may cause slight skin irritation.

Caution: ingestion may be harmful.

Caution: absorption through skin may be harmful.

Caution: slightly toxic, inhalation of the dry material or mists may cause respiratory irritation.

Fire may produce irritating or poisonous gases.

Runoff from fire or spills may cause pollution.

## SAFETY GUIDELINES

***Restricted Use Pesticide:*** Yes

***Solubility:*** Very slightly soluble in water

***Toxicity Category:*** Class II Eye Hazard, Class III Oral Hazard, Class III Inhalation Hazard, Class IV Dermal Hazard

***Restricted-entry Interval (REI):*** 24 hours

## HANDLING AND STORAGE PRECAUTIONS

***Use Precautions***

Avoid contact with susceptible crops and other desirable broadleaf plants.

Avoid spray drift.

Select and wear the proper PPE.

Mix ingredients in areas of sufficient ventilation or outdoors to minimize exposure. Avoid spilling material while mixing.

Wash with soap and water immediately after use or exposure. Follow good personal hygiene practices.

Do not reuse spray equipment for any purpose unless thoroughly cleaned with a suitable cleaner.

Reentry—do not enter treated areas without protective clothing until sprays have dried or after 24 hrs.

Notification of a pesticide application must be given to all workers who will be within 1/4 mile of a field during the application or before the REI expires. This notification may be either oral or written, e.g., a posted sign at the entrances to treated sites. For some products, both oral and written notifications are necessary.

### *Storage and Disposal Precautions*

The reader should consult Chapter 6, Part V, Sections A and B, pages 251-265 for recommendations on the proper storage and disposal precautions.

## PROTECTIVE CLOTHING

Protective eyewear such as goggles or face shield is strongly recommended. Chemical-resistant gloves and boots, long-sleeved shirt, long pants, and shoes and socks are strongly recommended. When handling undiluted product use goggles, impermeable gloves, splash apron, protective clothing and boots and, if dust or vapor is a problem, use approved air purifying respirators for pesticides.

## EMERGENCY GUIDELINES

### *Fire or Explosive Hazard*

Evacuate nonessential personnel from fire area.

The products are not considered flammable. They will support combustion and may decompose under fire conditions and give off toxic products.

Some of these materials may react violently with water.

Segregate from oxidizers and incompatible materials.

The use of self-contained breathing apparatus and chemical protective clothing (chemical-resistant gloves, boots, hood, and suit) is recommended.

The use of high-pressure water may cause spread of hazardous materials.

*Small fires:* Dry chemical, $CO_2$, Halon, water spray, or alcohol foam.

*Large fires:* Water spray, fog, or alcohol foam is recommended.

Move containers of chemicals from fire area if you can do so without risk.

Cool containers that are exposed to flames with water until fire is out and heat of container is reduced.

Stay upwind of fire.

### *Spills*

The reader should consult Chapter 6, Part V, Section C, pages 265-276 for the proper methods, adsorbent recommendations and directions to control, contain and clean-up spills, and for disposal of pesticide wastes.

### *First Aid*

The reader should consult Chapter 7, Section E, pages 312-314 for emergency first aid information.

# PESTICIDE GUIDE 61

## HEALTH HAZARDS

***Hazard Indicator: Caution***

Caution: may cause eye irritation.

Caution: contact may cause skin irritation.

Caution: ingestion may be harmful.

Caution: inhalation of the dry material or mists may cause respiratory irritation.

Caution: absorption through skin may be harmful.

Fire may produce irritating or poisonous gases.

Runoff from fire or spills may cause pollution.

## SAFETY GUIDELINES

***Restricted Use Pesticide:*** No

***Solubility:*** Soluble to insoluble in water

***Toxicity Category:*** Nonhazardous to Class III or IV Oral Hazard, Class III or IV Dermal Hazard, Class III or IV Inhalation Hazard

Few products in this group are flammable. Other products may contain one or more of the following chemicals: Aromatic 200, 1-butanol, calcium dodecylbenzene, chlorotoluene, dipropylene glycol monomethyl ether, ethylbenzene, isophorone, isopropanol, kerosene, methyl alcohol, methylene chloride, N-m-2-pyrrolidone, octylphenoxypolyethoxyethanol, oxirane, paraffinic oil, petroleum distillates, propylene glycol, solvent naphtha, tetrahydrofurfural alcohol, toluene, trimethylbenzenes, xylenes, and several other chemicals whose identities were withheld as trade secrets.

***Restricted-entry Interval (REI):*** 12 hours

## HANDLING AND STORAGE PRECAUTIONS

***Use Precautions***

Avoid contact with susceptible crops, desirable broadleaf plants, beneficial insects, and/or susceptible animals.

Avoid spray drift.

Select and wear the proper PPE.

Mix ingredients in areas of sufficient ventilation or outdoors to minimize exposure. Avoid spilling material while mixing.

Wash with soap and water immediately after use or exposure. Follow good personal hygiene practices.

Do not reuse spray equipment for any purpose unless thoroughly cleaned with a suitable cleaner.

Reentry—do not enter treated areas without protective clothing until sprays have dried or after 12 hrs.

Notification of a pesticide application must be given to all workers who will be within 1/4 mile of a field during the application or before the REI expires. This notification may be either oral or written, e.g., a posted sign at the entrances to treated sites. For some products, both oral and written notifications are necessary.

### *Storage and Disposal Precautions*

The reader should consult Chapter 6, Part V, Sections A and B, pages 251-265 for recommendations on the proper storage and disposal precautions.

## PROTECTIVE CLOTHING

Protective eyewear, chemical-resistant gloves and boots, long-sleeved shirt, long pants, and shoes and socks are strongly recommended. When mixing, use a splash apron and, if dust or vapor is a problem, use approved air purifying respirators for pesticides.

## EMERGENCY GUIDELINES

### *Fire or Explosive Hazard*

Evacuate nonessential personnel from fire area.

Most products are not considered flammable. They will support combustion and may decompose under fire conditions and give off toxic products. However, several products are flammable. For information on product flammability the reader should consult Table 1 Pesticides Listed Alphabetically by Trade Name under formulations (see footnotes). The reader should review Chapter 6, Section V, for information on storage, disposal, and handling of flammable products.

Some of these materials may react violently with water.

Segregate from oxidizers and incompatible materials.

The use of self-contained breathing apparatus and chemical protective clothing (chemical-resistant gloves, boots, hood, and suit) is recommended.

The use of high-pressure water may cause spread of hazardous materials.

*Small fires:* Dry chemical, $CO_2$, Halon, water spray, or standard foam.

*Large fires:* Water spray, fog, foam or, if required alcohol, foam is recommended.

Move containers of chemicals from fire area if you can do so without risk.

Cool containers that are exposed to flames with water until fire is out and heat of container is reduced.

Stay upwind of fire.

### *Spills*

The reader should consult Chapter 6, Part V, Section C, pages 265-276 for the proper methods, adsorbent recommendations and directions to control, contain and clean-up spills, and for disposal of pesticide wastes.

### *First Aid*

The reader should consult Chapter 7, Section E, pages 312-314 for emergency first aid information.

# CHEMICAL GUIDES

# CHEMICAL GUIDE 1

## HEALTH HAZARDS

***Hazard: Flammable, Corrosive Liquids***

Toxic if inhaled or ingested.

Contact can cause severe burns to skin and eyes.

Avoid any skin or eye contact.

Fire may produce irritating, corrosive or poisonous gases.

Vapors may cause dizziness or suffocation.

Runoff from fire control efforts or spills may cause pollution.

## SAFETY GUIDELINES

***Solubility:*** Soluble in water

***Hazard Rating-NFPA***

| | |
|---|---|
| Health | 2 |
| Flammability | 3 |
| Reactivity | 1 |

## HANDLING AND STORAGE PRECAUTIONS

The reader should consult Chapter 6, Part V, Sections A and B, pages 251-265 for recommendations on the proper storage and disposal precautions.

## PROTECTIVE CLOTHING

The use of goggles or face shield and chemical-resistant gloves is recommended.

When using large quantities or highly concentrated solutions the use of a splash hood and splash apron are highly recommended.

## EMERGENCY GUIDELINES

***Fire or Explosive Hazard***

Evacuate nonessential personnel from fire area.

Stay upwind and keep out of low areas.

The materials are considered flammable and may also support combustion and decompose under fire conditions, giving off toxic products.

Some of these materials may react violently with water.

The use a self-contained breathing apparatus and chemical protective clothing (impervious gloves, boots, hood and suit) is recommended.

The use of high-pressure water may cause the spread of the hazardous materials.

*Small fires*: The use of dry chemical, $CO_2$, Halon, water spray, or alcohol foam is recommended.

*Large fires:* The use of water spray, fog, or alcohol foam is recommended.

Remove containers from fire areas if you can do so without risk.

Use water to keep fire-exposed containers cool.

Dike or otherwise control water for later disposal.

### *Spills*

The reader should consult Chapter 6, Part V, Section C, pages 265-276 for the proper methods, adsorbent recommendations and directions to control, contain and clean-up spills, and for disposal of pesticide wastes.

### *First Aid*

The reader should consult Chapter 7, Section E, pages 312-314 for emergency first aid information.

# CHEMICAL GUIDE 2

## HEALTH HAZARDS

***Hazard: Corrosive and/or Toxic Chemicals***

May be harmful if inhaled or ingested.

Effects of contact or inhalation may be delayed.

Contact can cause severe burns to skin and eyes.

Avoid any skin or eye contact.

Fire may produce irritating, corrosive, or poisonous gases.

Vapors may cause dizziness or suffocation.

Runoff from fire control efforts or spills may cause pollution.

## SAFETY GUIDELINES

***Solubility:*** Soluble in water

***Hazard Rating-NFPA***

| | |
|---|---|
| Health | 2 |
| Flammability | 1 |
| Reactivity | 1 |

## HANDLING AND STORAGE PRECAUTIONS

The reader should consult Chapter 6, Part V, Sections A and B, pages 251-265 for recommendations on the proper storage and disposal precautions.

## PROTECTIVE CLOTHING

The use of goggles or face shield and chemical-resistant gloves is recommended.

When using large quantities or highly concentrated solutions the use of a splash hood and splash apron are recommended.

## EMERGENCY GUIDELINES

***Fire or Explosive Hazard***

Evacuate nonessential personnel from fire area.

Stay upwind and keep out of low areas.

The materials are not considered flammable or combustible but may decompose under fire conditions and give off toxic products.

Some of these materials may react violently with water.

Contact with metals may produce flammable hydrogen gas.

The use of self-contained breathing apparatus and chemical protective clothing (impervious gloves, boots, hood and suit) is recommended.

The use of high-pressure water may cause the spread of the hazardous materials.

*Small fires:* The use of dry chemical, $CO_2$, Halon, water spray, or alcohol foam is recommended.

*Large fires:* The use of water spray, fog, or alcohol foam is recommended.

Remove containers from fire areas if you can do so without risk.

Use water to keep fire-exposed containers cool.

Dike or otherwise control water for later disposal.

### *Spills*

The reader should consult Chapter 6, Part V, Section C, pages 265-276 for the proper methods, adsorbent recommendations and directions to control, contain and clean-up spills, and for disposal of pesticide wastes.

### *First Aid*

The reader should consult Chapter 7, Section E, pages 312-314 for emergency first aid information.

# CHEMICAL GUIDE 3

## HEALTH HAZARDS

***Hazard: Flammable, Polar-Water Miscible Liquids***

Caution; may be harmful if inhaled or absorbed through skin.

Ingestion may be harmful or fatal.

Contact may irritate or burn skin and eyes.

Vapors may cause dizziness or suffocation.

Fire may produce irritating, corrosive, and/or poisonous gases.

Runoff from fire control efforts or spills may cause pollution.

## SAFETY GUIDELINES

***Solubility:*** Slightly soluble to completely soluble in water

***Hazard Rating-NFPA***

| | |
|---|---|
| Health | 1 |
| Flammability | 3 |
| Reactivity | 0 |

**Handling and Storage Precautions**

The reader should consult Chapter 6, Part V, Sections A and B, pages 251-265 for recommendations on the proper storage and disposal precautions.

**Protective Clothing**

The use of goggles or face shield and chemical-resistant gloves is recommended.

When using large quantities of materials the use of a splash hood and/or splash apron is recommended.

## EMERGENCY GUIDELINES

***Fire or Explosive Hazard***

Remove all ignition sources.

Evacuate nonessential personnel from fire area.

Stay upwind and keep out of low areas.

The materials are flammable and will burn and decompose under fire conditions, giving off toxic products.

The use of self-contained breathing apparatus and fire-resistant, chemical protective clothing (impervious gloves, boots, hood and suit) is recommended.

The use of high-pressure water may cause the spread of the hazardous materials.

*Small fires:* The use of dry chemical, $CO_2$, Halon, water spray, or alcohol foam is recommended.

*Large fires:* The use of water spray, fog, or alcohol foam is recommended.

Remove containers from fire areas if you can do so without risk.

Use water to keep fire-exposed containers cool.

### *Spills*

The reader should consult Chapter 6, Part V, Section C, pages 265-276 for the proper methods, adsorbent recommendations and directions to control, contain and clean-up spills, and for disposal of pesticide wastes.

### *First Aid*

The reader should consult Chapter 7, Section E, pages 312-314 for emergency first aid information.

# CHEMICAL GUIDE 4

## HEALTH HAZARDS

***Hazard: Flammable, Non-Polar, Insoluble Liquids***

Caution; may be harmful if inhaled or absorbed through skin.

Ingestion may be harmful.

Contact may irritate or burn skin and eyes.

Vapors may cause dizziness or suffocation.

Fire may produce irritating, corrosive, and/or poisonous gases.

Runoff from fire control efforts or spills may cause pollution.

## SAFETY GUIDELINES

***Solubility:*** Insoluble in water

***Hazard Rating-NFPA***

| | |
|---|---|
| Health | 1 |
| Flammability | 2 |
| Reactivity | 0 |

## HANDLING AND STORAGE PRECAUTIONS

The reader should consult Chapter 6, Part V, Sections A and B, pages 251-265 for recommendations on the proper storage and disposal precautions.

## PROTECTIVE CLOTHING

The use of goggles or face shield and chemical-resistant gloves is recommended.

When using large quantities of materials the use of a splash hood and/or splash apron is recommended.

## EMERGENCY GUIDELINES

***Fire or Explosive Hazard***

Remove all ignition sources.

Evacuate nonessential personnel from fire area.

Stay upwind and keep out of low areas.

The materials are flammable and will burn and decompose under fire conditions, giving off toxic products.

The use of self-contained breathing apparatus and fire-resistant, chemical protective clothing (impervious gloves, boots, hood and suit) is recommended.

The use of high-pressure water may cause the spread of the hazardous materials.

*Small fires:* The use of dry chemical, $CO_2$, Halon, water spray, or alcohol foam is recommended.

*Large fires:* The use of water spray, fog, or alcohol foam is recommended.

Remove containers from fire areas if you can do so without risk.

Use water to keep fire-exposed containers cool.

### Spills

The reader should consult Chapter 6, Part V, Section C, pages 265-276 for the proper methods, adsorbent recommendations and directions to control, contain and clean-up spills, and for disposal of pesticide wastes.

### First Aid

The reader should consult Chapter 7, Section E, pages 312-314 for emergency first aid information.

# CHEMICAL GUIDE 5

## HEALTH HAZARDS

***Hazard: Flammable and Unstable Gases***

Caution; gas may be harmful if inhaled at high concentrations.

Contact with gas or liquefied gas may cause burns, severe injury, and/or frost-bite.

Vapors may cause dizziness or asphyxiation.

Fire may produce irritating or poisonous gases.

## SAFETY GUIDELINES

***Solubility:*** Insoluble in water

***Hazard Rating-NFPA***

| | |
|---|---|
| Health | 1 |
| Flammability | 4 |
| Reactivity | 3 |

## HANDLING AND STORAGE PRECAUTIONS

The reader should consult Chapter 6, Part V, Sections A and B, pages 251-265 for recommendations on the proper storage and disposal precautions.

## PROTECTIVE CLOTHING

The use of goggles or face shield and chemical-resistant gloves is recommended.

When using large quantities of materials the use of a splash hood and/or splash apron is recommended.

## EMERGENCY GUIDELINES

***Fire or Explosive Hazard***

Remove all ignition sources.

Evacuate nonessential personnel from fire area.

Stay upwind and keep out of low areas.

The materials are very flammable and will easily ignite and burn.

Will form explosive mixtures with air.

Vapors are heavier than air and will spread along ground.

Vapors may travel to source of ignition and flash back.

The use of self-contained breathing apparatus and fire-resistant, chemical protective clothing (impervious gloves, boots, hood and suit) is recommended.

The use of high-pressure water may cause the spread of the hazardous materials.

*Small fires:* Use dry chemical or $CO_2$.

*Large fires:* The use of water spray, fog, or standard foam is recommended.

Remove cylinders from fire areas if you can do so without risk.

Use water to keep fire exposed cylinders cool.

### *Spills*

The reader should consult Chapter 6, Part V, Section C, pages 265-276 for the proper methods, adsorbent recommendations and directions to control, contain and clean-up spills, and for disposal of pesticide wastes.

### *First Aid*

The reader should consult Chapter 7, Section E, pages 312-314 for emergency first aid information.

# CHEMICAL GUIDE 6

## HEALTH HAZARDS

***Hazard: Corrosive Gases***

Toxic; may be fatal if inhaled.

Vapors are irritating and corrosive.

Contact with gas or liquefied gas may cause burns, severe injury, and/or frost-bite.

Fire may produce irritating, corrosive, or poisonous gases.

Runoff from fire control efforts or spills may cause pollution.

## SAFETY GUIDELINES

***Solubility:*** Soluble in water

***Hazard Rating-NFPA***

| | Gas | Liquid |
|---|---|---|
| Health | 2 | 3 |
| Flammability | 1 | 1 |
| Reactivity | 0 | 0 |

## HANDLING AND STORAGE PRECAUTIONS

These products are compressed gases. The cylinders should be protected against physical damage.

The reader should consult Chapter 6, Part V, Sections A and B, pages 251-265 for recommendations on additional storage and disposal precautions.

## PROTECTIVE CLOTHING

The use of goggles or face shield and chemical-resistant gloves is recommended.

When using large quantities of materials the use of a splash hood and/or splash apron is recommended.

## EMERGENCY GUIDELINES

***Fire or Explosive Hazard***

Remove all ignition sources.

Evacuate nonessential personnel from fire area.

Stay upwind and keep out of low areas.

The materials are not considered flammable but may support combustion and decompose under fire conditions, giving off toxic products.

The use of self-contained breathing apparatus and fire-resistant, chemical protective clothing (fully encapsulating, vapor protective suit) is recommended.

The use of high-pressure water may cause the spread of the hazardous materials.

*Small fires:* Use dry chemical or $CO_2$.

*Large fires:* The use of water spray, fog, or standard foam is recommended.

Remove cylinders from fire areas if you can do so without risk.

Use water to keep fire-exposed cylinders cool.

### *Spills*

The reader should consult Chapter 6, Part V, Section C, pages 265-276 for the proper methods, adsorbent recommendations and directions to control, contain and clean-up spills, and for disposal of pesticide wastes.

### *First Aid*

The reader should consult Chapter 7, Section E, pages 312-314 for emergency first aid information.

# CHEMICAL GUIDE 7

## HEALTH HAZARDS

***Hazard: Low to Moderately Hazardous Substances***

Caution; may be harmful if inhaled.

Ingestion may be harmful.

Contact may irritate or burn skin and eyes.

Fire may produce irritating, corrosive, and/or poisonous gases.

Runoff from fire control efforts or spills may cause pollution.

## SAFETY GUIDELINES

***Solubility:*** Low to moderately soluble in water

***Hazard Rating-NFPA***

| | |
|---|---|
| Health | 2 |
| Flammability | 0 |
| Reactivity | 0 |

## HANDLING AND STORAGE PRECAUTIONS

The reader should consult Chapter 6, Part V, Sections A and B, pages 251-265 for recommendations on the proper storage and disposal precautions.

## PROTECTIVE CLOTHING

The use of goggles or face shield and chemical-resistant gloves is recommended.

When using large quantities of materials the use of a splash hood and/or splash apron is recommended.

## EMERGENCY GUIDELINES

***Fire or Explosive Hazard***

Remove all ignition sources.

Evacuate nonessential personnel from fire area.

Stay upwind and keep out of low areas.

The materials are not flammable and may burn and decompose under fire conditions, giving off toxic products.

The use of self-contained breathing apparatus and fire-resistant, chemical protective clothing (impervious gloves, boots, hood and suit) is recommended.

The use of high-pressure water may cause the spread of the hazardous materials.

*Small fires:* The use of dry chemical, $CO_2$, Halon, water spray, or foam is recommended.

*Large fires:* The use of water spray, fog, or foam is recommended.

Remove containers from fire areas if you can do so without risk.

Use water to keep fire-exposed containers cool.

### *Spills*

The reader should consult Chapter 6, Part V, Section C, pages 265-276 for the proper methods, adsorbent recommendations and directions to control, contain and clean-up spills, and for disposal of pesticide wastes.

### *First Aid*

The reader should consult Chapter 7, Section E, pages 312-314 for emergency first aid information.

# CHEMICAL GUIDE 8

## HEALTH HAZARDS

*Hazard: Oxidizers*

Inhalation, ingestion, or contact (skin and eyes) with vapors or solid material may cause injury, burns, or death.

Fire may produce irritating, corrosive, and/or poisonous gases.

Runoff from fire control efforts or spills may cause pollution.

## SAFETY GUIDELINES

*Solubility:* Soluble in water

*Hazard Rating-NFPA*

| | |
|---|---|
| Health | 0 |
| Flammability | 0 |
| Reactivity | 3 |

## HANDLING AND STORAGE PRECAUTIONS

The reader should consult Chapter 6, Part V, Sections A and B, pages 251-265 for recommendations on the proper storage and disposal precautions.

These are oxidizers and must be stored separately from flammable materials and avoid storage near combustible materials. Inside storage should be in a standard flammable liquids storage room or cabinet.

## PROTECTIVE CLOTHING

The use of goggles and gloves is recommended.

When using large quantities of materials the use of a splash hood and/or splash apron is recommended.

## EMERGENCY GUIDELINES

*Fire or Explosive Hazard*

Remove all ignition sources.

Evacuate nonessential personnel from fire area.

Stay upwind and keep out of low areas.

These materials are highly reactive oxidizing agents which will accelerate ignition. They are capable of undergoing detonation if heated under confinement that permits pressure build-up, or if subjected to strong shocks, such as an explosion.

The materials are oxidizers and in the presence flammable materials will burn and decompose under fire conditions and give off toxic products.

The use of self-contained breathing apparatus and fire-resistant, chemical protective clothing (impervious gloves, boots, hood and suit) is recommended.

The use of high-pressure water may cause the spread of the hazardous materials.

*Small fires:* Do not use dry chemical, $CO_2$, Halon, or foams. Use water spray only.

*Large fires:* Use water and water spray from a distance.

Remove containers from fire areas if you can do so without risk. Do not move container if exposed to heat.

Cool containers by flooding quantities of water until fire is out and heat of combustion has dissipated.

### *Spills*

The reader should consult Chapter 6, Part V, Section C, pages 265-276 for the proper methods, adsorbent recommendations and directions to control, contain and clean-up spills, and for disposal of pesticide wastes.

### *First Aid*

The reader should consult Chapter 7, Section E, pages 312-314 for emergency first aid information.

# CHEMICAL GUIDE 8

## HEALTH HAZARDS

***Hazard: Oxidizers***

Inhalation, ingestion, or contact (skin and eyes) with vapors or solid material may cause injury, burns, or death.

Fire may produce irritating, corrosive, and/or poisonous gases.

Runoff from fire control efforts or spills may cause pollution.

## SAFETY GUIDELINES

***Solubility:*** Soluble in water

***Hazard Rating-NFPA***

| | |
|---|---|
| Health | 0 |
| Flammability | 0 |
| Reactivity | 3 |

## HANDLING AND STORAGE PRECAUTIONS

The reader should consult Chapter 6, Part V, Sections A and B, pages 251-265 for recommendations on the proper storage and disposal precautions.

These are oxidizers and must be stored separately from flammable materials and avoid storage near combustible materials. Inside storage should be in a standard flammable liquids storage room or cabinet.

## PROTECTIVE CLOTHING

The use of goggles and gloves is recommended.

When using large quantities of materials the use of a splash hood and/or splash apron is recommended.

## EMERGENCY GUIDELINES

***Fire or Explosive Hazard***

Remove all ignition sources.

Evacuate nonessential personnel from fire area.

Stay upwind and keep out of low areas.

These materials are highly reactive oxidizing agents which will accelerate ignition. They are capable of undergoing detonation if heated under confinement that permits pressure build-up, or if subjected to strong shocks, such as an explosion.

The materials are oxidizers and in the presence flammable materials will burn and decompose under fire conditions and give off toxic products.

The use of self-contained breathing apparatus and fire-resistant, chemical protective clothing (impervious gloves, boots, hood and suit) is recommended.

The use of high-pressure water may cause the spread of the hazardous materials.

*Small fires:* Do not use dry chemical, $CO_2$, Halon, or foams. Use water spray only.

*Large fires:* Use water and water spray from a distance.

Remove containers from fire areas if you can do so without risk. Do not move container if exposed to heat.

Cool containers by flooding quantities of water until fire is out and heat of combustion has dissipated.

### *Spills*

The reader should consult Chapter 6, Part V, Section C, pages 265-276 for the proper methods, adsorbent recommendations and directions to control, contain and clean-up spills, and for disposal of pesticide wastes.

### *First Aid*

The reader should consult Chapter 7, Section E, pages 312-314 for emergency first aid information.

# CHEMICAL GUIDE 9

## HEALTH HAZARDS

***Hazard: Toxic and/or Corrosive (Non-Combustible or Water-Sensitive) and Slight Oxidizers***

Toxic; inhalation, ingestion, or contact (skin and eyes) with vapors or substance may cause severe injury, burns, or death.

Effects of contact or inhalation may be delayed.

Avoid any skin or eye contact.

Reacts with many materials to produce noxious or toxic vapors.

Fire may produce irritating, corrosive, or poisonous gases.

Runoff from fire control efforts or spills may cause pollution.

## SAFETY GUIDELINES

***Solubility:*** Soluble in water

***Hazard Rating-NFPA***

| | |
|---|---|
| Health | 3 |
| Flammability | 0 |
| Reactivity | 2 |

## HANDLING AND STORAGE PRECAUTIONS

The reader should consult Chapter 6, Part V, Sections A and B, pages 251-265 for recommendations on the proper storage and disposal precautions.

## PROTECTIVE CLOTHING

The use of goggles and/or face shield and chemical-resistant gloves is recommended.

When using large quantities or highly concentrated solutions the use of a splash hood and splash apron are recommended.

## EMERGENCY GUIDELINES

***Fire or Explosive Hazard***

Evacuate nonessential personnel from fire area.

Stay upwind and keep out of low areas.

The materials are considered not flammable, but they are highly reactive and capable of igniting finely divided combustible materials on contact.

Some of these materials may react violently with water.

Contact with metals may produce flammable hydrogen gas.

The use of self-contained breathing apparatus and chemical protective clothing (impervious gloves, boots, hood and suit) is recommended.

The use of high-pressure water may cause the spread of the hazardous materials.

*Small fires:* The use of dry chemical, $CO_2$, Halon, water spray, or alcohol foam is recommended.

*Large fires:* The use of water spray, fog, or alcohol foam is recommended.

Remove containers from fire areas if you can do so without risk.

Use water to keep fire-exposed containers cool.

Dike or otherwise control water for later disposal.

### *Spills*

The reader should consult Chapter 6, Part V, Section C, pages 265-276 for the proper methods, adsorbent recommendations and directions to control, contain and clean-up spills, and for disposal of pesticide wastes.

### *First Aid*

The reader should consult Chapter 7, Section E, pages 312-314 for emergency first aid information.

# CHEMICAL GUIDE 10

## HEALTH HAZARDS

***Hazard: Flammable, Non-Polar/Water Immiscible, Toxic Liquids***

May be toxic if inhaled or absorbed through skin.

Ingestion may be harmful.

Inhalation or contact may irritate or burn skin and eyes.

Vapors may cause dizziness or suffocation.

Fire may produce irritating, corrosive, and/or poisonous gases.

Runoff from fire control efforts or spills may cause pollution.

## SAFETY GUIDELINES

***Solubility:*** Insoluble in water

***Hazard Rating-NFPA***

| | |
|---|---|
| Health | 2 |
| Flammability | 3 |
| Reactivity | 0 |

## HANDLING AND STORAGE PRECAUTIONS

The reader should consult Chapter 6, Part V, Sections A and B, pages 251-265 for recommendations on the proper storage and disposal precautions.

## PROTECTIVE CLOTHING

The use of goggles and/or face shield and chemical-resistant gloves is recommended.

When using large quantities of materials the use of a splash hood and/or splash apron is recommended.

## EMERGENCY GUIDELINES

***Fire or Explosive Hazard***

Remove all ignition sources.

Evacuate nonessential personnel from fire area.

Stay upwind and keep out of low areas.

The materials are flammable and will burn and decompose under fire conditions, giving off toxic products.

The use of self-contained breathing apparatus and fire-resistant, chemical protective clothing (impervious gloves, boots, hood and suit) is recommended.

The use of high-pressure water may cause the spread of the hazardous materials.

*Small fires:* The use of dry chemical, $CO_2$, Halon, water spray, or regular foam is recommended.

*Large fires:* The use of water spray, fog, or regular foam is recommended.

Remove containers from fire areas if you can do so without risk.

Use water to keep fire-exposed containers cool.

### *Spills*

The reader should consult Chapter 6, Part V, Section C, pages 265-276 for the proper methods, adsorbent recommendations and directions to control, contain and clean-up spills, and for disposal of pesticide wastes.

### *First Aid*

The reader should consult Chapter 7, Section E, pages 312-314 for emergency first aid information.

# CHEMICAL GUIDE 11

## HEALTH HAZARDS

***Hazard: Solid Oxidizers***

Inhalation, ingestion, or contact (skin and eyes) with vapors or substance may cause severe injury, burns, or death.

Fire may produce irritating, corrosive, and/or poisonous gases.

Runoff from fire control efforts or spills may cause pollution.

***Solubility:*** Soluble in water

***Hazard Rating-NFPA***

| | |
|---|---|
| Health | 2 |
| Flammability | 0 |
| Reactivity | 2 |

## HANDLING AND STORAGE PRECAUTIONS

The reader should consult Chapter 6, Part V, Sections A and B, pages 251-265 for recommendations on the proper storage and disposal precautions.

## PROTECTIVE CLOTHING

The use of goggles and/or face shield and chemical-resistant gloves is recommended.

When using large quantities of materials the use of a splash hood and/or splash apron is recommended.

## EMERGENCY GUIDELINES

***Fire or Explosive Hazard***

Remove all ignition sources.

Evacuate nonessential personnel from fire area.

Stay upwind and keep out of low areas.

These materials are highly reactive oxidizing agents which will accelerate ignition. They are capable of undergoing detonation if heated under confinement that permits pressure build-up, or if subjected to strong shocks, such as an explosion.

The materials are oxidizers and in the presence of flammable materials will burn and decompose giving off toxic products.

The use of self-contained breathing apparatus and fire-resistant, chemical protective clothing (impervious gloves, boots, hood and suit) is recommended.

The use of high-pressure water may cause the spread of the hazardous materials.

*Small fires:* Do not use dry chemical, $CO_2$, Halon, or foams. Use water spray only.

*Large fires:* Use water and water spray from a distance.

Remove containers from fire areas if you can do so without risk. Do not move container if exposed to heat.

Use water to keep fire-exposed containers cool.

### *Spills*

The reader should consult Chapter 6, Part V, Section C, pages 265-276 for the proper methods, adsorbent recommendations and directions to control, contain and clean-up spills, and for disposal of pesticide wastes.

### *First Aid*

The reader should consult Chapter 7, Section E, pages 312-314 for emergency first aid information.

# CHEMICAL GUIDE 12

## HEALTH HAZARDS

***Hazard: Toxic and/or Corrosive Substances***

Toxic; inhalation, ingestion, or skin contact with substance can be fatal.

Effects of contact or inhalation may be delayed.

Contact can cause severe burns to skin and eyes.

Avoid any skin or eye contact.

Fire may produce irritating, corrosive, or poisonous gases.

Runoff from fire control efforts or spills may cause pollution.

## SAFETY GUIDELINES

***Solubility:*** Insoluble in water

***Hazard Rating-NFPA***

| | |
|---|---|
| Health | 3 |
| Flammability | 2 |
| Reactivity | 0 |

## HANDLING AND STORAGE PRECAUTIONS

The reader should consult Chapter 6, Part V, Sections A and B, pages 251-265 for recommendations on the proper storage and disposal precautions.

## PROTECTIVE CLOTHING

The use of goggles and/or face shield and chemical-resistant gloves is recommended.

When using large quantities or highly concentrated solutions, the use of a splash hood and splash apron is recommended.

## EMERGENCY GUIDELINES

***Fire or Explosive Hazard***

Evacuate nonessential personnel from fire area.

Stay upwind and keep out of low areas.

Combustible materials that may burn but do not ignite readily.

The use of self-contained breathing apparatus and chemical protective clothing (impervious gloves, boots, hood and suit) is recommended.

The use of high-pressure water may cause the spread of the hazardous materials.

*Small fires:* The use of dry chemical, $CO_2$, Halon, water spray, or alcohol foam is recommended.

*Large fires:* The use of water spray, fog, or alcohol foam is recommended.

Remove containers from fire areas if you can do so without risk.

Use water to keep fire-exposed containers cool.

Dike or otherwise control water for later disposal.

### *Spills*

The reader should consult Chapter 6, Part V, Section C, pages 265-276 for the proper methods, adsorbent recommendations and directions to control, contain and clean-up spills, and for disposal of pesticide wastes.

### *First Aid*

The reader should consult Chapter 7, Section E, pages 312-314 for emergency first aid information.

# CHEMICAL GUIDE 13

## HEALTH HAZARDS

***Hazard: Flammable Gases***

Caution; gas may be harmful if inhaled at high concentrations.

Contact with gas or liquefied gas may cause burns, severe injury, and/or frostbite.

Vapors may cause dizziness or asphyxiation.

Fire may produce irritating or poisonous gases.

## SAFETY GUIDELINES

***Solubility:*** Insoluble in water

***Hazard Rating-NFPA***

| | |
|---|---|
| Health | 1 |
| Flammability | 4 |
| Reactivity | 0 |

## HANDLING AND STORAGE PRECAUTIONS

These products may be in cylinders or aerosol cans; they should be protected against physical damage.

The reader should consult Chapter 6, Part V, Sections A and B, pages 251-265 for additional recommendations on the proper storage and disposal precautions.

## PROTECTIVE CLOTHING

The use of goggles and/or face shield and chemical-resistant gloves is recommended.

When using large quantities of materials the use of a splash hood and/or splash apron is recommended.

## EMERGENCY GUIDELINES

***Fire or Explosive Hazard***

Remove all ignition sources.

Evacuate nonessential personnel from fire area.

Stay upwind and keep out of low areas.

The materials are very flammable and will easily ignite and burn.

Will form explosive mixtures with air.

Many vapors are heavier than air and will spread along ground.

Vapors may travel to source of ignition and flash back.

The use of self-contained breathing apparatus and fire-resistant, chemical protective clothing (impervious gloves, boots, hood and suit) is recommended.

The use of high-pressure water may cause the spread of the hazardous materials.

*Small fires:* Use dry chemical or $CO_2$.

*Large fires:* The use of water spray or fog is recommended.

Remove cylinders from fire areas if you can do so without risk.

Use water to keep fire-exposed cylinders cool.

### *Spills*

The reader should consult Chapter 6, Part V, Section C, pages 265-276 for the proper methods, adsorbent recommendations and directions to control, contain and clean-up spills, and for disposal of pesticide wastes.

### *First Aid*

The reader should consult Chapter 7, Section E, pages 312-314 for emergency first aid information.

# CHEMICAL GUIDE 14

## HEALTH HAZARDS

***Hazard: Flammable and Toxic (Polar-Water Miscible) Liquids***

Toxic; may be harmful if inhaled or absorbed through skin.

Ingestion may be harmful or fatal.

Inhalation or contact may irritate or burn skin and eyes.

Vapors may cause dizziness or suffocation.

Fire may produce irritating, corrosive, and/or poisonous gases.

Runoff from fire control efforts or spills may cause pollution.

## SAFETY GUIDELINES

***Solubility:*** Miscible in water

***Hazard Rating-NFPA***

| | |
|---|---|
| Health | 1 |
| Flammability | 3 |
| Reactivity | 0 |

## HANDLING AND STORAGE PRECAUTIONS

The reader should consult Chapter 6, Part V, Sections A and B, pages 251-265 for recommendations on the proper storage and disposal precautions.

## PROTECTIVE CLOTHING

The use of goggles and/or face shield and chemical-resistant gloves is recommended.

When using large quantities of materials the use of a splash hood and/or splash apron is recommended.

## EMERGENCY GUIDELINES

***Fire or Explosive Hazard***

Remove all ignition sources.

Evacuate nonessential personnel from fire area.

Stay upwind and keep out of low areas.

The materials are flammable and will burn and decompose under fire conditions, giving off toxic products.

The use of self-contained breathing apparatus and fire-resistant, chemical protective clothing (impervious gloves, boots, hood and suit) is recommended.

The use of high-pressure water may cause the spread of the hazardous materials.

*Small fires:* The use of dry chemical, $CO_2$, Halon, water spray, or alcohol-resistant foam is recommended.

*Large fires:* The use of water spray, fog, or alcohol-resistant foam is recommended.

Remove containers from fire areas if you can do so without risk.

Use water to keep fire-exposed containers cool.

### *Spills*

The reader should consult Chapter 6, Part V, Section C, pages 265-276 for the proper methods, adsorbent recommendations and directions to control, contain and clean-up spills, and for disposal of pesticide wastes.

### *First Aid*

The reader should consult Chapter 7, Section E, pages 312-314 for emergency first aid information.

# CHEMICAL GUIDE 15

## HEALTH HAZARDS

***Hazard: Halogenated Solvents***

Toxic; may be harmful if inhaled or absorbed through skin.

Ingestion may be harmful or fatal.

Contact may irritate or burn skin and eyes.

Vapors may cause dizziness or suffocation.

Fire may produce irritating, corrosive, and/or poisonous gases.

Runoff from fire control efforts or spills may cause pollution.

## SAFETY GUIDELINES

***Solubility:*** Insoluble in water

***Hazard Rating-NFPA***

| | |
|---|---|
| Health | 2 |
| Flammability | 1 |
| Reactivity | 0 |

## HANDLING AND STORAGE PRECAUTIONS

The reader should consult Chapter 6, Part V, Sections A and B, pages 251-265 for recommendations on the proper storage and disposal precautions.

## PROTECTIVE CLOTHING

The use of goggles and/or face shield and chemical-resistant gloves is recommended.

When using large quantities of materials the use of a splash hood and/or splash apron is recommended.

## EMERGENCY GUIDELINES

***Fire or Explosive Hazard***

Remove all ignition sources.

Evacuate nonessential personnel from fire area.

Stay upwind and keep out of low areas.

The materials are not flammable but will burn and decompose under fire conditions, giving off toxic products.

The use of self-contained breathing apparatus and fire-resistant, chemical protective clothing (impervious gloves, boots, hood and suit) is recommended.

The use of high-pressure water may cause the spread of the hazardous materials.

*Small fires:* The use of dry chemical, $CO_2$, Halon, or water spray is recommended.

*Large fires:* The use of water spray, fog, or alcohol foam is recommended.

Remove containers from fire areas if you can do so without risk.

Use water to keep fire-exposed containers cool.

### *Spills*

The reader should consult Chapter 6, Part V, Section C, pages 265-276 for the proper methods, adsorbent recommendations and directions to control, contain and clean-up spills, and for disposal of pesticide wastes.

### *First Aid*

The reader should consult Chapter 7, Section E, pages 312-314 for emergency first aid information.

# CHEMICAL GUIDE 16

## HEALTH HAZARDS

***Hazard: Flammable Solids***

Ingestion may be harmful.

Inhalation or contact may irritate or burn skin and eyes.

Contact with molten material may cause severe burns to skin and eyes.

Fire may produce irritating and/or poisonous gases.

Runoff from fire control efforts or spills may cause pollution.

## SAFETY GUIDELINES

***Solubility:*** Insoluble in water

***Hazard Rating-NFPA***

| | |
|---|---|
| Health | 2 |
| Flammability | 2 |
| Reactivity | 0 |

## HANDLING AND STORAGE PRECAUTIONS

The reader should consult Chapter 6, Part V, Sections A and B, pages 251-265 for recommendations on the proper storage and disposal precautions.

## PROTECTIVE CLOTHING

The use of goggles and chemical-resistant gloves is recommended.

When using large quantities of materials the use of a splash hood and/or splash apron is recommended.

## EMERGENCY GUIDELINES

***Fire or Explosive Hazard***

Remove all ignition sources.

Evacuate nonessential personnel from fire area.

Stay upwind and keep out of low areas.

The materials are not flammable but will burn and decompose under fire conditions, giving off toxic products.

The use of self-contained breathing apparatus and fire-resistant, chemical protective clothing (impervious gloves, boots, hood and suit) is recommended.

The use of high-pressure water may cause the spread of the hazardous materials.

*Small fires:* The use of dry chemical, $CO_2$, Halon, water spray, or regular foam is recommended.

*Large fires:* The use of water spray, fog, or regular foam is recommended.

Remove containers from fire areas if you can do so without risk.

Use water to keep fire-exposed containers cool.

### *Spills*

The reader should consult Chapter 6, Part V, Section C, pages 265-276 for the proper methods, adsorbent recommendations and directions to control, contain and clean-up spills, and for disposal of pesticide wastes.

### *First Aid*

The reader should consult Chapter 7, Section E, pages 312-314 for emergency first aid information.

# CHEMICAL GUIDE 17

## HEALTH HAZARDS

***Hazard:* *Toxic, Solid Oxidizers***

Toxic: inhalation or ingestion can be fatal.

Contact may irritate or burn skin and eyes.

Fire may produce irritating, corrosive, and/or poisonous gases.

Runoff from fire control efforts or spills may cause pollution.

## SAFETY GUIDELINES

***Solubility:*** Soluble in water

***Hazard Rating-NFPA***

| | |
|---|---|
| Health | 2 |
| Flammability | 0 |
| Reactivity | 2 |

## HANDLING AND STORAGE PRECAUTIONS

The reader should consult Chapter 6, Part V, Sections A and B, pages 251-265 for recommendations on the proper storage and disposal precautions.

## PROTECTIVE CLOTHING

The use of goggles and/or face shield and chemical-resistant gloves is recommended.

When using large quantities of materials the use of a splash hood and/or splash apron is recommended.

## EMERGENCY GUIDELINES

***Fire or Explosive Hazard***

Evacuate nonessential personnel from fire area.

Stay upwind and keep out of low areas.

These materials are highly reactive oxidizing agents which will accelerate ignition. They are capable of undergoing detonation if heated under confinement that permits pressure build-up, or if subjected to strong shocks, such as an explosion.

The materials are oxidizers and in the presence flammable materials will burn and decompose giving off toxic products.

The use of self-contained breathing apparatus and fire-resistant, chemical protective clothing (impervious gloves, boots, hood and suit) is recommended.

Remove all ignition sources.

The use of high-pressure water may cause the spread of the hazardous materials.

*Small fires:* Do not use dry chemical, $CO_2$, Halon, or foams. Use water spray only.

*Large fires:* Use water and water spray from a distance.

Remove containers from fire areas if you can do so without risk. Do not move container if exposed to heat.

Cool containers with flooding quantities of water until fire is out and heat of combustion has dissipated.

### *Spills*

The reader should consult Chapter 6, Part V, Section C, pages 265-276 for the proper methods, adsorbent recommendations and directions to control, contain and clean-up spills, and for disposal of pesticide wastes.

### *First Aid*

The reader should consult Chapter 7, Section E, pages 312-314 for emergency first aid information.

# CHEMICAL GUIDE 18

## HEALTH HAZARDS

***Hazard: Flammable, Toxic Gases***

Toxic: may be fatal if inhaled or absorbed through skin.

Contact with gas or liquefied gas may cause burns, severe injury, and/or frostbite.

Fire may produce irritating or poisonous gases.

## SAFETY GUIDELINES

***Solubility:*** Soluble in water

***Hazard Rating-NFPA***

| | |
|---|---|
| Health | 2 |
| Flammability | 4 |
| Reactivity | 3 |

## HANDLING AND STORAGE PRECAUTIONS

These products are available in cylinders and aerosol cans. They should be protected against physical damage.

The reader should consult Chapter 6, Part V, Sections A and B, pages 251-265 for additional recommendations on the proper storage and disposal precautions.

## PROTECTIVE CLOTHING

The use of goggles and/or face shield and chemical-resistant gloves is recommended.

When using large quantities of materials the use of a splash hood and/or splash apron is recommended.

## EMERGENCY GUIDELINES

***Fire or Explosive Hazard***

Remove all ignition sources.

Evacuate nonessential personnel from fire area.

Stay upwind and keep out of low areas.

The materials are very flammable and will easily ignite and burn.

Will form explosive mixtures with air.

Many vapors are heavier than air and will spread along ground.

Vapors may travel to source of ignition and flash back.

The use of self-contained breathing apparatus and fire-resistant, chemical protective clothing (impervious gloves, boots, hood and suit) is recommended.

The use of high-pressure water may cause the spread of the hazardous materials.

*Small fires:* Use dry chemical, $CO_2$, water spray, or standard foam.

*Large fires:* The use of water spray, fog, or standard foam is recommended.

Remove cylinders from fire areas if you can do so without risk.

Use water to keep fire-exposed cylinders cool.

### *Spills*

The reader should consult Chapter 6, Part V, Section C, pages 265-276 for the proper methods, adsorbent recommendations and directions to control, contain and clean-up spills, and for disposal of pesticide wastes.

### *First Aid*

The reader should consult Chapter 7, Section E, pages 312-314 for emergency first aid information.

# GLOSSARY

**Abrasion**—The process of wearing away by rubbing.

**Absorption**—The process by which a chemical is taken into plants. animals, or minerals. Compare with adsorption.

**Acaricide**—A pesticide used to control mites and ticks. A miticide is an acaricide.

**Acetylcholine (ACh)**—Chemical transmitter of nerve and nerve-muscle impulses in animals and insects.

**Activator**—A chemical added to a pesticide to increase its activity.

**Active Ingredient (a.i.)**—The chemical or chemicals in a pesticide responsible for killing, poisoning, or repelling the pest. Listed separately in the ingredient statement.

**Acute toxicity**—The toxicity of a material determined at the end of 24 hours to cause injury or death from a single dose or exposure.

**Adherence**—Sticking to a surface.

**Adjuvant**—An ingredient that improves the properties of a pesticide formulation. Includes wetting agents, spreaders, emulsifiers, dispersing agents, foam suppressants, penetrants, and correctives.

**Adsorption**—The process by which chemicals are held or bound to a surface by physical or chemical attraction. Clay and high organic soils tend to adsorb pesticides.

**Adulterated pesticide**—A pesticide that does not conform to the professed standard or quality as documented on its label or labeling.

**Aerobic**—Living in the air or presence of oxygen. The opposite of anaerobic.

**Aerosol**—An extremely fine mist or fog consisting of solid or liquid particles suspended in air. Also, certain formulations used to produce a fine mist or smoke.

**Agitation**—The process of stirring or mixing in a sprayer or tank.

**Algicide**—Chemical used to control algae and aquatic weeds.

**Alkaloids**—Chemicals present in some plants. Some are used as pesticides.

**Anaerobic**—Living in the absence of air or oxygen. The opposite of aerobic.

**Annual**—Plant that completes its life cycle in one year, i.e., germinates from seed, produces seed, and dies in the same season.

**Antagonism**—(Opposite of synergism) Decreased activity arising from the effect of one chemical on another.

**Antibiotic**—Chemical substance produced by a microorganism and that is toxic to other microorganisms.

**Anticoagulant**—A chemical which prevents normal bloodclotting. The active ingredient in some rodenticides.

**Antidote**—A treatment used to counteract the effects of pesticide poisoning or some other poison in the body.

**Anti-Siphoning Device**—A device attached to the filling hose that prevents backflow or backsiphoning from a spray tank into a water source.

**Aqueous**—A term used to indicate the presence of water in a solution.

**Arachnid**—A wingless arthropod with two body regions and four pairs of jointed legs. Spiders, ticks, and mites are in the taxonomic class Arachnida.

**Aromatics**—Solvents containing benzene, or compounds derived from benzene.

**Arsenicals**—Pesticides containing arsenic.

**Arthropod**—An invertebrate animal characterized by a jointed body and limbs and usually a hard body covering that is molted at intervals. For example, insects, mites, and crayfish are in the phylum Arthropoda.

**Atropine (atropine sulfate)**—An antidote used to treat organophosphate and carbamate poisoning.

**Attractant**—A substance or device that will lure pests to a trap or poison bait.

**Auxin**—Substance found in plants, which stimulates cell growth in plant tissues.

**Avicide**—A pesticide used to kill or repel birds. Birds are in the class Aves.

**Bacteria**—Microscopic organisms, some of which are capable of producing diseases in plants and animals. Others are beneficial.

**Bactericide**—Chemical used to control bacteria.

**Bacteriostat**—Material used to prevent growth or multiplication of bacteria.

**Bait**—A food or other substance used to attract a pest to a pesticide or to a trap.

**Band Application**—Application of a pesticide in a strip alongside or around a structure, a portion of a structure, or any object.

**Barrier Application**—see band application.

**Beneficial Insect**—An insect that is useful or helpful to humans. Usually insect parasites, predators, pollinators, etc.

**Biennial**—Plant that completes its growth in 2 years. The first year it produces leaves and stores food; the second year it produces fruit and seeds.

**Biological Control**—Control of pests using predators, parasites, and disease-causing organisms. May be naturally occurring or introduced.

**Biological Control Agent**—Any biological agent that adversely affects pest species.

**Biomagnification**—The process where one organism accumulates chemical residues in higher concentrations from organisms they consume.

**Biota**—Animals and plants of a given habitat.

**Biotic Insecticide**—Usually microorganisms known as insect pathogens that are applied in the same manner as conventional insecticides to control pest species.

**Bipyridyliums**—A group of synthetic organic pesticides which includes the herbicide paraquat.

**Blight**—Common name for a number of different diseases on plants, especially when collapse is sudden, e.g., leaf blight, blossom blight, shoot blight.

**Botanical Pesticide**—A pesticide produced from chemicals found in plants. Examples are nicotine, pyrethrins, rotenone, and strychnine.

**Brand Name**—The name or designation of a specific pesticide product or device made by a manufacturer or formulator. A trade name or marketing name.

**Broadcast Application**—Application over an entire area rather than only on rows, beds, or middles.

**Broadleaf Weeds**—Plants with broad, rounded, or flattened leaves.

**Broad-Spectrum Insecticides**—Nonselective, having about the same toxicity to most insects.

**Brush Control**—Control of woody plants.

**Calibrate, Calibration of Equipment or Application Method**—The measurement of dispersal or output and adjustments made to control the rate of dispersal of pesticides.

**Canker**—A lesion on a stem.

**Carbamate Insecticide**—A synthetic organic pesticide containing carbon, hydrogen, nitrogen, and sulfur. They are a class of insecticides derived from carbamic acid.

**Carcinogen**—A substance that can cause cancer in living animals.

**Carcinogenic**—Can cause cancer.

**Carrier**—An inert liquid, solid, or gas added to an active ingredient to make a pesticide disperse effectively. A carrier is also the material, usually water or oil, used to dilute the formulated product for application.

**Causal Organism**—The organism (pathogen) that produces a specific disease.

**Certified Applicators**—Individuals who are certified to use or supervise the use of any restricted use pesticide covered by their certification.

**Chelating Agent**—Certain organic chemicals (i.e., ethylenediaminetetraacetic acid) that combine with metals or ions to form soluble chelates and prevent conversion to insoluble compounds.

**Chemical Control**—Pesticide application to kill pests.

**Chemical Name**—The scientific name of the active ingredient(s) found in the formulated product. This complex name is derived from the chemical structure of the active ingredient.

**Chemosterilant**—A chemical compound capable of preventing animal reproduction.

**Chemotherapy**—Treatment of a diseased organism, usually plants, with chemicals to destroy or inactivate a pathogen without seriously affecting the host.

**Chemtrec**—The Chemical Transportation Emergency Center has a toll-free number that provides 24-hour information for chemical emergencies such as a spill, leak, fire, or accident. 1-800-424-9300.

**Chlorinated Hydrocarbon**—A pesticide containing chlorine, carbon, and hydrogen. Many are persistent in the environment. Examples: chlordane, DDT, methoxychlor.

**Chlorosis**— The yellowing of a plant's green tissue.

**Cholinesterase (ChE), Acetylcholinesterase**—An enzyme of the body required for proper nerve function that is inhibited or inactivated by organophosphate or carbamate insecticides.

**Chronic Toxcity**—The ability of a material to cause injury or illness (beyond 24 hours following exposure) from repeated, prolonged exposure to small amounts (see also Acute Toxicity).

**Commercial Applicator**—A certified applicator who for compensation uses or supervises the use of any pesticide classified for restricted use for any purpose or on any property other than that producing an agricultural commodity.

**Common Name**—A name given to a pesticide's active ingredient by a recognized committee on pesticide nomenclature. Many pesticides are known by a number of trade or brand names but the active ingredient (s) has only one recognized common name. Example: The common name for Sevin insecticide is carbaryl.

**Community**—The different populations of animal species (or plants) that exist together in an ecosystem (see also Population and Ecosystem).

**Compatible (Compatibility)**—When two materials can be mixed together with neither affecting the action of the other.

**Competent**—Individuals properly qualified to perform functions associated with pesticide application. The degree of competency (capability) required is directly related to the nature of the activity and the associated responsibility.

**Concentration**—The amount of active ingredient in a given volume or weight of formulation. For example, lb/gallon or percent by weight.

**Contact Herbicide**—Phytotoxic by contact with plant tissue rather than as a result of translocation.

**Contact Pesticide**—A compound that causes death or injury to insects when it contacts them. It does not have to be ingested. Often used in reference to a spray applied directly on a pest.

**Contamination**—The presence of an unwanted pesticide or other material in or on a plant, animal, or their by-product (see residue).

**Cultural Control**—A pest control method that includes changing human habits, e.g., sanitation, changing work practices, changing cleaning and garbage pick-up schedules, etc.

**Cumulative Pesticides**—Those chemicals which tend to accumulate or build up in the tissues of animals or in the environment (soil, water).

**Curative Pesticide**—A pesticide which can inhibit or eradicate a disease-causing organism after it has become established in the plant or animal.

**Cutaneous Toxicity**—Same as dermal toxicity.

**Cuticle**—Outer covering of insects or leaves. Chemically they are quite different.

**Days-to-Harvest**—The least number of days between the last pesticide application and the harvest date, as set by law. Same as "harvest intervals."

**Deciduous Plants**—Perennial plants that lose their leaves during the winter.

**Decontaminate**—The removal or breakdown of any pesticide chemical from any surface or piece of equipment.

**Deflocculating Agent**—A material added to a suspension to prevent settling.

**Defoliant**—A chemical that initiates abscission.

**Degradation**—The process by which a chemical compound or pesticide is reduced to simpler compounds by the action of microorganisms, water, air, sunlight, or other agents. Degradation products are usually, but not always less toxic than the original compound.

**Deposit**—Quantity of a pesticide or other material deposited on a unit area.

**Dermal**—Of the skin: through or by the skin.

**Dermal Toxicity**—The ability of a pesticide to cause acute illness or injury to a human or animal when absorbed through the skin (see Exposure Route).

**Desiccant**—A type of pesticide that draws moisture or fluids from a pest causing it to die. Certain desiccant dusts destroy the waxy outer coating that holds moisture within an insect's body.

**Desiccation**—Accelerated drying of plant or plant parts.

**Detoxify**—To make an active ingredient in a pesticide or other poisonous chemical harmless and incapable of being toxic to plants and animals.

**Diagnosis**—The positive identification of a problem and its cause.

**Diatomaceous Earth**—A whitish powder prepared from deposits formed by the silicified skeletons of diatoms. Used as diluent in dust formulations.

**Diluent**—Any liquid or solid material used to dilute or carry an active ingredient.

**Dilute**—To make thinner by adding water, another liquid, or a solid.

**Disinfectant**—A chemical or other agent that kills or inactivates disease producing microorganisms in animals, seeds, or other plant parts. Also commonly referred to chemicals used to clean or surface-sterilize inanimate objects.

**DNA**—Deoxyribonucleic acid.

**Dormant Spray**—Chemical applied in winter or very early spring before treated plants have started active growth.

**Dose, Dosage**—Quantity, amount, or rate of pesticide applied to a given area or target.

**Drift**—The airborne movement of a pesticide spray or dust beyond the intended target area.

**Dust**—A finely ground, dry pesticide formulation containing a small amount of active ingredient and a large amount of inert carrier or diluent such as clay or talc.

**$EC_{50}$**—The median effective concentration (ppm or ppb) of the toxicant in the environment (usually water) which produces a designated effect in 50% of the test organisms exposed.

**Ecology**—A division of biology concerned with organisms and their relation to the environment.

**Ecosystem**—The interacting system of all the living organisms of an area and their nonliving environment.

**Ectoparasite**—A parasite feeding on a host from the exterior or outside.

**$ED_{50}$**—The median effective dose, expressed as mg/ kg of body weight, which produces a designated effect in 50% of the test organisms exposed.

**Emulsifiable Concentrate** —A pesticide formulation produced by mixing or suspending the active ingredient (the concentrate) and an emulsifying agent in a suitable carrier. When added to water, a milky emulsion is formed.

**Emulsifier**—Surface active substances used to stabilize suspensions of one liquid in another, for example, oil in water.

**Emulsion**—A mixture of two liquids which are not soluble in one another. One is suspended as very small droplets in the other with the aid of an emulsifying agent.

**Encapsulated Formulation**—A pesticide formulation with the active ingredient enclosed in capsules of polyvinyl or other materials; principally used for slow release. The enclosed active ingredient moves out to the capsule surface as pesticide on the surface is removed (volatilizes, rubs off, etc.).

**Endangered Species**—Groups of interbreeding plants or animals that have been reduced to the extent that they are near extinction and that have been designated by the EPA to be endangered.

**Endoparasite**—A parasite that enters host tissue and feeds from within.

**Entry Interval**—See Re-entry Interval.

**Environment**—Air, land, water, all plants, man, and other animals, and the interrelationships which exist among them.

**Environmental Protection Agency (EPA)**—The Federal agency responsible for pesticide rules and regulations and all pesticide registrations.

**EPA**—The U.S. Environmental Protection Agency.

**EPA Establishment Number**—A number assigned by EPA to each pesticide production plant. The number indicates the manufacturer's plant at which the pesticide product was produced and must appear on all labels of that product.

**EPA Registration Number**—A number assigned by EPA to a pesticide product when the product is registered by the manufacturer or his designated agent. The number must appear on all labels for a particular product.

**Eradicant**—Applies to fungicides in which a chemical is used to eliminate a pathogen from its host or environment.

**Eradication**—The complete elimination of a (pest) population from a designated area.

**Exposure Route or Common Exposure Route**—The manner (dermal, oral, or inhalation/respiratory) in which a pesticide may enter an organism.

**Exterminate**—Often used to imply the complete extinction of a species over a large continuous area such as an island or a continent.

**FEPCA**—The Federal Environmental Pesticide Control Act of 1972.

**FIFRA**—The Federal Insecticide, Fungicide, and Rodenticide Act of 1974. A federal law and its amendments that control pesticide registration and use.

**Filler**—Diluent in powder form.

**Fixed Coppers**—Insoluble copper fungicides where the copper is in a combined form. Usually finely divided, relatively insoluble powders.

**Flowable**—A pesticide formulation in which a very finely ground solid particle is suspended (not dissolved) in a liquid carrier.

**Foaming Agent**—A chemical which causes a pesticide preparation to produce a thick foam. This aids in reducing drift.

**Fog Treatment**—A fine mist of pesticide in aerosol-sized droplets (under 40 microns). Not a mist or gas. After propulsion, fog droplets fall to horizontal surfaces.

**Foliar Treatment**—Application of the pesticide to the foliage of plants.

**Food Chain**—Sequence of species within a community, each member of which serves as food for the species next higher in the chain.

**Formulation**—Way in which basic pesticide is prepared for practical use. Includes preparation as wettable powder, granular, emulsifiable concentrate, etc.

**Fumigant**—A pesticide formulation that volatilizes, forming a toxic vapor or gas that kills in the gaseous state. Usually, it penetrates voids to kill pests.

**Fungicide**—A chemical that kills fungi.

**Fungistatic**—Action of a chemical that inhibits the germination of fungus spores while in contact.

**Fungus (plural, Fungi)**—A group of small, often microscopic, organisms in the plant kingdom which cause rot, mold, and disease. Fungi need moisture or a damp environment (wood rots require at least 19% moisture). Fungi are extremely important in the diet of many insects.

**General Use Pesticide**—A pesticide which can be purchased and used by the general public without undue hazard to the applicator and environment as long as the instructions on the label are followed carefully (see Restricted Use Pesticide).

**Germicide**—A substance that kills germs (microorganisms).

**GPA**—Gallons per acre.

**GPM**—Gallons per minute.

**Granule**—A dry pesticide formulation. The active ingredient is either mixed with or coated onto an inert carrier to form a small, ready-to-use, low concentrate particle which normally does not present a drift hazard. Pellets differ from granules only in their precise uniformity, larger size, and shape.

**Groundwater**—Water sources located beneath the soil surface from which spring water, well water, etc. is obtained.

**Growth Regulator**—Organic substance effective in minute amounts for controlling or modifying (plant or insect) growth processes.

**Harvest Intervals**—Period between last application of a pesticide to a crop and the harvest as permitted by law.

**Hazard**—see Risk.

**Herbaceous Plant**—A plant that does not develop woody tissue.

**Herbicide**—A pesticide used to kill or inhibit plant growth.

**Hormone**—A product of living cells that circulates in the animal or plant fluids and that produces a specific effect on cell activity remote from its point of origin.

**Host**—Any animal or plant on or in which another lives for nourishment, development, or protection.

**Hydrogen-Ion Concentration**—A measure of acidity or alkalinity. Expressed in terms of the pH of the solution. For example, a pH of 7 is neutral, from 1 to 7 is acid, and from 7 to 14 is alkaline.

**Hydrolysis**—Chemical process of (in this case) pesticide breakdown or decomposition involving a splitting of the molecule and addition of a water molecule.

**IGR, Insect Growth Regulator, Juvenoid**—A pesticide constructed to mimic insect hormones that control molting and the development of some insect systems affecting the change from immature to adult (see Juvenile Hormone).

**Immune**—Not susceptible to a disease or poison.

**Impermeable**—Cannot be penetrated. Semipermeable means that some substances can pass through and others cannot.

**Incompatible**—Two or more materials which cannot be mixed or used together.

**Inert Ingredients**—The inactive materials in a pesticide formulation, which would not prevent damage or destroy pests if used alone.

**Ingest**—To eat or swallow.

**Ingredient Statement**—That portion of the label on a pesticide container which gives the name and amount of each active ingredient and the total amount of inert ingredients in the formulation.

**Inhalation**—Taking a substance in through the lungs; breathing in (see Exposure Route).

**Inhalation Toxicity**—Poisonous to man or animals when breathed into lungs.

**Insect Growth Regulator**—see IGR.

**Insect Pest Management**—The practical manipulation of insect (or mite) pest populations using any or all control methods in a sound ecological manner.

**Insecticide**—A pesticide used to manage or prevent damage caused by insects. Sometimes generalized to be synonymous with pesticide.

**Insects, Insecta**—A class in the phylum Arthropoda characterized by a body composed of three segments and three pairs of legs.

**Inspection**—To examine for pests, pest damage, other pest evidence, etc. (see Monitoring).

**Integrated Control**—The integration of the chemical and biological methods of pest control.

**Integrated Pest Management**—A management system that uses all suitable techniques and methods in as compatible a manner as possible to maintain pest populations at levels below those causing economic injury.

**Juvenile Hormone**—A hormone produced by an insect that inhibits change or molting. As long as juvenile hormone is present the insect does not develop into an adult but remains immature.

**Kg or Kilogram**—A unit of weight in the metric system equal to 2.2 pounds.

**Label**—All printed material attached to or part of the container.

**Labeling**—The pesticide product label and other accompanying materials that contain directions that pesticide users are legally required to follow.

**Larva (plural larvae)**—The developmental stage of insects that hatch from an egg with complete metamorphosis. A mature larva becomes a pupa. (Some other invertebrates have larvae, including crustaceans, and especially mites and ticks).

**$LC_{50}$**—Lethal concentration. The concentration of an active ingredient in air which is expected to cause death in 50 percent of the test animals treated. A means of expressing the toxicity of a compound present in air as dust, mist, gas, or vapor. It is generally expressed as micrograms per liter as a dust or mist but, in the case of a gas or vapor, as parts per million (ppm).

**$LD_{50}$**—Lethal dose. The dose of an active ingredient taken by mouth or absorbed by the skin which is expected to cause death in 50 percent of the test animals treated. If a chemical has an $LD_{50}$ of 10 milligrams per kilogram (mg/kg) it is more toxic than one having an $LD_{50}$ of 100 mg/kg.

**Leaching**—The movement of a pesticide chemical or other substance downward through soil as a result of water movement.

**Low Volume Spray**—Concentrate spray, applied to uniformly cover the crop, but not as a full coverage to the point of runoff.

**Metamorphosis**—A change in the shape, or form, of an animal. Usually used when referring to insect development.

**mg/kg (milligrams per kilogram)**—Used to designate the amount of toxicant required per kilogram of body weight of test organism to produce a designated effect, usually the amount necessary to kill 50% of the test animals.

**Microbial Degradation**—Breakdown of a chemical by microorganisms.

**Microbial Insecticide**—A microorganism applied in the same way as conventional insecticides to control an existing pest population.

**Microbial Pesticide**—Bacteria, viruses, fungi, and other microorganisms used to control pests. Also called biorationals.

**Microorganism**—An organism so small it can be seen only with the aid of a microscope.

**Mildew**—Fungus growth on a surface.

**Miscible Liquids**—Two or more liquids capable of being mixed in any proportions and that will remain mixed under normal conditions.

**Miticide**—A pesticide used to control mites (see Acaricide).

**Mode of Action**—The way in which a pesticide exerts a toxic effect on the target plant or animal.

**Molluscicide**—A chemical used to kill or control snails and slugs.

**Monitoring**—Ongoing surveillance. Monitoring includes inspection and recordkeeping. Monitoring records allows technicians to evaluate pest population suppression, identify infested or non-infested sites, and manage the progress of the management or control program.

**Mosaic**—Leaf pattern of yellow and green or light green and dark green produced by certain virus infections.

**Mutagen**—Substance causing genes in an organism to mutate or change.

**Mutagenic**—Can produce genetic change.

**Mycoplasma**—A microorganism intermediate in size between viruses and bacteria possessing many virus-like properties and not visible with a light microscope.

**Necrosis**—Death of plant or animal tissues which results in the formation of discolored, sunken, or necrotic (dead) areas.

**Necrotic**—Showing varying degrees of dead areas or spots.

**Negligible Residue**—A tolerance which is set on a food or feed crop permitting an ultra-small amount of pesticide at harvest as a result of indirect contact with the chemical.

**Nematicide**—Chemical used to kill nematodes.

**Nontarget Organism**—Any plant or animal other than the intended target(s) of a pesticide application.

**Nymph**—The developmental stage of insects that hatch from an egg with gradual metamorphosis. Nymphs become adults.

**Oncogenic**—The property to produce tumors (not necessarily cancerous) in living tissues (see carcinogenic.)

**Oral**—Pertaining to the mouth; through or by the mouth.

**Oral Toxicity**—The ability of a pesticide to cause injury or acute illness when taken by mouth. One of the common exposure routes.

**Organic Compounds**—Chemicals that contain carbon.

**Organochlorine Insecticide**—One of the many chlorinated insecticides, e.g., DDT, dieldrin, chlordane, BHC, lindane, etc.

**Organophosphate**—Class of insecticides (also one or two herbicides and fungicides) derived from phosphoric acid esters, e.g., parathion, malathion, diazinon, etc.

**Ovicide**—A chemical that destroys an organism's eggs.

**Parasite**—A plant, animal, or microorganism living in, on, or with another living organism for the purpose of obtaining all or part of its food.

**Pathogen**—Any disease-producing organism.

**Penetrant**—An additive or adjuvant which aids the pesticide in moving through the outer surface of plant tissues.

**Penetration**—The ability of a substance to enter a organism, structure, or garment..

**Perennial**—Plant that continues to live from year to year. Plants may be herbaceous or woody.

**Persistence**—The capability of an insecticide to persist as an effective residue due to its low volatility and chemical stability, e.g., certain organochlorine insecticides.

**Personal Protective Equipment (PPE)**—Devices and clothing intended to protect a person from exposure to pesticides. Includes such items as long-sleeved shirts, long trousers, coveralls, suitable hats, gloves, shoes, respirators, and other safety items as needed.

**Pest**—An undesirable organism: (1) any insect, rodent, nematode, fungus, weed, or (2) any other form of terrestrial or aquatic plant or animal life or virus, bacteria, or other micro-organism (except viruses, bacteria, or other micro-organisms on or in living man or other living animals) which the Administrator declares to be a pest under FIFRA, Section 25(c)(1).

**Pesticide**—An "economic poison" defined in most state and federal laws as any substance used for controlling, preventing, destroying, repelling, or mitigating any pest. Includes fungicides, herbicides, insecticides, nematicides, rodenticides, desiccants, defoliants, plant growth regulators, etc.

**pH**—A measure of the acidity/alkalinity of a liquid: acid below pH7; basic or alkaline above pH7 (up to 14).

**Pheromone**—A substance emitted by an animal to influence the behavior of other animals of the same species. Some are synthetically produced for use in insect traps.

**Photodegradation**—Breakdown of chemicals by the action of light.

**Physical Control**—Habitat alteration or changing the infested physical structure; e.g., caulking holes, cracks, tightening around doors, windows, moisture reduction, ventilation, etc.

**Physical Selectivity**—Refers to the use of broad spectrum insecticides in such ways as to obtain selective action. This may be accomplished by timing, dosage, formulation, etc.

**Physiological Selectivity**—Refers to insecticides which are inherently more toxic to some insects than to others.

**Phytotoxic**— Harmful to plants.

**Piscicide**—Chemical used to kill fish.

**Plant Regulator (Growth Regulator)**—A chemical which increases, decreases, or changes the normal growth or reproduction of a plant.

**Poison**—Any chemical or agent that can cause illness or death when eaten, absorbed through the skin, or inhaled by man or animals.

**Post-emergence**—After emergence of the specified weed or crop.

**ppb**—Parts per billion (parts in 109 parts) is the number of parts of toxicant per billion parts of the substance in question.

**ppm**—Part per million (parts in 106 parts) is the number of parts of toxicant per million parts of the substance in question. They may include residues in soil, water, or whole animals.

**Precipitate** —A solid substance that forms in a liquid and settles to the bottom of a container. A material that no longer remains in suspension.

**Predacide**—Chemical used to poison predators.

**Predator**—An animal that attacks, kills, and feeds on other animals. Examples of predaceous animals are hawks, owls, snakes, many insects, etc.

**Pre-planting Treatment**—Given before the crop is planted.

**Propellant**—The inert ingredient in pressurized products that forces the active ingredient from the container.

**Protectant**—Fungicide applied to plant surface before pathogen attack to prevent penetration and subsequent infection.

**Protective Clothing**—Clothing to be worn in pesticide-treated fields under certain conditions as required by federal law, e.g., reentry intervals.

**Protopam Chloride (2-PAM)**—An antidote for certain organophosphate pesticide poisoning, but not for carbamate poisoning.

**Pupa (plural Pupae)**—The developmental stage of insects with complete metamorphosis where major changes from the larval to the adult form occurs.

**Rate of Application**—The amount of pesticide applied to a plant, animal, unit area, or surface; usually measured as per acre, per 1,000 square feet, per linear feet, or per cubic feet.

**Raw Agricultural Commodity**—Any food in its raw and natural state, including fruits, vegetables, nuts, eggs, raw milk, and meats.

**Re-entry Intervals**—Waiting interval required by federal law between application of certain hazardous pesticides to crops and the entrance of workers into those crops areas without protective clothing.

**Registered Pesticides**—Pesticide products which have been approved by the Environmental Protection Agency for the uses listed on the label.

**Repellent**—A compound that keeps insects, rodents, birds, or other pests away from plants, domestic animals, buildings, or other treated areas.

**Residual**—Having a continued killing effect over a period of time.

**Residue**—The pesticide active ingredient or its breakdown product(s) which remains in or on the target after treatment.

**Resistance (insecticide)**—Natural or genetic ability of an organism to tolerate the poisonous effects of a toxicant.

**Restricted Use Pesticide**—A pesticide that can be purchased and used only by certified applicators or persons under their direct supervision. A pesticide classified for restricted use under FIFRA, Section 3(d)(1)(C).

**Risk**—A probability that a given pesticide will have an adverse effect on man or the environment in a given situation.

**Rodenticide**—Pesticide applied as a bait, dust, or fumigant to destroy or repel rodents and other animals, such as moles and rabbits.

**Runoff**—The movement of water and associated materials on the soil surface. Runoff usually proceeds to bodies of surface water.

**Rust**—A disease with symptoms that usually include reddish-brown or black pustules: a group of fungi in the Basidiomycetes.

**Safener**—Substance that prevents or reduces the phytotoxicity when two or more substances must be mixed which otherwise would not be compatible.

**Scientific Name**—The one name of a plant or animal used throughout the world by scientists and based on Latin and Greek.

**Secondary Pest**—A pest which usually does little if any damage but can become a serious pest under certain conditions, e.g., when insecticide applications destroy its predators and parasites.

**Seed Protectant**—A chemical applied to seed before planting to protect seeds and new seedlings from disease and insects.

**Selective Insecticide**—One which kills selected insects, but spares many or most of the other organisms, including beneficial species, either through differential toxic action or the manner in which insecticide is used.

**Selective Pesticide**—One which, while killing the pest individuals, spares much or most of the other fauna or flora, including beneficial species, either through differential toxic action or through the manner in which the pesticide is used (formulation, dosage, timing, placement, etc.).

**Senescence**—Process or state of growing old.

**Signal Word**—A required word which appears on every pesticide label to denote the relative toxicity of the product. The signal words are either

"Danger—poison" for highly toxic compounds, "Warning" for moderately toxic, or "Caution" for slightly toxic.

**Site**—Areas of actual pest infestation. Each site should be treated specifically or individually.

**Slurry**—Thin, watery mixture, such as liquid mud, cement, etc. Fungicides and some insecticides are applied to seeds as slurries to produce thick coating and reduce dustiness.

**Smut**—A fungus with sooty spore masses; a group of fungi in the Basidiomycetes.

**Soil Application**—Application of pesticide made primarily to soil surface rather than to vegetation.

**Soil Persistence**—Length of time that a pesticide application on or in soil remains effective.

**Soil Sterilant**—A chemical that prevents the growth of all plants and animals in the soil. Soil sterilization may be temporary or permanent, depending on the chemical.

**Soluble Powder**—A finely ground, solid material which will dissolve in water or some other liquid carrier.

**Solution**—A mixture of one or more substances in another substance (usually a liquid) in which all the ingredients are completely dissolved. Example: sugar in water.

**Solvent**—A liquid which will dissolve another substance (solid, liquid, or gas) to form a solution.

**Spore**—A single- to many-celled reproductive body in the fungi that can develop a new fungus colony.

**Spot Treatment**—Application to localized or restricted areas as differentiated from overall, broadcast, or complete coverage.

**Spreader**—Ingredient added to spray mixture to improve contact between pesticide and plant surface.

**Sterilize**—To treat with a chemical or other agent to kill every living thing in a certain area.

**Sticker**—Ingredient added to spray or dust to improve its adherence to plants.

**Stomach Poison**—A pesticide that must be eaten by an insect or other animal in order to kill or control the animal.

**Structural Pests**—Pests which attack and destroy buildings and other structures, clothing, stored food, and manufactured and processed goods. Examples: Termites, cockroaches, clothes moths. rats, dry-rot fungi.

**Surface Water**—Water on the earth's surface: rivers, lakes, ponds, streams, etc. (see Groundwater).

**Surfactant**—A chemical which increases the emulsifying, dispersing, spreading, and wetting properties of a pesticide product.

**Susceptible**—Capable of being diseased or poisoned; not immune.

**Susceptible Species**—A plant or animal that is poisoned by moderate amounts of a pesticide.

**Suspension**—A pesticide mixture consisting of fine particles dispersed or floating in a liquid, usually water or oil. Example: wettable powders in water.

**Swath**—The width of the area covered by a sprayer or duster making one sweep.

**Synergism**—Increased activity resulting from the effect of one chemical on another.

**Synthesize**—Production of a compound by joining various elements or simpler compounds.

**Systemic**—Compound that is absorbed and translocated throughout the plant or animal.

**Tank Mix**—Mixture of two or more pesticides in the spray tank at time of application. Such mixture must be cleared by EPA.

**Target Pest**—The pest at which a particular pesticide or other control method is directed.

**Technical Material** —The pesticide active ingredient in pure form, as it is manufactured by a chemical company. It is combined with inert ingredients or additives in formulations such as wettable powders, dusts, emulsifiable concentrates, or granules.

**Teratogenic**—Substance which causes physical birth defects in the offspring following exposure of the pregnant female.

**Tolerance**—(1) The ability of a living thing to withstand adverse conditions, such as pest attacks, weather extremes, or pesticides. (2) The amount of pesticide that may safely remain in or on raw farm products at time of sale.

**Tolerant**—Capable of withstanding effects.

**Topical Application**—Treatment of a localized surface site such as a single leaf blade, on an insect, etc., as opposed to oral application.

**Toxic**—Poisonous to living organisms.

**Toxicant**—A poisonous substance such as the active ingredient in pesticide formulations that can injure or kill plants, animals, or microorganisms.

**Toxin**—A naturally occurring poison produced by plants, animals, or microorganisms. Examples: the poison produced by the black widow spider, the venom produced by snakes, the botulism toxin.

**Trade Name (Brand Name)**—Name given a product by its manufacturer or formulator, distinguishing it as being produced or sold exclusively by that company.

**Translocation**—Transfer of food or other materials such as herbicides from one plant part to another.

**Trivial Name**—Name in general or common-place usage; for example, nicotine.

**Ultra Low Volume** (ULV)—Sprays that are applied at 0.5 gallon or less per acre or sprays applied as the undiluted formulation.

**Unclassified Pesticide**—See General Use Pesticide.

**Vapor Pressure**—The property which causes a chemical to evaporate. The higher the vapor pressure, the more volatile the chemical or the easier it will evaporate.

**Vector**—An organism, as an insect, that transmits pathogens to plants or animals.

**Vermin**—Pests; usually rats, mice, or insects.

**Vertebrate**—Animal characterized by a segmented backbone or spinal column.

**Viricide**—A substance that inactivates a virus completely and permanently.

**Virus**—Ultramicroscopic parasites composed of proteins. Viruses can only multiply in living tissues and cause many animal and plant diseases.

**Viscosity**—A property of liquids that determines whether they flow readily. Viscosity usually increases when temperature decreases.

**Volatile**—Evaporates at ordinary temperatures when exposed to air.

**Volatilize**—To vaporize.

**Water Table**—The upper level of the water saturated zone in the ground.

**Weed**—Plant growing where it is not desired.

**Wettable Powder**—Pesticide formulation of toxicant mixed with inert dust and a wetting agent which mixes readily with water and forms a short-term suspension (requires tank agitation).

**Wetting Agent**—Compound that causes spray solutions to contact plant surfaces more thoroughly.

# INDEX

## —A—



## —B—



## —C—

—D—



—E—



—F—

—M—



—N—



—O—



—P—